GRAPH THEORY
AND COMBINATORICS

GRAPH THEORY AND COMBINATORICS

Proceedings of the Cambridge Combinatorial Conference
in Honour of Paul Erdös

Edited by

BÉLA BOLLOBÁS

*Department of Pure Mathematics and
Mathematical Statistics, University of Cambridge
Cambridge, UK*

1984

ACADEMIC PRESS

(*Harcourt Brace Jovanovich Publishers*)
London   Orlando   San Diego   San Francisco   New York
Toronto   Montreal   Sydney   Tokyo   São Paulo

ACADEMIC PRESS INC. (LONDON) LTD.
24–28 Oval Road,
London NW1 7DX

United States Edition Published by
ACADEMIC PRESS INC.
(Harcourt Brace Jovanovich, Inc.)
Orlando, Florida 32887

British Library Cataloguing in Publication Data

Cambridge Combinatorial Conference (*1983*)
Graph theory and combinatorics
1. Graph theory
I. Title   II. Bollobás, Béla   III. Erdös, Paul
511′.5   QA166

ISBN 0-12-111760-X
LCCN 83-83432

Printed in Northern Ireland by The Universities Press (Belfast) Ltd.
Belfast BT6 9HF

*In honour of Professor Paul Erdös,
on the occasion of his seventieth birthday.*

Paul Erdös

Cold cast bronze by Gabriella Bollobás, 1976.

LIST OF CONTRIBUTORS

J-C. BERMOND, Laboratoire de Recherche en Informatique, ERA 452 du CNRS, Université de Paris-Sud, Centre d'Orsay, Bât. 490, F–91405 Orsay Cedex, France

A. BIALOSTOCKI, Department of Pure Mathematics, School of Mathematical Sciences, Tel-Aviv University, Tel-Aviv, Israel

B. BOLLOBÁS, Department of Pure Mathematics, University of Cambridge, 16 Mill Lane, Cambridge CB2 1SB, UK

J. BOND, Laboratoire de Recherche en Informatique, ERA 452 du CNRS, Université de Paris-Sud, Centre d'Orsay, Bât. 490 91405 Orsay Cedex, France

P. J. CAMERON, Merton College, Oxford OX1 4JD, UK

A. G. CHETWYND, Faculty of Mathematics, The Open University, Milton Keynes MK7 6AA, UK

D. E. DAYKIN, Department of Mathematics, University of Reading, Whiteknights, Reading RG6 2AX, UK

P. DONNELLY, Balliol College, Oxford, UK

P. DUCHET, Centre de Mathématique Sociale, Ecole des Hautes Etudes en Sciences Sociales, 54 Boulevard Raspail, 75270 Paris Cedex 06, France

P. ERDÖS, Mathematical Institute of the Hungarian Academy of Sciences, Reáltanoda utca 13–15, Budapest, Hungary H-1053

R. J. FAUDREE, Department of Mathematical Sciences, Memphis State University, Memphis, Tennessee 38152, USA

T. I. FENNER, Department of Computer Science, Birkbeck College, University of London, Malet Street, London WC1E 7HX, UK

P. FRANKL, CNRS, Université de Paris VII (UER), 2 Place Jussieu, 75005 Paris, France

A. M. FRIEZE, Department of Computer Science and Statistics, Queen Mary College, University of London, Mile End Road, London E1 4NS, UK

I. P. GOULDEN, Department of Combinatorics and Optimization, University of Waterloo, Waterloo, Ontario N2L 3G1, Canada (Present address: Department of Mathematics, Rm 2-336, Massachusetts Institute of Technology, 77 Massachusetts Avenue, Cambridge, Massachusetts 02139, USA)

M. GRÖTSCHEL, Mathematisches Institut, Universität Augsburg, Meminger Strasse 6, D-8900 Augsburg, West Germany

R. HÄGGKVIST, Department of Mathematics, University of Stockholm, Stockholm, Sweden

R. HALIN, Mathematisches Seminar, Universität Hamburg, Bundesstrasse 55, 2000 Hamburg 13, West Germany

A. J. HARRIS, Department of Pure Mathematics, University of Cambridge, 16 Mill Lane, Cambridge CB2 1SB, UK

S. T. HEDETNIEMI, Department of Computer Science, Clemson University, Clemson, South Carolina 29631, USA

A. J. W. HILTON, Department of Mathematics, The University of Reading, Whiteknights, Reading RG6 2AX, UK (Present address: Faculty of Mathematics, The Open University, Milton Keynes MK7 6AA, UK)

D. M. JACKSON, Department of Combinatorics and Optimization, University of Waterloo, Waterloo, Ontario N2L 3G1, Canada

B. KORTE, Institut für Ökonometrie und Operations Research, Rheinische Friedrich-Wilhelms-Universität, Nassestrasse 2, D-5300 Bonn 1, West Germany

M. LAS VERGNAS, Université Pierre et Marie Curie, UER 48-Mathématiques, 4 Place Jussieu, 75005 Paris, France

R. LASKAR Department of Mathematical Sciences, College of Sciences, Clemson University, Clemson, South Carolina 29631, USA

M. LESK, CNET, PAA/ATR, 38 Rue du General Leclere, 92131, Issy les Moulineaux, France

L. LOVÁSZ Bolyai Institute, Jósef Attila University, Szeged, Hungary

M. D. PLUMMER Department of Mathematics, Box 1543, Station B, Vanderbilt University, Nashville, Tennessee 37235, USA

W. R. PULLEYBANK Department of Combinatorics and Optimization, University of Waterloo, Waterloo, Ontario N2L 3G1, Canada

R. RADO, FRS, Department of Mathematics, University of Reading, Whiteknights, Reading RG6 2AX, UK

P. ROSENSTIEHL, Centre d'Analyse et de Mathématique Sociale, Ecole des Hautes Etudes en Sciences Sociales, 54 Boulevard Raspail, 75270 Paris Cedex 06, France

R. J. ROUSSEAU, Department of Mathematical Sciences, Memphis State University, Memphis, Tennessee 38152, USA

J. F. SACLE, Laboratoire de Recherche en Informatique, ERA 452 de CNRS, Université de Paris-Sud, Centre d'Orsay, Bât. 490, F–91405 Orsay Cedex, France

R. H. SCHELP, Department of Mathematical Sciences, Memphis State University, Memphis, Tennessee 38152, USA

A. SCHÖNHEIM, Department of Pure Mathematics, School of Mathematical Sciences, Tel-Aviv University, Tel-Aviv, Israel

J. SHEEHAN, Department of Mathematics, University of Aberdeen, Edward Wright Building, Dunbar Street, Aberdeen AB9 2TY, UK

I. SIMON, Instituto de Matemática e Estatistica, Universidade de São Paulo, Caixa Postal 20.570 (Ag. Iguatemi), 05508 São Paulo, Brazil

J. SPENCER, Department of Mathematics, State University of New York at Stony Brook, Stony Brook, New York 11790, USA

E. SZEMERÉDI, Mathematical Institute of the Hungarian Academy of Sciences, Reáltanoda utca 13–15, Budapest, Hungary H-1053

C. THOMASSEN, Matematisk Institut, Danmarks Tekniske Højskole, DK-2800 Lyngby, Denmark

W. T. TROTTER, Jʀ, Department of Mathematics and Statistics, University of South Carolina, Columbia, South Carolina 29208, USA

W. T. TUTTE, Department of Combinatorics and Optimization, University of Waterloo, Waterloo, Ontario N2L 3G1, Canada

W. F. DE LA VEGA, Laboratoire de Recherche en Informatique, Université de Paris-Sud, Centre d'Orsay, Bât. 490, F–91405 Orsay Cedex, France

D. J. A. WELSH, Merton College, Oxford OX1 4JD, UK

PREFACE

The Cambridge Combinatorial Conference was held at Trinity College from 21 to 25 March 1983, under the auspices of the Department of Pure Mathematics and Mathematical Statistics. The conference was in honour of Professor Paul Erdös on his seventieth birthday. Twenty-seven of the participants were invited to give talks. This volume consists of most of the papers they presented together with a few additional articles on closely related themes.

In his famous address on mathematical problems, Hilbert remarked that "the clearness and ease of comprehension insisted on for a mathematical theory I should still more demand for a mathematical problem, if it is to be perfect. For what is clear and easily comprehended attracts; the complicated repels us." Graph theory is fortunate to have, in the work of Paul Erdös, a rich source of problems which meet Hilbert's standard. During the past half-century his skills have contributed to number theory, logic, probability and function theory but, above all, they have made him a driving force behind the rapid development of graph theory and combinatorics. His problems and solutions now fill a list of publications of a length to rival Cayley's 967. In his many journeys through the world, Paul Erdös has come to know and to help a great army of mathematicians. They owe much to his encouragement and stimulus. Those of us at the conference were pleased to mark our own debt to him at a celebration dinner on 24 March 1983. Professor Richard Guy proposed a moving and memorable toast to Paul Erdös and received a witty and surprisingly cheerful reply.

Cambridge   Béla Bollobás
20 August 1983

CONTENTS

ON SOME PROBLEMS IN GRAPH THEORY, COMBINATORIAL ANALYSIS AND COMBINATORIAL NUMBER THEORY

Paul Erdös

ABSTRACT   In this paper I discuss some old problems, most of them due to myself and my collaborators. Many of these problems have been undeservedly neglected, even by myself. I also discuss some new problems.

I. Graph theory

1. $G(n)$ is a graph of n vertices and $G(n; e)$ is a graph of n vertices and e edges. Is it true that if every induced subgraph of a $G(10n)$ of $5n$ vertices has more than $2n^2$ edges then our $G(10n)$ contains a triangle? It is easy to show that if true this result is best possible. To see this let A_i, $i = 1$, $2, \ldots, 5$, be sets of $2n$ vertices, put $A_1 = A_6$ and join every vertex of A_i to every vertex of A_{i+1}. This $G(10n; 20n^2)$ has of course no triangle and every induced subgraph of $5n$ vertices contains at least $2n^2$ edges. Equality is of course possible: choose A_1, A_3 and half the vertices of A_4.

Simonovits pointed out to me that a graph of completely different structure also shows that the conjecture, if true, is best possible. Consider the Petersen graph, which is a $G(10; 15)$. Replace each vertex by a set of n vertices and replace every edge of the Petersen graph by the n^2 edges of a $K(n, n)$. This gives a $G(10n; 15n^2)$ and it is easy to see that every induced subgraph of $5n$ vertices has at least $2n^2$ edges.

The fact that two graphs of different structure are extremal perhaps indicates that the conjecture is either false or difficult to prove. I certainly hope that the latter is the case.

It is perhaps tempting to conjecture that my graph has the following extremal property. If a $G(10n)$ has no triangle and every induced subgraph of $5n$ vertices has at least $2n^2$ edges, then our graph can have at most $20n^2$ edges. Perhaps the graph of Simonovits has the smallest number of edges among all extremal graphs; perhaps in fact these two graphs are the only extremal graphs.

Many generalizations are possible; the triangle could be replaced by other graphs. Is it true that every $G((4h + 2)n)$, every induced subgraph

GRAPH THEORY AND COMBINATORICS
ISBN 0-12-111760-X

of $(2h+1)n$ vertices of which has more than $2n^2$ edges, contains an odd circuit of fewer than $2h+1$ edges?

Let $0 < \alpha < 1$. Determine the smallest c_α so that if every induced subgraph of $G(n)$ of $\lfloor \alpha n \rfloor$ vertices contains more than $c_\alpha n^2$ edges then $G(n)$ has a triangle. My conjecture was that $c_{1/2} = \frac{1}{50}$.

Observe that it is not difficult to prove that for $n > n_0(r)$ every $G(n; cn^2)$ contains a $K(r)$, provided that every induced subgraph of $n/2$ vertices has $(1 + o(1))(cn^2/4)$ edges.

2. Is it true that every graph of $5n$ vertices which contains no triangle can be made bipartite by the omission of at most n^2 edges? The same graph as used in paragraph 1 shows that the conjecture, if true, is best possible. (Replace each vertex of a pentagon by n vertices, etc.)

Is it true that a $G(5n)$ which has no triangle contains at most n^5 pentagons? Again the same graph shows that, if true, this is the best possible. Here also many generalizations are possible.

3. Let G be a bipartite graph of n white and $n^{2/3}$ black vertices. Is it true that if our graph has more than cn edges (where c is a sufficiently large constant), then it contains a C_6? It is easy to see that it must contain a C_8 but does not have to contain a C_4. Simonovits strongly disbelieves this conjecture and I have no real evidence for its truth. It is easy to see that this conjecture, if true, is best possible. To see this observe that Benson's graph has $cn^{4/3}$ edges and n black and n white vertices, thus a suitable subgraph has n white and $n^{2/3}$ black vertices and cn edges.

C. Benson, Minimal regular graphs of girth eight and twelve. *Canad. J. Math.* **18** (1966), 1091–1094.

4. Several hundred papers have been published on extremal graph theory and recently Bollobás published an excellent book on the subject. Here I state only a few problems which have not been thoroughly investigated. First a recent problem of Simonovits and myself.

Denote by $f(n; H)$ the smallest integer such that every graph $G(n; f(n; H))$ contains H as a subgraph. Turán, who started extremal graph theory, determined $f(n; K(r))$ for every r. In particular he proved that $f(n, K(3)) = \lfloor n^2/4 \rfloor + 1$.

Rademacher was the first to observe that every $G(n; \lfloor n^2/4 \rfloor + 1)$ contains at least $\lfloor n/2 \rfloor$ triangles. This result was generalized and extended by Bollobás, Lovász, Simonovits and myself. Simonovits and I asked: Is it true that every $G(n; f(n; C_4))$ contains at least $(1 + o(1)) \cdot n^{1/2}$ C_4s. We could not even prove that the number of these C_4s tends to infinity.

Brown, V. T. Sós, Rényi and I proved that $f(n; C_4) = (\frac{1}{2} + o(1))n^{3/2}$ and (p is a power of a prime) that

$$f(p^2 + p + 1, C_4) \geq \frac{1}{2}p(p+1)^2. \tag{1}$$

Is it true that there is equality in (1)? Füredi proved this for $p = 2^k$, and very recently he proved that there is always equality in (1). I observed that for every n,

$$f(n; C_4) \leq \frac{1}{2}n^{3/2} + (1 + o(1))\frac{n}{4}, \tag{2}$$

and conjectured that there is equality in (2).

Kővári, V. T. Sós and Turán proved that

$$f(n; K(r, r)) < c_r n^{2 - (1/r)}. \tag{3}$$

Very probably (3) is best possible for every r. This problem is still open for $r > 3$. Very recently Frankl proved that

$$f(n; K(r; 2^r)) > c_r' n^{2 - (1/r)}.$$

I conjectured some time ago that for every $\varepsilon > 0$ there is a c_ε such that every $G(n; \lfloor n^{1+\varepsilon} \rfloor)$ contains a non-planar subgraph of at most c_ε vertices.

Dirac asked in a conversation: Let m and n be fixed. How many edges can have a $G(n)$ if it does not contain a saturated planar subgraph of at least m vertices? Simonovits nearly completely solved this problem.

Denote by $f(n, m)$ the smallest integer such that every $G(n; f(n; m))$ contains all saturated planar graphs of m vertices. The four-colour theorem easily implies $f(n; m) = (1 + o(1))(n^2/3)$, but it might be of interest to get a sharper formula which would also show the dependence on m.

Simonovits and I posed the following problem: Is it true that

$$\lim f(n; H)/n^{3/2} = \infty \tag{4}$$

holds if and only if H contains a subgraph H', each vertex of which has degree greater than 2?

This attractive conjecture is very far from being settled. The following two graphs perhaps could give a counterexample. Define H_k as follows: The vertices of H_k are $x; y_1, \ldots, y_k, z_1, z_2, \ldots, z_{\binom{k}{2}}$. The vertex x is joined to all the y_is and each z_i is joined to two y_js (no two z_is are joined to the same pair). Is it true that

$$f(n; H_k) < cn^{3/2}? \tag{5}$$

I proved (5) for $k = 3$ but for $k > 3$ I do not know if (5) holds. The second graph H was once considered by V. T. Sós, Simonovits and myself. H has five black and five white vertices, x_1, x_2, a, x_3, x_4 and y_1, y_2, b, y_3, y_4

respectively; a is joined to every y_j and b to every x_i; a and b are not joined. $(x_1, x_2; y_1, y_2)$ and $(x_3, x_4; y_3, y_4)$ form a C_4. Is it true that $f(n; H) < cn^{3/2}$?

In a recent interesting paper Faudree and Simonovits obtain several striking new results in extremal graph theory. They hope that a further development of their method will lead to the construction of two graphs H_1 and H_2 for which

$$\lim_{m \to \infty} f(n; H_1; H_2)/\min\{f(n; H_1), f(n; H_2)\} = 0.$$

Here $f(n; H_1, H_2)$ is the smallest integer for which every $G(n; f(n; H_1, H_2))$ contains either H_1 or H_2.

B. Bollobás, *Extremal Graph Theory*, London Mathematical Society Monographs no. 11, Academic Press, London, 1978.

P. Erdös, On a theorem of Rademacher and Turán. *Illinois J. Math.* **6** (1962), 122–127.

B. Bollobás, Relations between sets of complete subgraphs. In *Fifth British Combinatorial Conference*, Utilitas Mathematicae, Winnipeg, 1976, pp. 161–170.

L. Lovász and M. Simonovits, On the number of complete subgraphs of a graph. In *Fifth British Combinatorial Conference*, Utilitas Mathematicae, Winnipeg, 1976, pp. 431–441.

P. Erdös, A. Rényi and V. T. Sós, On a problem of graph theory. *Studia Sci. Math. Hungar.* **1** (1966), 215–235.

W. Brown, On graphs that do not contain a Thomsen graph. *Canad. Math. Bull.* **9** (1966), 281–285.

Z. Füredi, Graphs without quadrilaterals. *J. Combinat. Theory* **34** (1983), 187–190.

T. Kővári, V. T. Sós and P. Turán, On a problem of K. Zarankiewicz. *Colloq. Math.* **3** (1959), 50–57.

M. Simonovits, On graphs not containing large saturated planar graphs. In *Infinite and Finite Sets*, Coll. Math. Soc. J. Bolyai no. 10, North-Holland, Amsterdam, 1975, pp. 1365–1386.

P. Erdös, On some extremal problems in graph theory. *Israel J. Math.* **3** (1965), 113–116.

R. Faudree and M. Simonovits, On a class of degenerate extremal graph problems. *Combinatorica* **3** (1983), 83–94.

P. Erdös and M. Simonovits, Supersaturated graphs. *Combinatorica* **3** (1983), in press.

5. Many papers and the excellent book of Graham, Rothschild and Spencer have recently been published on Ramsey theory. Here I only want to say a few words and mention a new problem of Hajnal and myself. Let G_1 and G_2 be two graphs, and denote by $r(G_1, G_2)$ the smallest integer m for which, if we colour the edges of $K(m)$ by two

colours, I and II, then either there is a G_1 all of whose edges have colour I or a G_2 all of whose edges have colour II. Graver and Yackel, Ajtai, Komlós and Szemerédi and I myself proved that

$$c_2 n^2/(\log n)^2 < r(K(n), K(3)) < c_1 n^2/\log n.$$

I always conjectured that

$$r(K(n), C_4) < n^{2-\varepsilon}. \tag{6}$$

Many colleagues doubt whether (6) holds. As far as I know everybody believes that

$$r(K(n), C_4)/r(K(n), K(3)) \to 0. \tag{7}$$

Szemerédi recently observed that

$$r(K(n), C_4) < cn/(\log n)^2 < cr(K(n), K(3)).$$

Denote by $R(G_1, G_2)$ the smallest integer m for which there is a $G(m)$ with the property that if we colour the edges of G with two colours I and II in an arbitrary way, then either there is an *induced* subgraph G_1 of G, all of whose edges have colour I, or an induced subgraph G_2, all of whose edges have colour II. The existence of $R(G_1, G_2)$ is not at all obvious. As far as I know this was first conjectured by Hansen and proved simultaneously by Deuber, Rödl, and Erdös, Hajnal and Pósa. Hajnal and I observed that if G_1 and G_2 have at most n vertices then

$$R(G_1, G_2) = m < 2^{2^{n^{1+\varepsilon}}}. \tag{8}$$

We have never published the not entirely trivial proof of (8) since Hajnal and I thought that perhaps

$$\max R(G_1, G_2) = r(K(n), K(n)). \tag{9}$$

Conjecture (9) is perhaps a little too optimistic, but we have no counterexample. Perhaps there is a better chance to prove $R(G_1, G_2) < 2^{cn}$.

R. L. Graham, B. L. Rothschild and J. H. Spencer, *Ramsey Theory*, John Wiley, New York, 1980.

W. Deuber, A generalization of Ramsey's theorem. In *Infinite and Finite Sets*, Coll. Math. Soc. J. Bolyai no. 10, North-Holland, Amsterdam, 1975, pp. 323–332.

P. Erdös, A. Hajnal and L. Pósa, Strong embeddings of graphs into coloured graphs. In *Infinite and Finite Sets*, Coll. Math. Soc. J. Bolyai no. 10, North-Holland, Amsterdam, 1975, pp. 585–596.

V. Rödl, A generalization of Ramsey's theorem. In *Hypergraphs and Block Systems*, Zielona Gora, 1976, pp. 211–220.

P. Erdös, Graph theory and probability II. *Canad. J. Math.* **13** (1961), 364–352.

J. E. Graver and I. Yackel, Some graph theoretic results associated with Ramsey's theorem. *J. Combinat. Theory* **4** (1968), 125–175.
M. Ajtai, J. Komlós and E. Szemerédi, A note on Ramsey numbers. *J. Combinat. Theory A* **29** (1980), 354–360.

6. During a recent visit to Prague, I learned of the following surprising result of Frankl and Rödl which settled an old conjecture of mine. Let $G_i^r(n_i ; e_i)$ be a sequence of r-graphs. I say that their density is α if α is the largest number for which there is a sequence of induced subgraphs $G_i^{(r)}(m_i ; e_i')$ of $G_i^{(r)}(n_i ; e_i)$ with $m_i \to \infty$ and

$$e_i' = (\alpha + o(1))\binom{m_i}{r} \quad \text{as} \quad i \to \infty.$$

In a slightly imprecise but more illuminating way we can say that a large r-graph $G^{(r)}(n ; e)$ has density α is α is the largest number for which there is a large subgraph $G(m_i ; e_i')$ (of m_i vertices and e_i' edges) for which

$$e_i' = (\alpha + o(1))\binom{m_i}{r}.$$

A well known theorem of Stone and myself asserts that for $r = 2$ the only possible values of the densities of an ordinary graph are $1 - 1/r$, $1 \leqslant r \leqslant \infty$. (See also some papers of Bollobás, Chvátal, Simonovits, Szemerédi and myself.)

I conjectured that for $r > 2$ the set of possible densities forms a well ordered set. This was disproved by Frankl and Rödl. At this moment it is not yet clear what are the possible values of the densities of r-graphs for $r > 2$. The results of Frankl and Rödl will be soon published in *Combinatorica*.

P. Erdös and A. H. Stone, On the structure of linear graphs. *Bull. Amer. Math. Soc.* **52** (1946), 1087–1091.
P. Erdös and M. Simonovits, A limit theorem in graph theory. *Studia Sci. Math. Hungar. Acad.* **1** (1966), 51–57.
B. Bollobás and P. Erdös, On the structure of edge graphs. *Bull. London Math. Soc.* **5** (1973), 317–321.
B. Bollobás, P. Erdös and M. Simonovits, On the structure of edge graphs II. *J. London Math. Soc.* **12** (1976), 219–224.
V. Chvátal and E. Szemerédi, On the Erdös–Stone theorem. *J. London Math. Soc.* **23** (1981), 193–384.
P. Erdös, On extremal problems of graphs and generalized graphs. *Israel J. Math.* **2** (1965), 183–190.

7. Frankl and Rödl also settled a very recent question of Nešetřil and myself. At the recent meeting at Poznan on random graphs Nešetřil and I

conjectured that for every $\varepsilon > 0$ there is a graph $G(n; e)$ which contains no $K(4)$ but every subgraph of $(\frac{1}{2} + \varepsilon)e$ edges of it contains a triangle. This conjecture was settled by Frankl and Rödl by the probability method early in September 1983. Many generalizations are possible which are not yet completely cleared up.

The graph of Frankl and Rödl has n vertices and $n^{3/2-\varepsilon}$ edges. Frankl and Rödl also proved that there is no $c > 0$ such that for every $\varepsilon > 0$ there is such a graph with at least cn^2 edges. Their method at present does not seem to work if $K(3)$ is replaced by $K(r)$, $K(4)$ by $K(r+1)$ and $(\frac{1}{2} + \varepsilon)$ by $1 - (1/r) + \varepsilon$.

8. With Nešetřil we posed the following problem: Is there a G whose every edge is contained in at most three triangles and for which $G \rightarrow (K(3), K(3))$ holds? (In other words, if we colour the edges of G by two colours then is there always a monochromatic triangle?) Observe that since $K(6) \rightarrow (K(3), K(3))$ "three" cannot be replaced by "four". Let $G(n) \rightarrow (K(3), K(3))$ critically (i.e. if we omit any edge of $G(n)$ then this property no longer holds). Nešetřil and Rödl proved that such graphs $G(n)$ exist. We thought that for every k there is an n_k so that if $n > n_k$ and $G(n) \rightarrow (K(3), K(3))$ critically then there is an edge of $G(n)$ which is contained in at least k triangles. Nešetřil has just proved that this is not so.

J. Nešetřil and V. Rödl, The structure of critical Ramsey graphs. *Acta Math. Acad. Sci. Hungar.* **32** (1978), 295–300.

9. Let $G(n)$ be a connected graph of n vertices. Denoted by $f(G)$ the smallest integer for which the vertices of G can be covered by $f(G)$ cliques and let $h(G)$ be the largest integer for which there are $h(G)$ edges no two of which belong to the same clique. Parthasaraty and Choudum conjectured that $h(G) \geq f(G)$. By probabilistic methods I disproved this conjecture. In fact I showed that there is a $G(n)$ for which

$$f(G(n)) > \frac{c_1 n}{(\log n)^3} h(G(n)). \tag{10}$$

I conjectured that (10) is best possible, i.e. that for every $G(n)$

$$f(G(n)) < \frac{c_2 n}{(\log n)^3} h(G(n)). \tag{11}$$

I had difficulties in proving (11), which as far as I know is still open. I think the proof of (11) will not be difficult and that I am perhaps overlooking a simple argument. I thought that if $h(G(n))$ is small then

$h(G(n)) = f(G(n))$ is true. This was known for $n = 3$ and Kostochka just informs me that if $h(G) \le 5$ then $f(G) = h(G)$, but there is a $G(n)$ for which $h(G(n)) = 6$ and $f(G(n)) > n^{1/17}$. Kostochka's nice result answers my question very satisfactorily; nevertheless many questions remain, e.g. is it true that if $f(G(n)) > n^{1-\varepsilon}$, $\varepsilon \to 0$, then $h(G(n)) \to \infty$?

An old conjecture of Hajós stated: Is it true that every k-chromatic graph contains a topological complete h-gon; i.e. it contains h vertices every two of which can be joined by paths every two of which are disjoint (except having common end-points)? This was disproved by Catlin and Kostochka. Fajtlowicz and I showed by probabilistic methods that if $K(G(n))$ is the chromatic number of G and $T(G)$ is the number of vertices of the largest topologically complete graph embedded in G, then for almost all graphs

$$K(G(n)) > c_1 \left(\frac{n^{1/2}}{\log n} \right) T(G). \tag{12}$$

Perhaps (12) is best possible, i.e.

$$K(G(n)) < c_2 \left(\frac{n^{1/2}}{\log n} \right) T(G).$$

Bollobás, Catlin and I, on the other hand, proved that Hadwiger's conjecture holds for almost all graphs $G(n)$.

P. Erdös, [Title unknown.] *J. Math. Res. Exposition* **2** (1982), 93–96.

P. Erdös and S. Fajtlowicz, On the conjecture of Hajós. *Combinatorica* **1** (1981), 141–143.

P. A. Catlin, Hajós's graph-colouring conjecture: variations and counterexamples. *J. Combinat. Theory B* **26** (1979), 268–274.

B. Bollobás, P. A. Catlin and P. Erdös, Hadwiger's conjecture is true for almost every graph. *Europ. J. Combinat.* **1** (1980), 195–199.

10. Denote by $C^{(n)}$ the graph of the n-dimensional cube. $C^{(n)}$ has 2^n vertices and $n2^{n-1}$ edges. How many edges of $C^{(n)}$ ensure the existence of a C_{2r}, a circuit of size $2r$? Perhaps for $r > 2$, $o(n2^n)$ edges suffice. Perhaps $(\frac{1}{2} + \varepsilon)n2^{n-1}$ edges suffice for a C_4. It is easy to see that $\frac{1}{2}n2^{n-1}$ do not suffice for a C_4 and Chung observed that $\frac{2}{3}n2^{n-1}$ suffice for a C_4. The following is an old and forgotten problem of Graham and myself. It is well known that $K(2^n)$ can be decomposed as the union of n bipartite graphs, but $K(2^n + 1)$ can not be so decomposed. Suppose we decompose $K(2^n + 1)$ as the union of n graphs. What is the least odd circuit which must occur in any of the decompositions? To be precise, denote by $f(n)$ the smallest integer such that if

$$K(2^n) = \bigcup_{i=1}^{n} G_i$$

then one of the G_is contains an odd circuit of order at most $f(n)$. How large is $f(n)$? As far as I know, nothing is known about this problem, which has been completely forgotten or at least neglected, perhaps undeservedly so.

P. Erdös and R. L. Graham, On partition theorems for finite graphs. In *Infinite and Finite Sets*, Coll. Math. Soc. J. Bolyai no. 10, North Holland, Amsterdam, 1975, pp. 515–528.

11. The following problem is due to Fajtlowicz and myself. A graph $G(n)$ is said to have property $I_{r,l}$ (respectively $I_{r,\infty}$) if every set of l independent vertices (respectively every set of independent vertices) has a common neighbour and $G(n)$ contains no $K(r)$. It is said to have property I_∞ if we only assume that every set of independent vertices has a common neighbour. $f(n; r, l)$ (respectively $f(n; r, \infty)$ or $f(n; \infty)$) is the largest integer such that every graph with property $I_{r,l}$ (respectively $I_{r,\infty}$ or I_∞) has a vertex of degree greater than or equal to $f(n; r, l)$ (respectively $f(n; r, \infty)$ or $f(n; \infty)$). Determine or estimate $f(n; r, l)$, $f(n; r, \infty)$ and $f(n; \infty)$ as well as possible. Pach proved that $f(n; 3, \infty) = (n + 1)/3$ and $f(n; 3, 3) < n^{1-c_{3,3}}$, where $c_{3,3}$ is a positive constant. Pach and I have just proved that $f(n; r, r) < n^{1-c_{r,r}}$ and by the probability method we established

$$f(n; \infty) = (1 + o(1)) \frac{n \log \log n}{\log n}.$$

We do not know if $f(n; 3, l) > cn$ holds for some $l > 3$ or $f(n; r, \infty) > cn$ is true for some $r > 3$.

The following slightly modified problem could also be considered. Assume that a graph $G(n)$ is such that every set of r vertices of $G(n)$ has a common neighbour. Then of course it is immediate that our $G(n)$ must contain a $K(r + 1)$. On the other hand it is easy to see by the probability method that there is such a graph which contains no $K(r + 2)$ and each vertex of which has degree $o(n)$.

J. Pach, Graphs whose every independent set has a common neighbour. *Discrete Math.* **37** (1981), 217–218.

12. A group is said to have property A_k if it has at most k elements which pairwise do not commute. About 10 years ago I asked: Determine or estimate the smallest $f(k)$ so that every group with property $A(k)$ is the union of $f(k)$ or fewer Abelian groups. This is a finite modification of a problem considered by Bernhard Neumann. Isaacs proved that

$$(1 + c)^k < f(k) < k!^{2+\varepsilon}.$$

13. Finally I state a few recent problems which have not been investigated carefully.

Let G be a graph. Denote by $f_b(G)$ the maximal number of edges of a bipartite subgraph contained in G and by $f_3(G)$ the maximal number of edges of a triangle-free graph contained in G. Trivially,

$$f_3(G) \geqslant f_b(G). \tag{13}$$

Can one characterize the graphs for which there is equality in (13)? Or at least can one give a fairly general class of graphs having this property? Clearly $K(n)$ has equality in (13) by Turán's theorem and Simonovits easily showed by the method of Zykov that all complete r-partite graphs also have equality in (13). V. T. Sós asked: Let G be such that every circuit has a diagonal (or perhaps only every odd circuit has a diagonal). Is it then true that G has equality in (13)?

Horak, Kratochvil and I considered the following questions. Let $G_1(n), \ldots, G_r(n)$ be edge-disjoint subgraphs of $K(n)$ such that every Hamiltonian circuit of $G(n)$ has an edge in common with the $G_i(n)$. Put

$$f(n; r) = \min \sum_{i=1}^{r} e(G_i(n)),$$

where $e(G)$ denotes the number of edges of G. It is easy to see that $f(n; r) = r(n-2)$ for $r \leqslant 3$ and $f(n; r) < 2^{c_i r}$ for $r > 3$. Can one determine the exact value of $f(n; r)$ for $r > 3$? What is the largest value $g(n)$ of r for which such graphs $G_i(n)$ exist? Similar questions can be asked for Hamiltonian paths or other subgraphs of $K(n)$.

Shortly after we posed these questions. Horák and Širáň succeeded in determining both $f(n; r)$ and $g(n)$. They proved that

$$g(n) = \left\lfloor 3 + \log_2 \frac{n-1}{3} \right\rfloor \quad \text{for} \quad n \geqslant 4.$$

Furthermore, setting $w(n, r) = 3 \cdot 2^{r-4}(2n - 3 \cdot 2^{r-3} - 1)$,

$$f(n; r) = w(n, r) \qquad \text{for} \quad 4 \leqslant r \leqslant 3 + \log_2 \frac{n + b_n}{9},$$

$$f(n; r) = w(n, r-1) + c_n \quad \text{for} \quad 3 + \log_2 \frac{n + b_n}{9} < r \leqslant g(n),$$

where $c_n = (n^2 - 1)/8$. $b_n = 1$ for odd values of n and $c_n = (n^2 + 2n - 8)/8$. $b_n = 4$ for even values of n.

Let G be a graph of e edges. Is it true that

$$r(G, G) < 2^{c_1 e^{1/2}}? \tag{14}$$

If true, (14) is easily seen to be best possible apart from the value of c_1. Probably $r(G)$ is maximal if G is as complete as possible.

V. T. Sós asked: A graph G is said to be Ramsey-critical if

$$r(G, G) > r(G - e, G - e) \qquad (15)$$

for every edge of G. Can one characterize the graphs which satisfy (15)? As far as I know this simple and interesting question has not yet been considered. It seems certain that $K(n)$ satisfies (15), but even this is not known and it is not clear to me that (15) holds for almost all graphs. A related question is: Can one characterize the graphs for which

$$r(G, G) < r(G + e, G + e), \qquad (16)$$

where e is any edge not in G which joins two vertices of G?

II. Set systems

I have published many problems on set systems, therefore I mention only some problems which are perhaps not too well known.

1. Rado and I once considered the following problem. Determine or estimate the largest integer $g(n)$ for which one can give $g(n)$ sets A_h, $|A_h| = n$, $1 \le h \le g(n)$, so that for every three of our A_hs the union of some two of them contains the third. Jean Larson showed $g(2n) \ge (n + 1)^2$ and Frankl and Pach proved that $g(2n) = (n + 1)^2$. More generally, let $g_t(n)$ denote the maximum number m such that there exists a family of n-sets $\{A_1, A_2, \ldots, A_m\}$ without t disjointly representable members (i.e. any t members of the family contains one which is covered by the remaining $t - 1$ members). Frankl and Pach conjecture that

$$g_t(n) = T(n + t - 1, t, t - 1),$$

where $T(n, k, l)$ denotes the Turán number, the maximum cardinality of a family of l-sets on an n-set without a complete subgraph with k points. They can prove that

$$T(n + t - 1, t, t - 1) < g_t(n) < \binom{n + t - 1}{t - 1}.$$

Let $h_t(n)$ denote the maximum cardinality of a (non-uniform) family on n points without t disjointly representable members. A theorem of Sauer implies that

$$h_t(n) \le \sum_{1 \le i \le t - 1} \binom{n}{i}.$$

Frankl gave a construction for $t = 3$ (unpublished) and Füredi and Quinn proved that equality holds here for all t. There are a large number of optimal families.

Frankl and Pach also conjecture that the following beautiful generalization of the Erdös–Ko–Rado theorem is true. Suppose that $\mathcal{H}$ is a system of k element subsets of an n element set and that

$$|\mathcal{H}| > \binom{n-1}{k-1}.$$

Then there exists a $H_0 \in \mathcal{H}$ such that for every subset $A \subseteq H_0$ one can choose $H \in \mathcal{H}$ satisfying $H \cap H_0 = A$. (We obtain the Erdös–Ko–Rado theorem in the special case $A = \varnothing$.) They can prove this for set-systems of cardinality greater than $\binom{n}{k-1}$ by using a linear algebraic approach.

P. Frankl and J. Pach, On the number of sets in a null-t design. *Europ. J. Combinat.* **4** (1983), 21–23.

P. Frankl and J. Pach, On disjointly representable sets. *Combinatorica* **4** (1984), in press.

Z. Füredi and F. Quinn, Traces of finite sets. *Ars Combinat.* in press.

N. Sauer, On density of families of sets. *J. Combinat. Theory A* **13** (1972), 145–147.

P. Erdös, Chao Ko and R. Rado, Intersection theorems for systems of finite sets. *Quart. J. Math. Oxford (2)* **12** (1961), 313–320.

2. A few years ago Frankl formulated the following interesting problem: Let $\{A_k\}, |A_k| = n$, be a two-chromatic clique, i.e. $|A_{k_1} \cap A_{k_2}| \neq \varnothing$ and $\bigcup_k A_k = S$ can be decomposed into the union of two disjoint sets S_1 and S_2 so that no A_k is contained in S_1 or S_2. Let $f(n)$ be the smallest integer for which there is a set B, $|B| = f(n)$, $A_i \not\subseteq B$, $B \cap A_i \neq \varnothing$. Frankl showed that $n + c\sqrt{n} < f(n) < n^2 \log n$. It would of course be very desirable to have a better estimate for $f(n)$.

3. Let $\{A_i : 1 \leq i \leq t_n\}$ be a family of n-sets which is a maximal clique, i.e. which is such that $A_i \cap A_j \neq \varnothing$ for all $1 \leq i < j \leq n$ and there is no n-set B for which $B \cap A_i \neq \varnothing$ for every $1 \leq i \leq t_n$. Füredi asked whether it is true that

$$\left| \bigcup_{i=1}^{t_n} A_i \right| \leq t_n ?$$

If true, this would be a generalization of Fischer's inequality.

Perhaps the conjecture remains true if $|A_i| = n$ is replaced by $|A_i| \leq n$.

This, if true, would be a generalization of the theorem of de Bruijn and myself.

N. G. de Bruijn and P. Erdös, On a combinatorial problem. *Indig. Math.* **10** (1948), 421–423.

4. The following problem is due to Duke and myself. Consider set systems $\{A_i : 1 \leqslant i \leqslant m\}$, $A_i \subset S$, $|S| = n$, $|A_i| = r \geqslant 3$, such that no t sets $A_{i_1}, A_{i_2}, \ldots, A_{i_t}$ intersect in the same element (i.e. the family $\{A_i : 1 \leqslant i \leqslant m\}$ contains no $\triangle$-systems of size t and kernel 1). Denote by $f(n; r, t)$ the maximal value of m. With Duke we proved that $f(n; r, t) < c(r, t)n^{r-2}$. Recently Chung and Frankl proved that

$$f(n; 3, t) = 2\binom{t}{2}(n - 2t) + 2\binom{t}{3}$$

for $n > n_0(t)$, t odd, and

$$f(n; 3, t) = (t(t - \tfrac{3}{2}) + O(1))(n - 2t + 1)$$

for t even.

Here is their construction for odd t: Consider two disjoint t element subsets X and Y of S. Our family consists of all three element subsets A_i of S which are disjoint from one of X and Y and intersect the other in at least two elements.

For $r \geqslant 4$ we can add to the above construction all r sets intersecting both X and Y in at least two elements.

Frankl and Füredi proved that for $n > n_0(r, t)$ this construction is best possible. They also proved that for every c_1 there is a c_2 such that if $r = 5$, $m > c_2 n^2$ then there are c_1 sets which form a $\triangle$-system of kernel 2. It would be desirable to determine c_2 as a function of c_1.

Here I want to remind the reader of one of my favourite problems, asked by Rado and myself more than 20 years ago. Is it true that there is an absolute constant C so that any family of C^n n-sets contains a $\triangle$-system of three sets? I offer US$1000 for a proof or disproof.

R. Duke and P. Erdös, Systems of finite sets having a common intersection. In *Proceedings of the Southeastern Conference on Combinatorics, etc.*, Congressus Num. XIX, Utilitas Mathematicae, Winnipeg, 1977, pp. 247–252.
P. Erdös and R. Rado, Intersection theorems for systems of sets. *J. London Math. Soc.* **35** (1960), 85–90.

5. Let $|S| = 2n$, $A_k \subset S$, $|A_k| = n$, $1 \leqslant k \leqslant T_n$ be a system of subsets of S. Assume that the number of pairs A_i, A_j satisfying $A_i \cap A_j = \varnothing$ is greater

than or equal to 2^{2n}. I conjectured a few years ago that this implies

$$T_n > (1 - \varepsilon)2^{n+1}.$$

Frankl recently proved my conjecture. Frankl conjectures that if $T_n > 2^n n^c$ then the number of disjoint pairs is at most $\varepsilon(c)T_n^2$, where $\varepsilon(c) \to 0$ as $c \to \infty$.
Frankl proved that if $2^{1/(s+1)} < a \leq 2^{1/s}$ and $T_n = a^n$ then there are at most

$$\frac{s-1+\varepsilon}{s}\binom{T}{2^n}$$

disjoint pairs. It is easy to see that this result is best possible. In some sense this is a phenomenon similar to the Erdös–Stone theorem.
I hope Frankl will soon publish his interesting results in detail.

6. Let $A_1, \ldots, A_m$, $|A_i| = r$, $1 \leq i \leq m$, $A_i \cap A_j \neq \emptyset$, $1 \leq i < j \leq m$. Assume that for every A_i and every proper subset B of A_i there is an A_j satisfying $A_j \cap B = \emptyset$ (i.e. the family $\{A_i\}$ is a clique but this property gets destroyed if we replace any of the sets by a smaller one). Denote by $f(r)$ the largest possible value of

$$\left|\bigcup_{i=1}^{m} A_i\right|.$$

Calczinska-Karlowitz showed that $f(r)$ is finite. After some results of Ehrenfeucht and Mycielski, we proved with Lovász that

$$\frac{1}{2}\binom{2r-1}{n-1} + 2r - 2 \leq f(r) \leq \frac{r}{2}\binom{2r-1}{r}. \tag{17}$$

By making use of a theorem of Bollobás, Tuza improved (17) to

$$2\binom{2r-4}{r-2} + 2r - 4 \leq f(r) \leq \binom{2r-1}{r-1} + \binom{2r-4}{r-2}. \tag{18}$$

Tuza conjectures that the lower bound in (18) is exact.

M. Calczinska-Karlowitz, Theorem on families of finite sets. *Bull. Acad. Polon. Sci. Ser. Math. Astr. Phys.* **12** (1964), 87–89.
A. Ehrenfeucht and J. Mycielski, Interpolation of functions over a measure space and conjectures about memory. *J. Approximation Theory* **9** (1973), 218–236.
P. Erdös and L. Lovász, Problems and results on three-chromatic hypergraphs and some related questions. In *Infinite and Finite Sets*, Coll. Math. Soc. J. Bolyai no. 10, North Holland, Amsterdam, 1975, pp. 609–627.

B. Bollobás, On generalized graphs. *Acta Math. Acad. Sci. Hungar.* **16** (1965), 447–452.

Z. Tuza, Critical hypergraphs and intersecting set-pair systems. *Discrete Math.* to appear.

7. J. Larson and I asked: Is there an absolute constant C so that for every n there is a partially balanced block design $\{A_i\}$ on a set S of n elements so that

$$|A_i| > n^{1/2} - C \tag{19}$$

holds for every i? Our guess was that such a C does not exist for all n.

I just heard that Shrikande and Singhi have proved that if our guess is wrong, i.e. if a block design exists for every n satisfying (19), then there is a finite geometry whose order is not of the form $p^2 + p + 1$, where p is a power of a prime.

As far as I know it is not impossible that there is a sequence $n_1 < n_2 < \ldots$ which is such that for every i we have $n_{i+1} - n_i < c$ and there exists a finite geometry $n_i^2 + n_i + 1$ elements. It is easy to see that in this case there is a block design for every n satisfying (19).

Is there a partially balanced block design on n elements, say $\{A_i : 1 \le i \le t_n\}$, so that for every r the number of indices i for which $|A_i| = r$ is less than $cn^{1/2}$? It is easy to see that, if true, then apart from the value of c this result is best possible, though perhaps $cn^{1/2}$ could be replaced by $cr^{1/2}$.

P. Erdös and J. Larson, On pairwise balanced block designs with the sizes of blocks as uniform as possible. *Ann. Discrete Math.* **15** (1983), 129–134.

8. Is it true that in every finite geometry of $p^2 + p + 1$ elements there is a blocking set which meets every line in fewer than C points? Bruen and Freeman showed that such geometries exist for infinitely many p. I have been informed that T. Evans stated this problem several years before me.

9. Let $\{A_k\}$ be a family of subsets of a set with n elements. When is there a set S such that

$$1 \le |A_k \cap S| \le C \tag{20}$$

for every k? Perhaps there is an S satisfying (20) provided that $|A_k| > c_1 n^{1/2}$ and $|A_{k_1} \cap A_{k_2}| \le 1$ for all k, k_1 and k_2 with $k_1 \ne k_2$. Maybe the last condition can even be weakened to $|A_{k_1} \cap A_{k_2}| \le c_2$ with $c = c(c_1, c_2)$.

The following is an old question posed by Grünbaum and myself. Let $\{A_k : 1 \le k \le T_n\}$ be a partially balanced block design on n elements. Define a graph whose vertices are the A_is. Join two A_is if they have a non-empty intersection. Is it true that the chromatic number of this graph is at most n?

To conclude this section I wish to call attention to a forthcoming book on extremal problems on set systems being written by Frankl, Füredi and Katona. I await the appearance of this book with great interest.

III. Combinatorial number theory

1. Let $a_1 < a_2 < \ldots < a_n$ be a B_2 sequence of Sidon, i.e. a sequence such that the sums $a_i + a_j$ are all distinct. Can this sequence be embedded into a perfect difference set? In other words, does there exist a sequence $b_1 < b_2 < \ldots < b_{p+1}$ such that all the differences are incongruent modulo $p^2 + p + 1$ and the a_is all occur among the b_js? (A slightly stronger requirement would be $a_i = b_i$, $1 \leq i \leq n$.) For the applications I have in mind it would suffice if there is an X and a B_2 sequence $1 \leq b_1 < \ldots < b_t \leq X$ such that $t = X^{1/2}(1 + o(1))$ and all the a_is occur among the b_js. Unfortunately I could make no progress with these problems.

2. During the International Congress in Warsaw, Pisier asked me the following question. Call a sequence $a_1 < \ldots < a_n$ *independent* if the sums

$$\sum_{i=1}^{n} \varepsilon_i a_i, \qquad \varepsilon_i = 0 \quad \text{or} \quad 1$$

are all distinct. Now let $b_1 < b_2 < \ldots$ be an infinite sequence of integers and assume that there is a $\delta > 0$ such that for every n every subsequence of n terms of the b_is contains an independent subsequence of δn terms. Is it then true that our sequence $b_1 < b_2 < \ldots$ is the union of finitely many independent sequences? (Of course, their number should depend only on δ.) Pisier is not a combinatorialist but an outstanding analyst; he would need this lemma to characterize Sidon sets. As the reader can see "all roads lead to Rome". Unfortunately, so far I have been able to make no contribution to this very interesting question.

One of my oldest problems deals with independent sequences. Let $1 \leq a_1 < a_2 < \ldots < a_t \leq n$ be independent. Is it true that

$$\max t = \frac{\log n}{\log 2} + O(1)? \tag{21}$$

Leo Moser and I proved 30 years ago by using the second moment method that

$$\max t \leq \frac{\log n}{\log 2} + \frac{\log \log n}{2 \log 2} + O(1). \tag{22}$$

As far as I know, (22) is still the best known upper bound for max t. I offer a reward of US\$500 for a proof or disproof of (21).

Here is another additive Pisier-type problem. Assume that the infinite sequence $B = \{b_1 < b_2 < \ldots\}$ has the property that for every n every subsequence of n terms of our sequence has a subsequence of δn terms which is a B_2 sequence. Is it then true that B is the union of $C_\delta B_2$ sequences? Many generalizations and extensions are possible but I have no non-trivial results.

3. The problem posed by Pisier led me to several related questions. Let G be an infinite graph. Assume that there is a $\delta > 0$ such that every set of e edges of G contains a subgraph of δe edges which do not contain a C_4. Is it then true that G is the union of C_δ graphs which do not contain a C_4? The answer is expected to be negative. C_4 could of course be replaced by any other graph H, but it is uncertain whether any non-trivial positive results can be obtained.

4. I conjectured that every sequence of integers $a_1 < \ldots < a_n$ contains a subsequence $a_{i_1} < a_{i_2} < \ldots < a_{i_r}$ which is a B_2 sequence and $r = (1 + o(1))n^{1/2}$. Komlós, Sulyok and Szemerédi proved this with $r < cn^{1/2}$.

I proved that there is a $B_2^{(2)}$ sequence $a_1 < \ldots < a_n$ (i.e. the number of solutions of $a_i + a_j = t$ is greater than or equal to 2), such that the largest B_2 subsequence of it has at most $cn^{2/3}$ terms. Is this best possible? Or can $\frac{2}{3}$ be replaced by $\frac{1}{2} + \varepsilon$, or even by $\frac{1}{2}$?

J. Komlós, M. Sulyok and E. Szemerédi, Linear problems in combinatorial number theory. *Acta Math. Acad. Sci. Hungar.* **26** (1975), 113–121.

Clearly many further problems of the Pisier type can be asked, but it is not clear whether there ever are any non-trivial positive results.

2

LARGE HYPERGRAPHS OF DIAMETER 1

J. C. Bermond, J. Bond and J. F. Saclé

ABSTRACT The design of interconnection networks leads to numerous combinatorial problems, some of which are discussed in this paper. The main problem studied here concerns the diameter of hypergraphs with bounds on the degree and the size of the edges.

1. Introduction

The problem considered here is motivated by the study of combinatorial properties of interconnection networks (which can be telecommunication networks, microprocessor networks, etc.). In most of the classical networks the connections are from point to point and, therefore, a network is modelled by a graph in which the edges represent the communication links. Now, there is some interest in considering networks in which computers share a communication medium such as a bus. (Our interest in that problem came from work with researchers at the French National Telecommunication Research Centre (CNET) on a project called REBUS.) In this case the network is well modelled by a hypergraph whose vertices correspond to the computer centres, and the edges to the buses. In the design and study of these networks several parameters are of importance: for example, message delay, message traffic density, reliability or fault tolerance, cost, etc. For technical reasons a computer must not belong to too many buses. In other words, every vertex of the hypergraph has a bounded degree (in practice 2, 3, or 4). Furthermore, the traffic load on each bus is limited. Since for general networks without symmetry properties, the traffic per bus is not easily calculated, we will restrict our study to networks with some uniformity and, in a first approximation, replace the constraint on traffic load by a limitation of the number of processors connected to each bus. This means that the size of the edges of the associated hypergraph is bounded.

We will focus our attention on the problem of minimizing the message delay: it is important that the length of the longest path that a message must travel from a vertex to another in the network (that is, the distance between the vertices) be as small as possible. In other words, the diameter of the hypergraph must be small (in practice no more than 4). The

GRAPH THEORY AND COMBINATORICS
ISBN 0-12-111760-X

problem is to construct or design an interconnection network satisfying the two constraints above (on the degree of a vertex and on the size of the edges), with a small diameter D and having the maximum number of vertices.

The first case to study is the case of diameter 1, and we will see that it gives rise to interesting problems of hypergraphs or set systems. For more details on the problem and the results obtained on it, we refer to the recent survey by Bermond, Bond, Paoli and Peyrat [2]. The case of diameter $D > 1$ will be considered by the authors in a later article.

2. Notation and definitions

We shall follow the terminology of Berge's book [1]. $H = (X, E)$ will denote a *hypergraph* with vertex set X and edge set E. The degree of a vertex is the number of edges containing it. Δ will denote the *maximum degree* of all vertices of H. r will denote the *maximal size of an edge*; if all the edges have size r, the hypergraph is r-uniform (also sometimes called an r-graph in the literature). A *path* in a hypergraph connecting x and y is a sequence $x = x_1, E_1, x_2, \ldots, x_i, E_i, x_{i+1}, \ldots, E_p, x_{p+1} = y$ with $\{x_i, x_{i+1}\} \subset E_i$. Its length will be p, the number of its edges. The distance between x and y is the length of a shortest path between them. The *diameter* of the hypergraph is the maximum of the distances over all pair of vertices. Therefore, *a hypergraph is of diameter 1 if any pair of vertices belongs to at least one edge.*

3. The $(\Delta, 1, r)$-hypergraph problem

Let us call a hypergraph of diameter D, maximum degree Δ and maximum size of the edges r, a (Δ, D, r)-*hypergraph*. The problem considered in the introduction is to determine the maximum number $n(\Delta, D, r)$ of vertices of a (Δ, D, r)-hypergraph. Here we restrict our attention to $D = 1$; therefore we have

MAIN PROBLEM   Determine the maximum number $n(\Delta, 1, r)$ of vertices of a hypergraph with maximum degree Δ, maximal edge-size r and such that every pair of vertices belongs to at least one edge.

PROPOSITION 1 $n(\Delta, 1, r) \leq 1 + \Delta(r - 1)$.

PROOF   Choose a vertex x and consider the edges containing it; they must contain all the other vertices. There are at most Δ such edges and each contains x and at most $(r-1)$ other vertices. $\square$

The reader may have seen the similarity of our problem to problems in design theory. Let us recall that a (v, r, λ)-BIBD (balanced incomplete block design) is a collection of subsets, called blocks, of a given set on v vertices such that every block contains exactly r vertices and every pair of vertices belongs to exactly λ blocks. Furthermore, in such a BIBD, every vertex belongs to the same number of blocks:

$$\Delta = \lambda(v-1)/(r-1).$$

PROPOSITION 2 $n(\Delta, 1, r) = 1 + \Delta(r-1)$ *if and only if there exists a* $(v, r, 1)$-*design with* $v = 1 + \Delta(r-1)$.

PROOF   One can consider the blocks of the design as the edges of a hypergraph and take $\lambda = 1$. Then a $(v, r, 1)$-design with $v = 1 + \Delta(r-1)$ is a $(\Delta, 1, r)$-hypergraph. Conversely, in view of the proof of Proposition 1, if $n(\Delta, 1, r) = 1 + \Delta(r-1)$, then every pair of vertices belongs to exactly one edge, and the size of each edge is exactly r. $\square$

REMARK   Clearly, many results of design theory will help to obtain examples of $(\Delta, 1, r)$-hypergraphs which have the maximum number of vertices. For these results we refer the reader to Hanani [4] or Doyen and Rosa [3].

Note also that by Fisher's inequality we always have $\Delta \geqslant r$. However, in practice Δ is often at most 4 and r can be of the order of 15–20; as a consequence, results concerning designs will not be readily applicable.

Before we proceed further, we pause to give a dual formulation of the problem, in terms of intersecting families, a subject considered by Erdös.

DUAL FORMULATION   Consider the dual hypergraph $H^* = (E^*, X^*)$ of H where the vertices of H^* represent the edges of H and the edges of H^* represent the vertices of H and where $e_j^* \in X_i^*$ if and only if $x_i \in E_j$ in H. Then the maximum degree of H^* is equal to r, and the maximum size of an edge of H^* to Δ. Furthermore, the property that H is of diameter 1 corresponds to the fact that in H^* any pair of edges intersects in at least one point; a family of edges having this property is usually called an intersecting family. Therefore our problem can be expressed as follows:

Find the maximum number of edges of an intersecting family of edges, with each edge of size at most Δ and each vertex of degree at most r.

Note that without the restriction on the maximum degree, but with a fixed number of vertices, this problem has been extensively studied.

4. The case of r small

Let us recall that in the case $\Delta = r$ the problem of existence of a $(v, r, 1)$-design with $v = 1 + \Delta(r-1) = (r-1)^2 + (r-1) + 1$ is equivalent to the existence of a projective plane of $q^2 + q + 1$ points $(q = r - 1)$. It is well known that such a plane exists when q is a power of a prime. Furthermore, it is known that such a plane does not exist when q is congruent to 1 or 2 (mod 4) and cannot be written as the sum of two squares of integers – for example $q = 6$ or $q = 14$.

In the case $r = 2$ we have, in fact, a graph of diameter 1, namely the complete graph, and the bound is always attained.

By using results on coverings we will solve the case $r = 3$, and quite completely the case $r = 4$.

PROPOSITION 3

$$n(\Delta, 1, 3) = 2\Delta + 1 \, if \, \Delta \equiv 0 \; or \; 1 \pmod 3$$

and

$$n(\Delta, 1, 3) = 2\Delta \; if \; \Delta \equiv 2 \pmod 3.$$

PROOF By Proposition 2 we have seen that the value $1 + 2\Delta$ can be reached if and only if a $(v, 3, 1)$-design exists with $v = 1 + 2\Delta$. Such a design is known as a Steiner triple system. A Steiner triple system exists if and only if $v \equiv 1, 3 \pmod 6$, which means that $\Delta \equiv 0, 1 \pmod 3$ (see any textbook on combinatorics). It remains to consider the case $\Delta \equiv 2 \pmod 3$, where the value $1 + 2\Delta$ cannot be achieved, so we have $n(\Delta, 1, 3) \leqslant 2\Delta$ for $\Delta \equiv 2 \pmod 3$. This value can easily be reached from an $n(\Delta - 1, 1, 3)$ system, having $2(\Delta - 1) + 1 = 2\Delta - 1$ points. Add to this system a new vertex 2Δ and the edges $(2i - 1, 2i, 2\Delta)$, for $1 \leqslant i \leqslant \Delta - 1$, and $(2\Delta - 1, 2\Delta)$. □

PROPOSITION 4

   (i) $n(\Delta, 1, 4) = 3\Delta + 1$ *if* $\Delta \equiv 0, 1 \pmod 4$;
   (ii) $n(\Delta, 1, 4) = 3\Delta$ *if* $\Delta \equiv 2 \pmod 4$;
   (iii) $n(\Delta, 1, 4) = 3\Delta$ *or* $3\Delta - 1$ *if* $\Delta \equiv 3 \pmod 4$.

PROOF By Proposition 2, we have $n(\Delta, 1, 4) = 3\Delta + 1$ if and only if there exists an $(n, 4, 2)$-BIBD; it is known (see [4]) that such systems exist if and only if $n \equiv 1$ or $4 \pmod{12}$, in which case $\Delta \equiv 0$ or $1 \pmod{4}$. Therefore we have part (i) of the proposition, and that if $\Delta \equiv 2$ or 3 $\pmod{4}$, $n(\Delta, 1, 4) \leq 3\Delta$.

The existence, for $\Delta \equiv 2 \pmod{4}$, of a $(\Delta, 1, 4)$-hypergraph with 3Δ vertices will follow from a more general result on coverings. Recall that a covering of the pairs by r-tuples consists of a family of r-tuples such that every pair of vertices belongs to at least one of them. A theoretical bound for the number of r-tuples is $m \geq \lceil (n/r) \lceil (n-1)/(r-1) \rceil \rceil$. This bound is obtained by considering that each vertex belongs to at least $\lceil (n-1)/(r-1) \rceil$ r-tuples. If $(n/r) \lceil (n-1)/(r-1) \rceil$ is an integer, and if there exists a covering with this number of r-tuples, then necessarily each vertex belongs to exactly $\lceil (n-1)/(r-1) \rceil$ r-tuples. Therefore, in this case, we have a $(\Delta, 1, r)$-hypergraph on n vertices with $\Delta = \lceil (n-1)/(r-1) \rceil$. In the case $r = 4$, we know by the work of various authors (see [5]) that there exists a covering of pairs with $\lceil (n/r) \lceil (n-1)/(r-1) \rceil \rceil$ quadruples.

If we take $n = 12t + 6$, then $((12t + 6)/4) \lceil (12t + 5)/3 \rceil$ is an integer and there exists a $(4t + 2, 1, 4)$-hypergraph with $12t + 6 = 3\Delta$ vertices (therefore proving (ii)).

If we take $n = 12t + 8$, then $((12t + 8)/4) \lceil (12t + 7)/3 \rceil$ is an integer and there exists a $(4t + 3, 1, 4)$-hypergraph with $12t + 8 = 3\Delta - 1$ vertices, therefore proving (iii). $\square$

REMARK We shall see later that $n(3, 1, 4) = 8$, which is $3\Delta - 1$. However, one of us has recently proved that $n(\Delta, 1, 4) = 3\Delta$ for $\Delta \equiv 7 \pmod{12}$, by using Kirkman triple systems (see Bond's thesis), and it might be possible that $n(\Delta, 1, 4) = 3\Delta$ for $\Delta \equiv 3 \pmod{4}$, $\Delta > 3$.

5. The case of Δ small

It is this case which is the most interesting in practice. First, as we have seen in Proposition 2, if $\Delta = r$ then $n(\Delta, 1, \Delta) = \Delta^2 - \Delta + 1$ if and only if there exists a $(\Delta^2 - \Delta + 1, 1, \Delta)$ design, which is in fact a projective plane of order q, where $\Delta = q + 1$. It is known that such a projective plane exists if q is a prime power. From this projective plane one can easily construct $(\Delta, 1, r)$-hypergraphs for $r \geq \Delta$, which will turn out, at least for $\Delta \leq 4$, to have the maximum number of vertices.

PROPOSITION 5 *If there exists a projective plane of order q, then for $\Delta = q + 1$ there exists a $(\Delta, 1, r)$-hypergraph for $r \geq \Delta$, with $f(\Delta) = \Delta r - (\Delta - 1) \lceil r/\Delta \rceil$ vertices, that is $n(\Delta, 1, r) \geq \Delta r - (\Delta - 1) \lceil r/\Delta \rceil$.*

PROOF   For $r = \Delta$ let us simply consider the projective plane with $\Delta^2 - \Delta + 1$ vertices. Otherwise let $r = t\Delta - \alpha$, with $t \geq 2$ and $\alpha \leq \Delta - 1$. Let x_0 be any vertex of the projective plane and let $L_1, L_2, \ldots, L_\alpha$ be any α lines containing x_0. Replace each vertex x by a set S_x of cardinality $t - \alpha$ if $x = x_0$, $t - 1$ if x is a vertex of L_i ($1 \leq i \leq \alpha$) different from x_0, and t otherwise. Replace a line L containing the vertices $x_1, x_2, \ldots, x_\Delta$ by an edge containing the sets $S_{x_1}, S_{x_2}, \ldots, S_{x_\Delta}$. If the line L contains x_0, then the edge obtained is at most of size $t\Delta - \alpha$. Otherwise the line L intersects each of the α lines L_i ($1 \leq i \leq \alpha$) and again its size is at most $t\Delta - \alpha$.

The degree of any vertex of S_x is that of x, that is, Δ. Furthermore, if we consider two vertices, either they belong to the same set S_x and therefore to many edges, or they belong to two different sets S_x and S_y, and, therefore, to the edge associated with the line containing x and y. The total number of vertices is

$$t(\Delta^2 - \Delta + 1) - \alpha - \alpha(\Delta - 1) = \Delta(t\Delta - \alpha) - (\Delta - 1)t = \Delta r - (\Delta - 1)\lceil r/\Delta \rceil.$$

Thus we have constructed a hypergraph with edge-size at most r having $f(\Delta)$ vertices. (This hypergraph always has $\Delta^2 - \Delta + 1$ edges.) $\square$

PROPOSITION 6   *If $\Delta \leq 4$ then $n(\Delta, 1, r) \leq \Delta r - (\Delta - 1)\lceil r/\Delta \rceil = f(\Delta)$.*

Before embarking on the proof, which will be somewhat long, in particular for $\Delta = 4$, let us remark that this bound is different from the bound of Proposition 1 and that, with Proposition 5, we have $(1, 2, 3$ being prime)

THEOREM 1

   (i) $n(2, 1, r) = 2r - \lceil r/2 \rceil$;
   (ii) $n(3, 1, 4) = 3r - 2\lceil r/3 \rceil$;
   (iii) $n(4, 1, r) = 4r - 3\lceil r/4 \rceil$.

PROOF OF PROPOSITION 6   The proof is somewhat similar for the cases $\Delta = 2, 3, 4$ and therefore let us give first some lemmas concerning the cardinality of the intersection of k edges.

Let M_k denote the maximum cardinality of the intersection of any k edges of a $(\Delta, 1, r)$-hypergraph. Then $r \geq M_1 \geq M_2 > \ldots \geq M_\Delta \geq 1$. In what follows let $f(\Delta) = \Delta r - (\Delta - 1)\lceil r/\Delta \rceil$. Let n be the number of vertices of the hypergraph.

LEMMA 1 $M_{k+1} \geq \lceil M_k/\Delta \rceil$; *in particular $M_2 \geq \lceil r/\Delta \rceil$.*

PROOF   Let $E_1, E_2, \ldots, E_k$ be k edges whose intersection A is of cardinality M_k and let x be a vertex not belonging to any E_i. Then x is joined to the points of A by at most Δ edges and one of these edges intersects A in at least $|A|/\Delta$ vertices. Therefore $M_{k+1} \geq |A|/\Delta = M_k/\Delta$. $\square$

LEMMA 2   If $M_\Delta \geq \lceil r/\Delta \rceil$, then $n \leq f(\Delta)$.

PROOF   Let us consider Δ edges intersecting in M_Δ vertices, any vertex of the intersection having a degree Δ. These edges cover all the vertices of the hypergraph. Therefore $n \leq \Delta r - (\Delta - 1)M_\Delta \leq \Delta r - (\Delta - 1)\lceil r/\Delta \rceil = f(\Delta)$. $\square$

COROLLARY 1   For $\Delta = 2$, $n(2, 1, r) \leq f(2) = 2r - \lceil r/2 \rceil$. $\square$

Indeed, by Lemma 1, $M_2 \geq \lceil r/2 \rceil$ and, by Lemma 2, $n \leq f(2)$.

LEMMA 3   If $M_2 \geq \lceil (\Delta - 1)r/\Delta \rceil$, then $n \leq f(\Delta)$.

PROOF   Consider two edges E_1 and E_2 intersecting in M_2 vertices and consider a given vertex x in the intersection. Then the edges E_1, E_2 and the $\Delta - 2$ other edges containing x cover all the vertices. Therefore $n \leq \Delta r - M_2 - (\Delta - 2) \leq \Delta r - \lceil (\Delta - 1)r/\Delta \rceil - (\Delta - 2) \leq f(\Delta)$. $\square$

LEMMA 4   Let $E_1, E_2, \ldots, E_k$ be k edges intersecting in M_k vertices and let C_k be the set of vertices not belonging to any of the E_i. If $M_{k+1} < M_k$, then $|C_k| \leq (\Delta - k)^2 M_2$.

PROOF   Let B_k be the intersection of E_i $(i = 1, \ldots, k)$, and let x be a vertex in B_k; x belongs to at most $(\Delta - k)$ other edges which intersect C_k. Therefore, one of them, say F, intersects C_k in at least $|C_k|/(\Delta - k)$ vertices. If $M_k > M_{k+1}$, $B_k \not\subseteq F$, then there exists a vertex y in $B_k - F$ and an edge F' containing y which also intersects C_k in at least $|C_k|/(\Delta - k)$ vertices. But then $|F \cap F'| \geq |C_k|/(\Delta - k)^2$ and $M_2(\Delta - k)^2 \geq |C_k|$. $\square$

LEMMA 5   If $M_{\Delta - 1} \geq \lceil r/\Delta \rceil$, then $n \leq f(\Delta)$.

PROOF   By Lemma 2 we can suppose $M_\Delta < \lceil r/\Delta \rceil$. Therefore $M_\Delta < M_{\Delta - 1}$ and, by Lemma 4, $|C_{\Delta - 1}| \leq M_2$. By Lemma 3, $M_2 \leq \lceil (\Delta - 1)r/\Delta \rceil - 1$. Let $E_1, E_2, \ldots, E_{\Delta - 1}$ be $\Delta - 1$ edges intersecting in $M_{\Delta - 1}$ vertices. Then

$$n \leq |E_1 \cup E_2 \cup \ldots \cup E_{\Delta - 1}| + |C_{\Delta - 1}|,$$

and we have

$$n \leq (\Delta - 1)r - (\Delta - 2)M_{\Delta - 1} + |C_{\Delta - 1}|.$$

so

$$n \leqslant (\Delta - 1)r - (\Delta - 2)\lceil r/\Delta \rceil + \lceil (\Delta - 1)r/\Delta \rceil - 1,$$

and

$$n \leqslant f(\Delta) - r + \lceil r/\Delta \rceil + \lceil (\Delta - 1)r/\Delta \rceil - 1 = f(\Delta). \quad \square$$

COROLLARY 2   For $\Delta = 3$, $n(3, 1, r) \leqslant f(3) = 3r - 2\lceil r/3 \rceil$.

PROOF   By Lemma 1, $M_2 \geqslant \lceil r/\Delta \rceil$ and, by Lemma 2, $n \leqslant f(3)$. $\quad \square$

Now we prove the case $\Delta = 4$.

By Lemma 1, $M_2 \geqslant \lceil r/4 \rceil$. By Lemma 3 and Lemma 2 we can suppose (otherwise the proof is finished) that $M_2 < \lceil 3r/4 \rceil$ and that $M_3 < \lceil r/4 \rceil$. Let E_1 and E_2 be two edges intersecting in M_2 points. Let $A = E_1 \cup E_2$ and $B = E_1 \cap E_2$ and let $C = X - A$. By Lemma 4, we have

$$|C| \leqslant 4M_2. \tag{1}$$

Furthermore, $n \leqslant |A| + |C| \leqslant 2r - M_2 + |C|$; that is,

$$n \leqslant 2r + 3M_2, \tag{2}$$

which can be expressed as

$$M_2 \geqslant \lceil (n - 2r)/3 \rceil. \tag{2'}$$

Let F_1 be an edge intersecting C and B chosen in such a way that $F_1 \cap B$ is as large as possible. Let $B_1 = F_1 \cap B$; $C_1 = f_1 \cap C$; $\bar{C}_1 = C - C_1$. If $x \in B_1$, then the fourth edge containing x must contain $\bar{C}_1$. If there exist two distinct edges intersecting B_1 and $\bar{C}_1$, then $|\bar{C}_1| \leqslant M_2$ and $n \leqslant |A| + |F_1| - |B_1| + |\bar{C}_1| \leqslant 2r - M_2 + r - 1 + M_2 = 3r - 1 \leqslant f(4)$.

Therefore, we can suppose (otherwise the proof is finished) that there exists one edge F_1' containing B_1 and $\bar{C}_1$. By the maximality of B_1, $F_1' \cap B = B_1$. Let $C_1' = F_1' \cap C$, then $\bar{C}_1 \subset C_1'$. As $B_1 \neq B$, let F_2 be an edge intersecting $B - B_1$ and C, and chosen in such a way that $F_2 \cap B$ is as large as possible. Let $B_2 = F_2 \cap B$. As above there exists another edge F_2' containing B_2.

We can construct a partition of B into sets B_i, $2 \leqslant i \leqslant k$, where $k \leqslant 4$ (because the degree of a vertex is at most 4). Let F_i and F_i' be such that $B_i = F_i \cap B = F_i' \cap B$. Let $C_i = F_i \cap C$; $C_i' = F_i' \cap C$; $C_i \cup C_i' = C$. Therefore

$$|C| \leqslant 2r - 2|B_i| \leqslant 2r - 2|B_1|. \tag{3}$$

To finish the proof we consider the three possible values of k.

Case 1: $k = 4$ The four edges containing a vertex z of C intersect B. One of them, say F, intersects $A = E_1 \cup E_2$ in at least $|A|/4 = (n - |C|)/4$ vertices. Let x be a vertex of $B \cap F$. By considering the four edges containing x, we obtain $n \leqslant 2r - M_2 + r - (n - |C|)/4 + r - 1 = 4r - n/4 - 1 + (|C|/4 - M_2)$ and, by (1), $n \leqslant 4r - n/4 - 1$ or $n \leqslant \lfloor (16r - 4)/5 \rfloor$. A calculation shows that $\lfloor (16r - 4)/5 \rfloor \leqslant f(4)$, except if $r = 5$ or 9.

If $r = 5$, $f(4) = 14$ and we cannot have $n = 15$ because $M_2 \leqslant \lceil 15/4 \rceil - 1 = 3$ and, therefore, we cannot partition B into four disjoint sets.

If $r = 9$, $f(4) = 27$, suppose that $n = 28$. Then, by (2'), $M_2 \geqslant 4$ and $|C| \geqslant n - |A| \geqslant 14$. But, as above, $|C| \leqslant 2r - (n - |C|)/4 - 1 = 10 + |C|/4$. Therefore $|C| \leqslant 40/3$. Since $|C| \leqslant 13$ is not possible, $n \neq 28$.

Case 2: $k = 3$ In this case $|B_1| \geqslant M_2/3$. Any edge intersecting B and C intersects C in at most $2M_2$ vertices, otherwise one of the other two edges intersecting B intersects it in $M_2 + 1$ vertices. Without loss of generality, let us suppose that $|C_1| \geqslant |C|/2$ (otherwise interchange F_1 and F_1') and $|C_2| \geqslant |C|/2$. Then there exists $z \in C_1 \cap C_2$ (if $|C_1| = |C|/2$ and $C_1 \cap C_2 = \varnothing$ choose C_1' instead of C_1). The third edge containing z and B_3 has at least $|C| - 2M_2$ vertices in C. Let $A_i = F_i \cap A$; then $|A_1| + |A_2| + |A_3| \leqslant 2(r - |C|/2) + r - |C| + 2M_2 = 3r - 2|C| + 2M_2$. The fourth edge intersects A in at most $2M_2$ vertices. Therefore $|A| \leqslant 3r - 2|C| + 4M_2$ and

$$n \leqslant 3r - |C| + 4M_2. \qquad (4)$$

On the other hand, as F_i and F_i' cover C, $|C| \leqslant 2r - |A_i| + |B_i|$ for every i and $3|C| \leqslant 6r - (|A_1| + |A_2| + |A_3|) - M_2$. Combining with $|A| \leqslant |A_1| + |A_2| + |A_3| + 2M_2$ and $|A| + |C| = n$ we obtain $n + 2|C| \leqslant 6r + M_2$, or

$$n \leqslant 6r - 2|C| + M_2. \qquad (5)$$

Recall that

$$n = |A| + |C| \leqslant 2r - M_2 + |C|. \qquad (6)$$

Then (4) and (6) give $2n \leqslant 5r + 3M_2$, (5) and (6) give $3n \leqslant 10r - M_2$ and, therefore, $11n \leqslant 35r$, or $n \leqslant \lfloor 35r/11 \rfloor$. In general $\lfloor 35r/11 \rfloor \leqslant f(4)$, except for $r = 5, 6, 9, 13, 17$. These cases can be dealt with directly by looking more closely at the structure.

Case 3: $k = 2$ Now $|B_1| \geqslant M_2/2$ and $|C| \leqslant 2r - M_2$ by (3). Therefore, $n = |A| + |C| \leqslant 4r - 2M_2$. Combining with (2), $n \leqslant 2r + 3M_2$, we obtain $5n \leqslant 16r$, or $n \leqslant \lfloor 16r/5 \rfloor$. In general, $\lfloor 16r/5 \rfloor \leqslant f(4)$, except for $r = 5, 6, 9, 10, 13, 17, 21, 25$. There, again, a careful examination of the structure proved that the values $n = f(4) + 1$ or $f(4) + 2$ are impossible. $\square$

6. Another construction

We can construct $(\Delta, 1, r)$-hypergraphs having a large number of vertices by using composition methods, as in design theory. Let us give an example. Suppose there exists a $(t, 1, r)$-hypergraph on p vertices; we will construct a $(\Delta, 1, r)$-hypergraph on pk vertices as follows.

Let us partition the pk vertices into k sets X_i, $i = 1, \ldots, k$, each of size p. Consider a covering of the pairs of vertices (x_i, x_j) with $x_i \in X_i$, $x_j \in X_j$ and $i \neq j$ with sets of size r in such a way that each vertex belongs to at most h such sets. The union of these sets and of the edges of k $(t, 1, r)$-hypergraphs constructed on each X_i forms a $(\Delta, 1, r)$-hypergraph on pk vertices with $\Delta = h + t$.

As a particular case, let $k = r$. In this case we know that there exists a partition of the pairs of vertices x_i, x_j with $x_i \in X_i$, $x_j \in X_j$, $i \neq j$, with sets of size r if and only if there exist $r - 2$ orthogonal Latin squares of order p. In such a case every vertex belongs to exactly p edges (i.e. $h = p$) and we have

PROPOSITION 7   *If there exists a $(t, 1, r)$-hypergraph on p vertices and $(r - 2)$ orthogonal Latin squares of order p, then there exists a $(t + p, 1, r)$-hypergraph on pr vertices.* $\square$

For example, this can be applied with $r = 4$, $p \neq 2, 6$, because there exist two orthogonal Latin squares of any order $p \neq 2, 6$. If $r = 5$, we can apply Proposition 7 with $p \neq 2, 3, 6, 10, 14$, as these are the only values of p for which three orthogonal Latin squares do not exist.

References

1. C. Berge, *Graphs and hypergraphs*, North Holland, Amsterdam, 1973.
2. J. C. Bermond, J. Bond, M. Paoli and C. Peyrat, Graphs and interconnection networks: diameter and vulnerability. In *Surveys in Combinatorics*, London Mathematical Society Lecture Notes No. 82, Cambridge University Press, Cambridge, 1983, pp. 1–30.
3. J. Doyen and H. Rosa, Updated bibliography and survey of Steiner systems. *Ann. Discrete Math.* **7** (1980), 317–349.
4. H. Hanani, Balanced incomplete block designs and related designs. *Discrete Math.* **11** (1975), 255–369.
5. W. H. Mills, Covering designs I: Coverings by a small number of subsets. *Ars Combinat.* **8** (1979), 199–315.

Note added in proof

We conjectured that $f(\Delta) = \Delta r - (\Delta - 1)\lceil r/\Delta \rceil$. P. Frankl has drawn our attention to the following result of Z. Füredi: Maximum degree and fractional matchings in uniform hypergraphs, *Combinatorica* **1**, (1981), 155–162 (Corollary 3). In our notation, this result can be stated as: $n(\Delta, 1, r) \leq r(\Delta^2 - \Delta + 1)/\Delta$ and if the hypergraph does not contain a projective plane of order $\Delta - 1$, then $n(\Delta, 1, r) \leq (\Delta - 1)r$; that seems to imply the conjecture at least for $r \geq (\Delta - 1)^2$.

ON SOME TURÁN AND RAMSEY NUMBERS FOR C_4

A. Bialostocki and J. Schönheim

ABSTRACT It is shown that the maximum of n, for which the edges of K_n can be partitioned into three classes, none of which contain a circuit of length 4, is 10. The result is obtained by establishing, on the one hand, that the maximum number of edges of a graph having 11 vertices and not containing any circuit of length 4 is 18, hence smaller than the biggest class in any three-colouring of the edges of K_{11}; and on the other hand, giving a required partition into three classes of the edges of K_{10}.

1. Definitions and notation

We shall use the symbols $T(n, G)$ and $R_k(G)$ for Turán and Ramsey numbers respectively. $T(n, G)$ is the maximum number of edges which a graph with n vertices can have and not contain any subgraph isomorphic to G. $R_k(G)$ is the smallest number n, such that colouring the edges of K_n arbitrarily with k colours, some colour must contain a subgraph isomorphic to G. The number of edges of a graph G will be denoted by $e(G)$.

2. Introduction

The circuit of length 4, C_4, being a small graph, has been considered for a long time in extremal graph theory and Ramsey theory as well. Extremal graph theorists [1, 3] have shown that $T(n, C_4)$ is asymptotically $\frac{1}{2}n^{3/2}$; while in Ramsey theory it is known [2] that $R_2(C_4) = 6$. No exact results are known in generalized Ramsey theory, i.e. for $k > 2$, $R_k(C_4)$ is not known. However, one does know the bounds [4, 6]

$$k^2 - k + 2 \leqslant R_k(C_4) \leqslant k^2 + k + 1,$$

where in the first inequality $k - 1$ is supposed to be a prime power. For $k = 3$, this gives $8 \leqslant R_3(C_4) \leqslant 13$.

We shall prove in this paper that $R_3(C_4) = 11$. The proof will be given by, on the one hand, exhibiting a partition of the edges of K_{10} into three C_4-free classes, showing that $R_3(C_4) > 10$; on the other hand, by proving the upper bound to be 11.

GRAPH THEORY AND COMBINATORICS
ISBN 0-12-111760-X

The attempt to prove this upper bound led us to investigate the exact value of $T(11, C_4)$. We found that the size of K_{11} is 55, so in any three-partition there is a class containing at least $\lceil 55/3 \rceil = 19$ edges. So $R_3(C_4) \leqslant 11$ would be established if $T(11, C_4) < 19$. In fact, this turns out to be true, as a consequence of $T(10, C_4) = 16$, which we shall also prove.

The above argument can be regarded as an application of a slightly stronger theorem than Theorem 3 of Section 1 in [5].

3. On $T(10, C_4)$ and $T(11, C_4)$

PROPOSITION 1   *If a graph G with 10 vertices not containing any C_4 as a subgraph has more than 15 edges, then the maximal degree Δ of its vertices is 4.*

PROOF $\Delta < 4$ implies $e(G) \leqslant 15$. If $\Delta = 4 + \alpha$, $0 < \alpha \leqslant 9$, let v be a vertex with degree Δ. Let A be the induced subgraph on v and its neighbours, C the induced subgraph on the remaining vertices and B the bipartite graph (A, C). Then, as a consequence of C_4-freeness,

$$e(A) \leqslant 4 + \alpha + \left\lfloor \frac{4 + \alpha}{2} \right\rfloor = 6 + \alpha + \left\lfloor \frac{\alpha}{a} \right\rfloor,$$

$$e(B) \leqslant 5 - \alpha,$$

$$e(C) \leqslant \beta(\alpha),$$

where $\beta(\alpha) = 4, 3, 1, 0, 0$ for $\alpha = 1, 2, 3, 4, 5$. Therefore,

$$e(G) \leqslant 11 + \left\lfloor \frac{\alpha}{2} \right\rfloor + \beta(\alpha) \leqslant 15. \quad \square$$

PROPOSITION 2   *If a graph with 10 vertices not containing any C_4 as a subgraph has maximal degree $\Delta = 4$, then,*
   (i) *It has at most 16 edges.*
   (ii) *The only extremal graphs are*

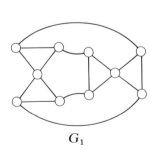 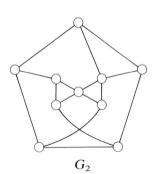

and

G_1 G_2

PROOF Admitting in the argument of the proof of Proposition 1 the value $\alpha = 0$, then $\beta(0) = 6$ and therefore $e(G) \leqslant 17$. But the value 17 cannot be attained since this could occur only if $e(A)$, $e(B)$, $e(C)$ are all maximal, and this is impossible since $e(C) = 6$ implies $e(B) < 5$. This proves (i).

For (ii), notice that, by the above implication, the only ways to have $e(G) = 16$ are $e(A) = 6$, $e(B) = 4$, $e(C) = 6$ and $e(A) = 6$, $e(B) = 5$, $e(C) = 5$. The first possibility leads uniquely to the graph G_1, while the second leads to G_2. The last assertion follows from the fact that in this case C must be C_5, otherwise $e(B) < 5$. □

THEOREM 1 (i) $T(10, C_4) = 16$.
(ii) *The set of extremal graphs is* $\{G_1, G_2\}$.

PROOF The theorem is a corollary of Proposition 1 and 2. □

THEOREM 2 $T(11, C_4) = 18$.

PROOF Clearly $T(11, C_4) \geqslant 18$, since one can add a vertex of degree 2 to G_1 or G_2 without having a C_4. We shall prove that $T(11, C_4) < 19$.

Suppose there is a graph G not containing any C_4 as a subgraph with $e(G) \geqslant 19$. Then deleting some edges we get $H \subset G$, $e(H) = 19$. Such a graph must have a vertex of degree 3. Indeed, if all edges have degree greater than 3, then $e(H) \geqslant 22$. So, there are vertices with degree of at most 3. But all those vertices v have to have degree 3. Otherwise, $H - v$ would contradict Theorem 1.

Thus, let v have degree 3 and consider $H \backslash v$. This graph, having 10 vertices and 16 edges, can be only G_1 or G_2 of Proposition 1.

We accomplish the proof by showing that it is impossible to add to G_i, $i = 1, 2$, a vertex of degree 3 and not to have a C_4.

Observe that at least two edges incident to v have to have the second end-point in A or C. For G_1, both A and C are isomorphical to

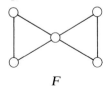

F

For G_2, A is as above and C is C_5. Adding a vertex of degree 2 to F one obtains a C_4; hence two edges of v end up in C_5, the end-points being necessarily neighbours in C_5. Then the third edge emerging from v creates a C_4. □

4. On $R_3(C_4)$

THEOREM 3 $R_3(C_4) = 11$.

PROOF   The three graphs below are edge disjoint, their union is K_{10} and none contains a C_4. Therefore $R_3(C_4) > 10$.

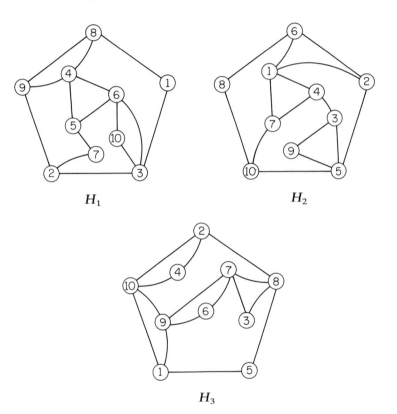

H_1 H_2

H_3

On the other hand, such a partition is impossible for K_{11}, since in any partition of the edges of K_{11} into three classes, at least one class contains at least 19 edges, and this class has to contain a C_4, by Theorem 2.   □

REMARK   Notice that H_1, H_2, H_3 are not isomorphic. We were unable to decide whether such a partition is possible with isomorphic parts. However, Christopher Clapham has exhibited such a partition since the Conference.

References

1. B. Bollobás, *Extremal Graph Theory*, Academic Press, London, 1978, p. 314.
2. B. Bollobás, *Graph Theory – An Introductory Text*, Springer Verlag, New York, Berlin and Heidelberg, 1979, p. 120.
3. P. Erdös, On sequences of integers no one of which divides the product of the others and some related problems. *Mitt. Forschungsinst. Math. Mech. Tomsk* **2** (1938), 74–82.
4. R. K. Chung and R. L. Graham, On multicolor Ramsey numbers for complete bipartite graphs. *J. Combinat. Theory B* **18**, 164–169.
5. R. L. Graham, B. L. Rothschild and J. H. Spencer, *Ramsey Theory*, John Wiley, New York, 1980.
6. R. W. Irving, Generalized Ramsey numbers for small graphs. *Discrete Math.* **9** (1974), 251–264.

4

THE EVOLUTION OF SPARSE GRAPHS

Béla Bollobás

ABSTRACT The paper contains a detailed study of random graphs with not more than $cn \log n$ edges, where c is a constant. Among other things, it is shown that almost every graph process is such that the first edge increasing the minimum degree to 2 ensures a Hamilton cycle. The diameter and connectivity are also examined in detail.

0. Introduction

The theory of random graphs was founded by Erdös and Rényi in a series of papers [9–14]. In order to recall some of the classical results due to Erdös and Rényi we introduce some definitions. A *graph process* is a nested sequence of graphs $G_0 \subset G_1 \subset \ldots \subset G_N$ with vertex set $V = \{1, 2, \ldots, n\}$ such that G_t has precisely t edges. $\left(\text{We shall follow the notation and terminology in [2] and [3]; in particular, } N \text{ stands for } \binom{n}{2}.\right)$ We shall write $\tilde{G} = (G_t)_0^N$ for a particular graph process; G_t is the *state* of $\tilde{G}$ at time t or simply the *graph* at time t. The collection of graph processes is $\tilde{\mathscr{G}}$; note that $|\tilde{\mathscr{G}}| = N!$.

We can view $\tilde{\mathscr{G}}$ as a probability space by giving all members of $\tilde{\mathscr{G}}$ the same probability. A *random graph process* refers to a member of this probability space. We say that *almost every* (a.e.) graph process has a property Q is the probability that $\tilde{G} \in \tilde{\mathscr{G}}$ has Q tends to 1 as $n \to \infty$.

The space $\tilde{\mathscr{G}}$ contains all the information about the spaces $\mathscr{G}_M = \mathscr{G}(n, M)$ of random graphs with vertex set V and M edges: the map $\tilde{\mathscr{G}} \to \mathscr{G}(n, M)$ given by $\tilde{G} = (G_t)_0^N \to G_M$ is measure preserving.

Following Erdös and Rényi we can think of a random graph process $\tilde{G}$ as a living organism which at time 0 is just the empty graph with vertex set $V = \{1, 2, \ldots, n\}$ and which develops by acquiring edges in a random fashion. Erdös and Rényi proved that a great variety of graph properties appear rather suddenly: for example if $\omega(n) \to \infty$ and

$$M_i = M_i(n) = \left\lfloor \frac{n}{2} (\log n + (-1)^i \omega(n)) \right\rfloor$$

then a.e. $\tilde{G} = (G_t)_0^N$ is such that G_{M_1} is disconnected but G_{M_2} is connected.

GRAPH THEORY AND COMBINATORICS
ISBN 0-12-111760-X

Erdős and Rényi [**10**, **11**] proved that the global structure of a random graph changes suddenly around time $t = \lfloor n/2 \rfloor$. Suppose $M = \lfloor cn \rfloor$, where c is a positive constant. If $0 < c < \frac{1}{2}$ then a.e. G_M is such that the order of a largest component is $\log n$ but if $c > \frac{1}{2}$ then a.e. G_M has a unique largest component (called a *giant* component) with about $\alpha(c)n$ vertices, where $\alpha(c) > 0$. Finally, if $c = \frac{1}{2}$ then the order of a largest component of G_M is $n^{2/3}$.

The results of Erdős and Rényi left open a number of questions concerning the emergence of the giant component. Is there a range of s ($s = o(n)$) in which almost every $G_{n/2+s}$ has a unique largest component and that component has about $f(n, s)$ vertices? What is this function $f(n, s)$? How large is a second largest component? These questions were answered in [**7**]. Let us state some of the results from [**7**]. Denote by $L_j(G)$ the order of the jth largest component of a graph G.

THEOREM 1   *A.e. graph process $\tilde{G} = (G_t)_0^N$ is such that for every $t \geq n/2 + 2(\log n)^{1/2} n^{2/3}$ the graph G_t has a unique component of order at least $n^{2/3}$ and all other components have at most $n^{2/3}/2$ vertices each.* $\Box$

THEOREM 2   *Let $t = n/2 + s$. If $2(\log n)^{1/2} n^{2/3} \leq s = o(n)$, then a.e. G_t is such that*

$$L_1(G_t) = 4s + o(n^2/s^2)$$

and

$$L_2(G_t) \leq (\log n)n^2/s^2.$$

If $n^{1-\gamma_0} \leq s = o(n/\log n)$ for some $\gamma_0 < \frac{1}{3}$ and $\omega(n) \to \infty$ then a.e. G_t satisfies

$$|L_2(G_t) - 4(6 \log s - 4 \log n - \log \log n)n^2/s^2| \leq \omega(n)n^2/s^2.$$ $\Box$

THEOREM 3   *Let $1 < c_0 < c_1$ be fixed, $\omega(n) \to \infty$ and set $\tau_0 = \lfloor c_0 n/2 \rfloor$, $\tau_1 = \lfloor c_1 n/2 \rfloor$. For $t \geq \tau_0$ define $c = c(t) = 2t/n$, $\alpha = \alpha(t) = c - 1 - \log c$ and*

$$a(t) = \frac{1}{\alpha}(\log n - \tfrac{3}{2} \log \log n + \omega(n)).$$

Then a.e. $\tilde{G}$ is such that for $\tau_0 \leq t \leq \tau_1$ the graph G_t has no small component of order at least $a(t)$. $\Box$

The basis of the proofs of these results is the existence of a function $t_0(n)$ only slightly greater than $n/2$ such that a.e. $\tilde{G} = (G_t)_0^N$ has the property that for $t \geq t_0(n)$ the graph G_t has no component whose order is between $n^{2/3}/2$ and $n^{2/3}$. This gap in the sequence of the orders of components ensures that the number of components of order greater than

$n^{2/3}$ is decreasing after time $t_0(n)$: if $t \geq t_0(n)$ and C is a component of G_{t+1} with $|C| > n^{2/3}$, then C contains a component C' of G_t with $|C'| > n^{2/3}$. From this it follows that once all components of $G_{t_0(n)}$ having more than $n^{2/3}$ vertices merge into a single component of some G_{t_1}, this graph G_{t_1} (and every G_t for $t \geq t_1$) has a unique component of order greater than $n^{2/3}$ and all other components have order less than $n^{2/3}/2$.

The proof of the emergence of the gap mentioned above is based on a uniform estimate of the number of labelled sparse graphs. The aim of Section 1 is to prove this uniform bound.

Another problem left open by Erdös and Rényi concerned Hamilton cycles. For what $M = M(n)$ is it true that a.e. G_M is Hamiltonian? If $\omega(n) \to \infty$ and $M_1 = \lfloor (n/2)(\log n - \omega(n)) \rfloor$ then a.e. G_{M_1} is disconnected so, rather trivially, we must have $M \geq M_1$. Erdös and Rényi conjectured that this trivial bound is essentially best possible, namely that there is a constant c such that for $M = \lfloor cn \log n \rfloor$ a.e. G_M is Hamiltonian. This conjecture was proved by Pósa [32] and Korshunov [28, 29]. Recently Komlós and Szemerédi [25] proved a much more precise result about Hamilton cycles in random graphs: if

$$M = M(n) = \lfloor (n/2)(\log n + \log \log n + c) \rfloor,$$

where c is a constant, then

$$P(G_M \text{ is Hamiltonian}) \to e^{-e^{-c}}.$$

The quantity $e^{-e^{-c}}$ is in fact the limit of the probability that the minimum degree of G_M is at least 2: as Komlós and Szemerédi proved, the main obstruction to a random graph being Hamiltonian is the existence of vertices of degree less than 2.

Erdös and Spencer conjectured (see [8]) that vertices of degree less than 2 are practically the only obstruction to Hamilton cycles in random graphs: a.e. graph process is such that the very first edge which increases the minimum degree to 2 also makes the graph Hamiltonian. This conjecture was reiterated by Komlós and Szemerédi [25]. The aim of Section 2 is to prove this conjecture.

The diameter of random graphs was investigated by Moon and Moser [31], Korshunov [27], Klee and Larman [23], Bollobás [5] and Klee, Larman and Wright [24]. The results in the papers above concern fairly dense graphs, namely graphs with considerably more than $n \log n$ edges. Among others the following natural question was left open: what is the diameter of the first connected graph in a random graph process? Equivalently, if $p = (\log n - \omega(n))/n$, where $\omega(n)$ tends to ∞ rather slowly, what is the distribution of the diameter of G_p conditional on G_p being connected. In Section 3 we shall answer this question.

The last substantial section of the paper, Section 4, concerns the connectivity of a random graph shortly after the time the graph becomes connected. We shall also study the size of the largest k-connected subgraph shortly before the graph becomes connected.

To conclude the paper we solve a problem which Paul Erdös posed at the conference.

1. The number of connected sparse graphs

Denote by $C(n, n + k)$ the number of connected labelled graphs with n vertices and $n + k$ edges. Thus $C(n, n + k) = 0$ if $k \leq -2$, $C(n, n - 1)$ is the number of labelled trees of order n, so $C(n, n - 1) = n^{n-2}$ and $C(n, n)$ is the number of connected unicyclic graphs. The function $C(n, n)$ was determined by Katz [21] and Rényi [33], who made use of the fact that if $F(r, s)$ denotes the number of forests on $\{1, 2, \ldots, r\}$ which have s components and in which the vertices $1, 2, \ldots, s$ belong to distinct components, then

$$F(r, s) = sr^{r-1-s}. \tag{1}$$

For $k \geq 1$ Wright [37–39] proved several results about $C(n, n + k)$. Among other things, he showed that for $k = o(n^{1/3})$

$$C(n, n + k) = f_k n^{n+(3k-1)/2}\{1 + O(k^{3/2}/n)\},$$

where f_k depends only on k. Furthermore, $f_0 = (\pi/8)^{1/2}$, $f_1 = 5/24$ and $f_2 = 5\sqrt{2}/128$. The proof of Theorem 1 (published in [7]) is based on the following uniform bound on $C(n, n + k)$:

THEOREM 4   *There is an absolute constant c such that for $1 \leq k \leq n$ we have*

$$C(n, n + k) \leq (c/k)^{k/2} n^{n+(3k-1)/2}.$$

PROOF   Denote by $\mathscr{C}(n, n + k)$ the set of connected graphs with vertex set $V = \{1, 2, \ldots, n\}$ which have $n + k$ edges. Every graph G in $\mathscr{C}(n, n + k)$ contains at least two cycles, so G contains a unique maximal connected subgraph H with minimum degree at least 2. Let W be the vertex set of H. Clearly $e(H) = w + k$, where $w = |W|$. Furthermore, G is obtained from H by adding to it a forest with vertex set W, in which each component contains exactly one vertex of W. Hence, by relation (1), wn^{n-1-w} graphs $G \in \mathscr{C}(n, n + k)$ give the same graph H.

Denote by T the set of branch vertices of H and by $U = W - T$ the set

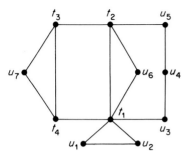 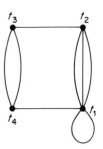

FIG. 1. A graph H and a multigraph M with $T = \{t_1, \ldots, t_4\}$ and $U = \{u_1, \ldots, u_7\}$.

of vertices of degree 2. Set $t = |T|$ and $u = |U| = w - t$. Clearly H is obtained from a multigraph M with vertex set T by subdividing the edges with the vertices from U (see Fig. 1). The multigraph M has $t + k$ edges and loops, so there are at most

$$u! \binom{u + t + k - 1}{t + k - 1}$$

ways of inserting the u vertices in order to obtain H from M. Since every vertex of M has degree at least 3 (a loop contributing 2 to the degree of the vertex incident with it), $3t \le 2(t + k)$; that is $t \le 2k$. Consequently,

$$C(n, n+k) \le \sum_{t=1}^{2k} \binom{n}{t} \psi(t, t+k) \sum_{u=0}^{n-1} \binom{n-t}{u} u! \binom{u + t + k - 1}{t + k - 1}(t + u)n^{n-1-t-u}, \tag{2}$$

where $\psi(t, t + k)$ is the number of labelled multigraphs of order t in which every vertex has degree at least 3. Rather crudely,

$$\psi(t, t+k) \le B\left(\binom{t}{2} + t + (t + k - 1), t + k - 1\right), \tag{3}$$

since a multigraph M is determined by the partition of its $t + k$ edges and loops into $\binom{t}{2} + t$ classes, $\binom{t}{2}$ classes for (multiple) edges and t classes for (multiple) loops.

To prove the theorem we shall simply estimate the bound for $f(n, n + k)$ implied by (2) and (3). We shall write $c_1, c_2, \ldots$ for positive absolute constants. Note first that if $t \le 2k$ then

$$\binom{t^2/2 + 3t/2 + k - 1}{t + k - 1} \le (c_1 k)^{k + t - 1}$$

and

$$\binom{n}{t}\binom{n-t}{u}u! = \frac{1}{t!}(n)_{t+u} \leq \frac{n^{t+u}}{t!}\exp(-u^2/2n).$$

Therefore

$$C(n, n+k) \leq n^{n-1}\sum_{t=1}^{2k}\sum_{u=0}^{n-k}A(t, u),$$

where

$$A(t, u) = \frac{1}{t!}(c_1 k)^{k+t-1}\exp(-u^2/2n)\binom{u+t+k-1}{t+k-1}(t+u).$$

Let us split the double sum above into two, according to the size of u. Clearly

$$\sum_{t=1}^{2k}\sum_{u=1}^{k}A(t, u) \leq \sum_{t=1}^{2k}\sum_{u=1}^{k}(c_1 k)^{t+k-1}\exp(-u^2/2n)2^{t+k}(3k)$$

$$\leq n^{1/2}\sum_{t=1}^{2k}(c_2 k)^{t+k-1}/t!$$

$$\leq n^{1/2}(c_3 k)^{k-3/2} \leq (c_3/k)^{k/2}n^{3k/2-1}. \tag{4}$$

The case in which u is large is also easily dealt with:

$$\sum_{t=1}^{2k}\sum_{u=k+1}^{n-t}A(t, u) \leq \sum_{t=1}^{2k}\sum_{u=k+1}^{n-t}\frac{1}{t!}(c_4 k)^{t+k-1}\exp(-u^2/2n)\frac{u^{t+k}}{(t+k-1)!}.$$

By making use of the standard inequality

$$\int_0^\infty x^{t+k}e^{-x^2/2}\,dx \leq ((t+k)!)^{1/2},$$

we find that

$$\sum_{t=1}^{2k}\sum_{u=k+1}^{n-t}A(t, u) \leq \sum_{t=1}^{2k}(c_5 k)^{t+k-1}\frac{((t+k)!)^{1/2}}{t!(t+k-1)!}n^{(t+k+1)/2}$$

$$< n^{(k+1)/2}\sum_{t=1}^{2k}\frac{(c_6 k)^{t+k}}{t!((t+k)!)^{1/2}}n^{t/2}$$

$$< n^{(k+1)/2}\sum_{t=1}^{2k}(c_3 k)^{t/2+k/2}n^{t/2}/t!$$

$$< n^{(k+1)/2}(c_8/k)^{k/2}n^k. \tag{5}$$

The assertion of our theorem follows from inequalities (4) and (5):

$$C(n, n+k) \leq n^{n-1}\{(c_3/k)^{k/2}n^{3k/2-1}+(c_8/k)^{k/2}n^{(3k+1)/2}\}$$

$$\leq (c/k)^{k/2}n^{n+(3k-1)/2}. \quad \square$$

If k is close to n, then the bound in the theorem above is worse than the bound, valid for all $k \geq -1$, given by the total number of graphs of order n and size $n + k$:

$$C(n, n+k) \leq \binom{N}{n+k} \leq \frac{en^{n+k}}{2}. \tag{6}$$

Inequality (6) and Theorem 4 have the following immediate consequence.

COROLLARY 5. $C(n, n+k) \leq c_0 2^{-k} n^{n+(3k-1)/2}$ for some absolute constant c_0 and every k, $-1 \leq k \leq N - n$. $\square$

2. Hamilton cycles

Let $Q = Q_n$ be a non-trivial monotone increasing property of graphs with vertex set $V = \{1, 2, \ldots, n\}$. Thus if G_1, G_2 are graphs on V, $G_1 \subset G_2$ and $G_1 \in Q$ then $G_2 \in Q$. Furthermore, $E^n \notin Q$ and $K^n \in Q$. The *hitting time* of the property Q is a function $\tau(\cdot; Q)$ on $\tilde{\mathcal{G}}$: for $\tilde{G} = (G_t)_0^N \in \tilde{\mathcal{G}}$,

$$\tau(\tilde{G}; Q) = \min\{t: G_t \in Q\}.$$

In other words, $\tau(\tilde{G}; Q) = t_0$ if the t_0th edge is the first edge after whose addition the graph has Q.

In particular, we shall write $\tau(\tilde{G}; \delta(G) \geq k)$ for the hitting time of the property of having minimum degree at least k and $\tau(\tilde{G}; \text{Ham})$ for the hitting time of the property of being Hamiltonian. Trivially,

$$\tau(\tilde{G}; \text{Ham}) \geq \tau(\tilde{G}; \delta(G) \geq 2) \tag{7}$$

for every graph process $\tilde{G}$, since the minimum degree of a Hamiltonian graph is at least 2. It was conjectured by Erdös and Spencer that in (7) equality holds for a.e. $\tilde{G}$: almost every graph process is such that the first edge which increases the minimum degree to 2 also makes the graph Hamiltonian. This conjecture was also stated by Komlós and Szemerédi [25] without any indication of a proof. The main aim of this section is to prove this conjecture.

The proof relies heavily on a colouring method introduced by Fenner and Frieze [15] and applied to Hamilton cycles in regular graphs by Bollobás [6] and Fenner and Frieze [16]. The existence of such a proof was mentioned in [6]. Another ingredient of the proof is a lemma concerning a somewhat artificial probability space of graphs, which enables one to relate a hitting time $\tau(\tilde{G}; Q)$ to $\tau(\tilde{G}; \delta(G) \geq k)$. We start with the introduction of this probability space.

Let $k \in \mathbb{N}$ be fixed and let $\alpha(n)$ be such that $\alpha(n) \to \infty$, $0 \leq \alpha(n) = o(\log \log \log n)$ and

$$M(n) = \frac{n}{2}(\log n + (k-1)\log \log n - \alpha(n))$$

is an integer for every n. To construct an element of $\mathcal{G}(n, M; \geq k)$ we start with a random graph G_M and colour its edges *blue*. Suppose G_M has g vertices of degree less than k, say $x_1, x_2, \ldots, x_g$. For each x_i select a random oriented edge with initial vertex x_i. Give the obtained multigraph, g of whose edges are oriented, probability

$$\binom{N}{M}^{-1}(n-1)^{-g}.$$

Let $\vec{\mathcal{G}}(n, M; \geq k)$ be the probability space of these multigraphs, some of whose edges are oriented.

Let $\mathcal{G}_c(n, M; \geq k)$ be the space of blue–green coloured graphs obtained from elements of $\vec{\mathcal{G}}(n, M; \geq k)$ by forgetting the orientation of the edges, replacing multiple edges by simple edges and colouring *green* the new edges, i.e. those edges which have not been coloured blue. Finally, to obtain the space $\mathcal{G}(n, M; \geq k)$, forget the colouring of the edges.

The following lemma justifies the introduction of the rather artificial probability space $\mathcal{G}(n, M; \geq k)$.

LEMMA 6   *Let Q be a monotone (increasing) property of graphs. Suppose a.e. element of $\mathcal{G}(n, M; \geq k)$ has Q. Then*

$$\tau(\tilde{G}; Q) \leq \tau(\tilde{G}; \delta(G) \geq k)$$

for a.e. graph process $\tilde{G}$.

PROOF   Note that a.e. graph process $\tilde{G}$ is such that (i) $\delta(G_M) = k-1$; (ii) G_M has $g \leq 2e^{\alpha(n)} \leq \log \log n$ vertices of degree $k-1$, say $x_1, x_2, \ldots, x_g$; (iii) if $x_i y_i$ is the first edge incident with x_i after time M then $y_i \notin \{x_1, \ldots, x_g\}$; (iv) a.e. extension of G_M (as a blue subgraph) to a graph in $\mathcal{G}(n, M; \geq k)$ has property Q.

Indeed, the first two assertions follow from standard results on degree sequences (see [**4**]); (iii) holds since

$$P(y_i \in \{x_1, \ldots, x_g\}) \leq \frac{g}{n-k}$$

for every i, and (iv) holds by assumption.

Since for $g = o(n^{1/2})$ we have $(n-g)^g \sim (n-1)^g$, and all g-sets $\{x_1 y_1, x_2 y_2, \ldots, x_g y_g\}$ are equally likely to appear, the lemma follows.   □

In order to find Hamilton cycles, we shall rely on a number of properties of most graphs in $\mathscr{G}(n, M; \geq 2)$. The following lemma is more than adequate for our purpose.

LEMMA   7   *Let* $M(n) = (n/2)(\log n + \log \log n - \alpha(n)) \in \mathbb{N}$, $\alpha(n) \to \infty$, $\alpha(n) = o(\log \log \log n)$ *and   let $k \geq 1$ be   fixed. Then   a.e.   graph $G \in$ $\mathscr{G}(n, M; \geq 2)$ is such that*

(i) $\delta(G) = 2$, $\Delta(G) \leq 3 \log n$ *and G is connected.*
(ii) *The graph G has at most $g_0 = \lfloor \log \log n \rfloor$ green edges.*
(iii) *There are $o(n/(\log n))$ vertices of degree at most $\frac{1}{2} \log n$.*
(iv) *No two vertices of degree at most $\frac{1}{10} \log n$ are within distance k of each other.*
(v) *Every set of $l \leq k$ vertices spans at most l edges.*
(vi) *Every set of $l \leq 5n/(\log n)^2$ vertices spans at most $10l$ edges.*
(vii) *For $n/(\log n)^2 \leq s \leq n/5$ every set of s vertices has at least $3s$ neighbours different from these vertices.*

PROOF   A.e. G_M is connected and has $g \leq \log \log n$ vertices of degree 1. A.e. extension of such a graph is such that the green edges are independent and no green edge joins two of these g vertices. Hence (i) and (ii) hold.

To prove (iii) it suffices to show that in G_M the expected number of vertices of degree at most $d_0 = \lfloor \frac{1}{2} \log n \rfloor$ is $o(n/(\log n))$. Since the property of having at most x such vertices is a monotone property, by Theorem 8 [3, p. 133] we may replace $\mathscr{G}_M$ by $\mathscr{G}_p$, where $p = M/N = (\log n + \log \log n - \alpha(n))/n$. The expectation in $\mathscr{G}_p$ is

$$n \sum_{d=0}^{d_0} \binom{n-1}{d} p^d (1-p)^{n-1-d} \leq n \sum_{d=0}^{d_0} \left(\frac{enp}{d} \right)^d n^{-1} = o(n^{0.9}).$$

A.e. graph G_M is such that $\Delta(G_M) \leq 3 \log n$ and there are at most $n^{1/2}$ vertices of degree at most $\frac{1}{10} \log n$. Consequently a.e. $G_c \in \mathscr{G}_c(n, M; \geq 2)$ is such that in the blue graph there are at most $2n^{1/2}(3 \log n)^{k+1} < n^{2/3}$ vertices within distance k from a vertex of degree at most $(\log n)/10$ or from an end-vertex of a green edge. Since $g(2n^{2/3}/n) = o(1)$ for a.e. G_M, arguing as in the proof of Lemma 6 we find that a.e. G_c is such that no two green edges are within distance k of each other or of a vertex of degree at most $\frac{1}{10} \log n$. Hence (iv) follows if we show that a.e. G_M is such that no two vertices of degree at most $d_1 = \lfloor \frac{1}{10} \log n \rfloor$ are within distance k of each other.

The expected number of such pairs of vertices is at most

$$\sum_{l=1}^{k} n^{l+1} \sum_{d=0}^{2d_1} \binom{2n}{d}\binom{N-2n+3}{M-l-d}\Big/\binom{N}{M} \leqslant 2\sum_{l=1}^{k} n^{l+1}\binom{2n}{2d_1}\binom{N-2n}{M-l-2d_1}\Big/\binom{N}{M}$$

$$\leqslant 3n^{k+1}\binom{2n}{2d_1}\binom{N-2n}{M-2d_1-k}\Big/\binom{N}{M}$$

$$= 3n^{k+1}\binom{2n}{2d_1}\frac{(N-M)_{2n-2d_1-k}(M)_{2d_1+k}}{(N)_{2n}}$$

$$\leqslant 4n^{k+1}\left(\frac{en}{d_1}\right)^{2d_1} e^{-2nM/N}\left(\frac{M}{N}\right)^{2d_1+k}$$

$$\leqslant n^{k+1}(10e)^{(\log n)/5}n^{-2}\left(\frac{\log n}{n}\right)^{k} = o(1).$$

Hence (iv) is proved.

Using the fact that a.e. G_c is such that $g \leqslant \log\log n$ and in the blue graph at most $n^{1/4}$ vertices are within distance k of a vertex, one can easily show that a.e. G_c is such that no cycle of length at most k contains a green edge and no cycle of length at most k is within distance k of a green edge. Hence (v) follows if we show that a.e. G_M is such that every set of $l \leqslant k$ vertices spans at most l edges. Since this property is monotone, by Theorem 8 [3, p. 133], it suffices to note that a.e. G_p has this property, where

$$p = \frac{(\log n + \log\log n - \alpha(n))}{n}.$$

This holds, for the threshold probability for a graph of order l and size $l+1$ is $n^{-l/(l+1)}$.

As a.e. G_c is such that the green edges are independent, (vi) follows if we show that a.e. G_M is such that every set of $l \leqslant n_0 = \lfloor 5n/(\log n)^2 \rfloor$ vertices spans at most $9l$ edges. This assertion does hold since in G_p (where as before $p = M/N = (\log n + \log\log n - \alpha(n))/n$) the expected number of sets violating the property is at most

$$\sum_{l=19}^{n_0} \binom{n}{l}\binom{\binom{l}{2}}{9l}p^{9l} \leqslant \sum_{l=19}^{n_0} \left(\frac{en}{l}\right)^{l}\left(\frac{el}{18}\right)^{9l}p^{9l}$$

$$\leqslant \sum_{l=19}^{n_0}\left(\frac{e^{10}}{18^9}pn(lp)^8\right)^{l}$$

$$\leqslant \sum_{l=19}^{n_0}\left\{(\log n)\left(\frac{l\log n}{n}\right)^8\right\}^{l} = o(1).$$

Finally, to show (vii) it suffices to prove that a.e. G_p is such that for $n_0 = \lfloor n/(\log n)^2 \rfloor \leqslant s \leqslant n_1 = \lfloor n/5 \rfloor$ every set of s vertices has at least $3s$ neighbours. Once again it suffices to give an upper bound for the expectation of the number of sets violating the condition:

$$\sum_{s=n_0}^{n_1} \binom{n}{4s}\binom{4s}{s}(1-p)^{s(n-4s)} \leqslant \sum_{s=n_0}^{n_1} \left(\frac{en}{4s}\right)^{4s} (4e)^{s_n - s(n-4s)}$$

$$\leqslant \sum_{s=n_0}^{n_1} \left(\frac{e^5 n^4}{64 s^4}\right)^s n^{-s/5}$$

$$\leqslant \sum_{s=n_0}^{n_1} (2 \log n)^{8s} n^{-s/5} = o(1).$$

This completes the proof of Lemma 7. □

We are ready to prove the main result of the section.

THEOREM 8 *A.e. graph process $\tilde{G}$ is such that $\tau(\tilde{G}; \mathrm{Ham}) = \tau(\tilde{G}; \delta(G) \geqslant 2)$.*

PROOF The proof is based on a lemma due to Pósa [32] (see also [3, p. 79]) which is used in most proofs of results concerning Hamilton cycles in random graphs, a colouring method due to Fenner and Frieze [15], and Lemma 7 above. In fact, we shall need the following simple consequence of Pósa's lemma (see [3, Lemma 1]).

LEMMA 9 *Suppose $k, u \in \mathbb{N}$ and a graph B is such that its longest path has length k but it does not contain a cycle of length $k + 1$. Suppose furthermore that for every set S of $s \leqslant u$ vertices we have*

$$|S \cup \Gamma_B(S)| \geqslant 3s. \tag{8}$$

Then there are u distinct vertices $y_1, y_2, \ldots, y_u$ and u not necessarily distinct subsets $Y_1, Y_2, \ldots, Y_u$ of $V(B)$ such that $|Y_i| = u$, $y_i \notin Y_i$ and no vertex in Y_i is adjacent to y_i, $i = 1, 2, \ldots, u$. Furthermore, the addition of any of the $y_i - Y_i$ edges creates a cycle of length $k + 1$. □

Let $M = (n/2)(\log n + \log \log n - \alpha(n))$ be as in Lemma 7 and denote by $\mathscr{F}_c = \mathscr{F}_c(n, M; \geqslant 2)$ the collection of graphs in $\mathscr{G}_c(n, M; \geqslant 2)$ which are non-Hamiltonian and satisfy the conditions of Lemma 7 with $k = 6$. For $G_c \in \mathscr{F}_c$ let $L(G_c)$ be a path of maximal length, say of length k. Note that since G_c is connected, it does not contain a cycle of length $k + 1$.

Let us assume that Theorem 8 is false. Since $\tau(\tilde{G}; \mathrm{Ham}) \geqslant \tau(\tilde{G}; \delta(G) \geqslant 2)$, by Lemmas 6 and 9 there are $\alpha > 0$ and an infinite set

$\Lambda \subset \mathbb{N}$ such that for $n \in \Lambda$ we have

$$P_c(\mathscr{F}_c) \geq \alpha. \tag{9}$$

Let us define a *coloured pattern* or simply a *pattern* as a blue, green and red coloured graph obtained from a graph $G_c \in \mathscr{F}_c$ by recolouring $R = \lfloor (\log n)^2 \rfloor$ blue edges of G_c *red* in such a way that (1) no edge of $L(G_c)$ is coloured red; (2) no edge incident with a vertex of degree at most $\frac{1}{2}\log n$ is coloured red; (3) the R red edges are independent.

How many patterns can be obtained from a graph $G_c \in \mathscr{F}_c$?

Denote by M_0 the number of blue edges which are on the path $L(G_c)$ or are incident with a vertex of degree at most $\frac{1}{10}\log n$. By assumption $M_0 \leq 2n$ if n is sufficiently large. There are $\binom{M-M_0}{R}$ colourings which satisfy conditions (1) and (2). What is the probability that if we choose one of these colourings at random then the colouring does not satisfy condition (3)? This probability is at most

$$n\binom{3\log n}{2}\binom{M-M_0-2}{R-2}\binom{M-M_0}{R}^{-1} = O(n(\log n)^2 R^2 M^{-2}) = o(1).$$

Hence, if n is large enough, at least

$$\frac{1}{2}\binom{M-M_0}{R} \geq \frac{1}{2}\binom{M-2n}{R}$$

patterns can be obtained from G_c. Consequently if (9) holds then the number of patterns is at least

$$\frac{\alpha}{2}\binom{N}{M}\binom{M-2n}{R}.$$

Let B be the subgraph of $G = G_c$ obtained after the omission of the red edges of a pattern constructed from G_c. We claim that the graph B satisfies the conditions of Lemma 9 with $u = \lfloor n/5 \rfloor$. Since G is connected and non-Hamiltonian, G and, *a fortiori*, B, cannot contain a cycle of length $k+1$. It remains to check (8). This is clear if $n(\log n)^{-2} \leq s \leq u = \lfloor n/5 \rfloor$ since then

$$|S \cup \Gamma_G(S)| \geq 4s \tag{10}$$

and the red edges are independent.

Now suppose that $|S| = s \leq n(\log n)^{-2}$ and

$$|S \cup \Gamma_B(S)| < 3|S|. \tag{11}$$

Set $S_1 = \{x \in S: d_G(x) \leq \frac{1}{10}\log n\}$, $S_2 = S \backslash S_1$ and

$$T = S_2 \cup \Gamma_G(S_2) \backslash (S_1 \cup \Gamma_G(S_1)).$$

Since for $x \in S_1$ we have $d_B(x_1) = d_G(x_1) \geq 2$ and in G no two vertices of S_1 are within distance 6 of each other,

$$|S_1 \cup \Gamma_B(S_1)| = |S_1 \cup \Gamma_G(S_1)| = \sum_{x \in S_1} (d_G(x) + 1) \geq 3 |S_1|. \tag{12}$$

Inequalities (11) and (12) imply that

$$|(S_2 \cup \Gamma_B(S_2)) \backslash (S_1 \cup \Gamma_B(S_1))| < 3 |S_2|. \tag{13}$$

Since the red edges of G are independent, by (13) we have

$$|T| < 4 |S_2| \leq \frac{4n}{(\log n)^2}.$$

By Lemma 7 (iv) and (v) each vertex $y \in S_2$ sends at least $d = \lfloor \frac{1}{10} \log n \rfloor - 2$ edges to T. Hence the graph $G[T]$ has at least $(d/2) |S_2|$ edges, contradicting assertion (vi) of Lemma 7. This shows that B does satisfy inequality (8).

Let $y_1, \ldots, y_u$ be the vertices and $Y_1, \ldots, Y_n$ be the sets guaranteed by Lemma 9. The set $\{y_i z : z \in Y_i, i = 1, \ldots, u\}$ of unordered pairs has at least $u^2/2$ elements and by Lemma 9 none of these pairs is a red edge. Consequently any blue–green graph B is the blue–green subgraph of at most

$$\binom{N - \lfloor u^2/2 \rfloor}{R}$$

patterns. How many choices do we have for B? Since B has $M - R$ blue edges and at most g_0 green ones, there are at most

$$\binom{N}{M - R} n^{2g_0}$$

choices for B. Therefore there are at most

$$\binom{N - \lfloor u^2/2 \rfloor}{R} \binom{N}{M - R} n^{2g_0}$$

patterns. Hence

$$\frac{\alpha}{2} \binom{N}{M} \binom{M - 2n}{R} \leq \binom{N - \lfloor u^2/2 \rfloor}{R} \binom{N}{M - R} n^{2g_0}. \tag{14}$$

Inequality (14) leads to a contradiction. Indeed, since $N - \lfloor u^2/2 \rfloor \leq \frac{25}{26} N$, inequality (14) implies that

$$\frac{\alpha}{2} (N - M)^M (M - 3n)^R \leq M^R (N - \lfloor u^2/2 \rfloor)^R N^{M-R} n^{2g_0},$$

SO

$$0 < \frac{\alpha}{2} \leq \left(1 - \frac{M}{N}\right)^{-M} n^{2g_0} \left(\frac{26}{27}\right)^R .$$

This is impossible since the right-hand side tends to 0 as $n \to \infty$, $n \in \Lambda$. The proof of Theorem 8 is complete. $\square$

3. The diameter

In this section we shall examine the diameter of a sparse random graph. Let τ_0 be the hitting time of connectedness, i.e. given a graph process $\tilde{G} = (G_t)_0^N$, set $\tau_0(\tilde{G}) = \min\{t: G_t \text{ is connected}\}$. How large is the diameter of G_{τ_0}?

THEOREM 10   *Almost every graph process $\tilde{G}$ is such that*

$$\left\lfloor \frac{\log n - \log 2}{\log \log n} \right\rfloor + 1 \leq \text{diam}(G_{\tau_0}) \leq \left\lceil \frac{\log n + 6}{\log \log n} \right\rceil + 3.$$

PROOF   Denote by Q the probability that the diameter is at most $d = \lceil (\log n + 6)/\log \log n \rceil + 3$. The main assertion of the theorem is the second inequality which states that a.e. graph process is such that G_{τ_0} has Q. This will follow if we show that

$$\tau(\tilde{G}; Q) \leq \tau(\tilde{G}; \delta(G) \geq 1). \tag{15}$$

To prove this we proceed as in the proof of Theorem 8, i.e. we make use of the model $\mathcal{G}(n, M; \geq 1)$ and Lemma 6, where $M = (n/2)(\log n - \alpha(n))$, $\alpha(n) \to \infty$ and $\alpha(n) = o(\log \log \log n)$. Set $p = (\log n - \alpha(n))/n$. We know that the models $\mathcal{G}_M$ and $\mathcal{G}_p$ are very close to each other; in fact if R is a monotone property of graphs then, rather crudely,

$$P(G_M \text{ has } R) \leq 3P(G_p \text{ has } R).$$

In what follows we shall often use this inequality to estimate the probability in the blue subgraph of a graph in $\mathcal{G}(n, M; \geq 1)$.

We shall show that a.e. graph in $\mathcal{G}(n, M; \geq 1)$ is such that for $k \leq d/2$ every vertex has many vertices at distance k from it. For a graph G and vertex $x \in V$ set

$$\Gamma_k(x) = \{y \in V: d_G(x, y) = k\} \quad \text{and} \quad d_k(x) = |\Gamma_k(x)|.$$

We claim that a.e. $G \in \mathcal{G}(n, M; \geq k)$ is such that for every vertex x we have

$$\tfrac{1}{9}(\log n)^2 \leq d_3(x) \leq 9(\log n)^3 \tag{16}$$

and if $4 \leq k \leq d/2$ then

$$\tfrac{1}{9}(\log n)^{k-1}\left(1 - \frac{\alpha(n)}{\log n}\right)^k \leq d_k(x) \leq 9(\log n)^k\left(1 + \frac{\alpha(n)}{\log n}\right)^k. \tag{17}$$

To see (16), note that in a.e. graph there are few green edges and few vertices of degree at most $\tfrac{1}{2}\log n$. Hence a.e. graph is such that if xy is a green edge then

$$|\Gamma_1(x) \cup \Gamma_1(y)| \geq \lceil \tfrac{1}{3}\log n\rceil + 1 = m_0. \tag{18}$$

Furthermore, the probability that some blue edge xy fails to satisfy (18) is at most

$$3\binom{n}{2}p \sum_{m=0}^{m_0}\binom{n-2}{m}(1-(1-p)^2)^m (1-p)^{2(n-m-2)}$$

$$\leq 3n(\log n)\frac{n^{m_0}}{m_0!}(2p)^{m_0}n^{-2}e^{2\alpha(n)} \leq n^{-1}(\log n)(6e)^{(\log n)/3} = o(1).$$

A similar inequality shows that with probability $1 - o(n^{-1})$ a given set of m vertices have at least $(m/2)\log n$ and at most $(3m/2)\log n$ neighbours in the blue graph, provided that $m_0/10 \leq m \leq (\log n)^3$. Since a.e. graph has maximum degree at most $3\log n$, this implies (16). Furthermore, we find that

$$\max\left\{(\tfrac{1}{3}\log n)^2, \left(\frac{\log n}{2}\right)d_2(x)\right\} \leq d_3(x)$$

in a.e. graph.

Relation (17) follows by induction on k from the following simple assertion. Let $A \subset B \subset V$, $b = |B| \leq 2a = 2|A|$ and

$$\tfrac{1}{9}(\log n)^2 \leq a \leq (\log n)^3 n^{1/2}.$$

Set $\delta = (\log\log n)/\log n$ and

$$\tilde{\Gamma}(A) = \{y: y \in V\backslash B, \exists x \in A, xy \in E(G)\}.$$

Then

$$P_p(|\;|\tilde{\Gamma}(A)| - a\log n| \geq \delta a\log n) = o(n^{-2}). \tag{19}$$

To prove (19) note that the probability that a given vertex $y \in V\backslash B$ does not belong to $\tilde{\Gamma}(A)$ is $\bar{q} = (1-p)^a$, so $|\tilde{\Gamma}(a)|$ has binomial distribution with parameters $n - b$ and $\bar{p} = 1 - \bar{q}$. Since

$$ap(1 - ap) \leq 1 - e^{-ap} \leq \bar{p} = 1 - (1-p)^a \leq pa,$$

$$|(n-b)\bar{p} - a\log n| \leq \left(\frac{\delta}{10}\right)a\log n.$$

Consequently, by a standard result concerning large deviations of the binomial distribution, the left-hand side of (19) is at most

$$\exp\{-\delta^2(a \log n)/4\} = o(n^{-\log \log n}) = o(n^{-2}).$$

This completes the proof of (17).

Now to see that a.e. graph in $\mathcal{G}(n, M; \geqslant 1)$ has diameter at most $d = \lceil(\log n + 6)/\log \log n\rceil + 3$, set $d_1 = \lfloor d/2 \rfloor$ and $d_2 = d - d_1 - 1$. Then a.e. graph is such that for any two vertices x and y there are at least $a_1 = \lfloor\frac{1}{10}(\log n)^{d_1-1}\rfloor$ vertices at distance d_1 from x and there are at least $a_2 = \lfloor\frac{1}{10}(\log n)^{d_2-1}\rfloor$ vertices at distance d_2 from y.

Let A_1 and A_2 be two given disjoint sets of vertices with $|A_1| = a_1$, $|A_2| = a_2$. The probability that no blue edge joins A_1 to A_2 is at most

$$3(1-p)^{a_1 a_2} \leqslant \exp\left\{-\frac{1}{101n}(\log n)^{d-2}\right\} \leqslant \exp\{-\tfrac{5}{2}\log n\} = o(n^{-2}).$$

This shows that a.e. graph is such that any two of its vertices are within distance $d_1 + d_2 + 1 = d$ of each other, completing the proof of (15).

The proof of the first inequality in Theorem 10 is very similar. We take $M = M(n) = (n/2)(\log n + \alpha(n))$, $\alpha(n) = o(\log \log \log n)$, $\alpha(n) \to \infty$. Then a.e. G_M is connected so the first inequality holds if a.e. G_M has diameter at least $\lfloor(\log n - \log 2)/\log \log n\rfloor + 1$. This follows from the analogues of inequalities (16) and (17). $\square$

With a little more work the bounds in Theorem 10 could be tightened. Furthermore, a slight variant of the proof above shows that as a graph process progresses, the diameter stays almost determined.

THEOREM 11 *Suppose that* $M = M(n) = (n/2)d(n) \in \mathbb{N}$, $d(n) - \log n \to \infty$ *and* $d(n) \leqslant n - 1$. *Then a.e.* G_M *is such that*

$$\left\lfloor\frac{\log n - \log 2}{\log d(n)}\right\rfloor + 1 \leqslant \operatorname{diam}(G_m) \leqslant \left\lceil\frac{\log n + 6}{\log d(n)}\right\rceil + 3. \quad \square$$

In fact, by some arguments similar to those in the proof above one can show that if $M = \lfloor cn/2 \rfloor$, where $c > 1$, then the diameter of the giant component of a G_M is $O(\log n)$ with probability tending to 1. If c is large enough then the asymptotic order of the diameter of the giant component is almost determined.

THEOREM 12 *Let* $\varepsilon > 0$. *If* c *is sufficiently large and* $M = \lfloor cn/2 \rfloor$, *then a.e.* G_M *consists of a giant component* H *and components of order at most* $(\log n)/(c - 1 - \log c - \varepsilon)$; *furthermore, the diameter of the giant component satisfies*

$$(1-\varepsilon)\frac{\log n}{\log c} \leq \operatorname{diam} H \leq (1+\varepsilon)\frac{\log n}{\log c}. \quad \square$$

4. The connectivity

The connectedness of random graphs has been studied by a great many authors: Erdös and Rényi [9, 12], Gilbert [17], Austin, Fagen, Penney and Riordan [1], Stepanov [34–36], Kelmans [22], Ivhenko [18, 19], Ivchenko and Medvedev [20], Kordecki [26], Kovalenko [30], etc. The aim of this section is to examine an aspect of connectedness which is somewhat different from those studied so far.

We know that if $p = (\log n - \omega(n))/n$, where $\omega(n) \to \infty$, then a.e. G_p is disconnected. Our first result shows that if p is only slightly larger than this value then the probability that G_p is not k-connected is rather small:

THEOREM 13 *Suppose that* $p = (1+c)(\log n)/n$, *where* $c > 0$ *is a constant. Then for* $\omega(n) \to \infty$ *and* $k \geq 1$ *we have*

$$P(G_p \text{ is not } k\text{-connected}) = O(\omega(n)(\log n)^{k+1}n^{-c}).$$

PROOF Suppose G_p is not k-connected. Then either there is a vertex of degree less than k or else G_p contains a set R of r vertices and a set S of $k-1$ vertices such that $2 \leq r \leq \lfloor (n-k+1)/2 \rfloor = n_0$, $R \cap S = \varnothing$, every vertex in R has degree at least k and no edge joins R to $V - R \cup S$. Denote by X_l the number of vertices of degree l and by Y_r the number of pairs (R, S) described above. To prove our theorem it suffices to show that

$$\sum_{l=0}^{k-1} E(X_l) + \sum_{r=2}^{n_0} E(Y_r) = O((\log n)^{k-1}n^{-c}).$$

Clearly

$$\sum_{l=0}^{k-1} E(X_l) = \sum_{l=0}^{k-1} n\binom{n-1}{l}p^l(1-p)^{n-1-l}$$

$$\leq 2n\frac{n^{k-1}}{(k-1)!}\left(\frac{(1+c)\log n}{n}\right)^{k-1}n^{-1-c}$$

$$= \frac{2(1+c)^{k-1}}{(k-1)!}(\log n)^{k-1}n^{-c}.$$

In order to estimate $E(Y_r)$ we distinguish between two cases: $2 \leq r \leq r_0 = \lfloor n^{1/4} \rfloor$ and $r_0 \leq r \leq n_0$. In the first case, note that the set $R \cup S$ spans at least $s = \lceil kr/2 \rceil$ edges and the probability that a given set of $r+k-1$ vertices spans at least s edges is at most

$$\binom{(r+k-1)^2/2}{s}p^s \leq 2^{2k^2}\binom{r^2}{s}p^s$$

$$\leq 2^{2k^2}\left(\frac{er^2p}{s}\right)^s$$

$$\leq 2^{2k^2}\left(\frac{6r(1+c)\log n}{kn}\right)^{rk/2}$$

$$\leq n^{-rk/3}.$$

Consequently

$$E(Y_2) \leq \binom{n}{k-1}\binom{n-k+1}{2}2^{2k^2}\left(\frac{12(1+c)\log n}{kn}\right)^k(1-p)^{2(n-k-1)}$$

$$= O(n^{k+1}(\log n)^k n^{-k} n^{-2(1+c)})$$

$$= o(n^{-2c})$$

and

$$\sum_{r=3}^{r_0} E(Y_r) \leq \sum_{r=3}^{r_0} n^{k+r-1}n^{-rk/2}(1-p)^{rn}$$

$$= \sum_{r=3}^{r_0} n^{k+r-1}n^{-rk/3}n^{-r(1+c)} = o(n^{-3c}).$$

When $r_0 \leq r \leq n_0$, all we use is that the graph contains no $R - (V\backslash(R\cup S))$ edges:

$$\sum_{r=r_0}^{n_0} E(Y_r) \leq \sum_{r=r_0}^{n_0}\binom{n}{k-1}\binom{n-k+1}{r}(1-p)^{r(n-k+1-r)}$$

$$\leq n^{k-1}\sum_{r=r_0}^{n_0}\frac{n^r}{r!}n^{-r(1+c)(n-k+1-r)/n}$$

$$= n^{k-1}\sum_{r=r_0}^{n_0} A_r.$$

Since for $r_0 \leq r < n_0$ we have $A_{r+1}/A_r < \frac{1}{2}$,

$$\sum_{r=r_0+1}^{n_0} E(Y_r) \leq 2n^{k-1}A_{r_0} \leq 2n^{k-1}\left(\frac{enn^{-1-c/2}}{r_0}\right)^{r_0} = o(n^{-1}),$$

completing our proof. □

How large a k-connected subgraph can we expect in G_p, where $p = c/n$

and c is constant? We know that for a large c the order of the giant component is about $(1 - e^{-c})n$, so for $k \geq 1$ we cannot hope for more than $(1 - e^{-c})n$ vertices in a k-connected subgraph. In fact, taking into account the trivial fact that the minimum degree of a k-connected graph is at least k, we see that at least $(c^{k-1}/(k-1)!)e^{-c}n$ must be expected to be outside a k-connected subgraph. The next result shows that this crude bound is essentially best possible. For the sake of convenience no attempt will be made to obtain the best bounds which the method could prove.

THEOREM 14   *Suppose that $8 \leq k + 3 \leq c/2$, $c > 67$ and $p = c/n$. Then with probability $1 - O(n^{-1})$ a random graph G_p contains a k-connected subgraph with at least*

$$\left\{ 1 - 5 \frac{c^{k+3}}{(k+3)!} e^{-c} \right\} n \geq \left\{ 1 - 5 \left(\frac{ec}{k+3} \right)^{k+3} e^{-c} \right\} n \geq \left(1 - 5 \left(\frac{2}{e} \right)^{c/2} \right) n$$

vertices.

PROOF   The k-*core* of a graph is the unique maximal subgraph with minimum degree at least k. We shall prove two assertions, which clearly imply the theorem:

(a) With probability $1 - O(n^{-1})$ the k-core has at least $n - 5(c^{k+3}/(k+3)!)e^{-c}n$ vertices.
(b) With probability at least $1 - O(n^{-1})$ the k-core is k-connected.

Assertion (a) is rather crude, but assertion (b) is essentially best possible.

To prove (a) we shall construct a sequence of sets $U_0, U_1, \ldots$ such that $V - \bigcup_{i=0}^{\infty} U_i$ is contained in the k-core. Let

$$U_0 = \{ x \in V : d(x) \leq k + 3 \}.$$

Having constructed $U_1, U_2, \ldots, U_l$, set

$$U_{l+1} = \left\{ x \in V - \bigcup_{i=0}^{l} U_i : x \text{ sends at least four edges to } \bigcup_{i=0}^{l} U_i \right\}$$

and

$$U = \bigcup_{i=0}^{\infty} U_i, \qquad W = V \backslash U.$$

The minimum degree of $G_p[W]$ is at least k, so it suffices to prove that

$$P(|U| \geq 2u_0) = O(n^{-1}),$$

where

$$u_0 = \left\lfloor \frac{5}{2} \left(\frac{c^{k+3}}{(k+3)!} \right) e^{-c} n \right\rfloor.$$

Clearly

$$E(|U_0|) = n \sum_{j=0}^{k+3} \binom{n-1}{j} p^i (1-p)^{n-1-j} \leq 2\left(\frac{c^{k+3}}{(k+3)!}\right) e^{-c} n.$$

Furthermore, it is straightforward to check that the variance of $|U_0|$ is $O(n)$. Consequently

$$P(|U_0| \geq u_0) = O(n^{-1}). \tag{20}$$

By the definition of the sets U_i, the subgraph $G[U]$ of G has at least $4\sum_{i=1}^{\infty} |U_i|$ edges. Moreover, if $2|U_0| \leq 2u_0 \leq |U|$ then G_p has a subgraph of order $2u_0$ with at least $4u_0$ edges. Hence, by (20), assertion (a) follows if we show that the probability that some $2u_0$ vertices span at least $4u_0$ edges is $O(n^{-1})$. This does hold with plenty to spare, for $c > 67$ implies that

$$\binom{n}{2u_0}\binom{2u_0^2}{4u_0} p^{4u_0} \leq \left\{\frac{en}{2u_0}\left(\frac{eu_0}{2}\right)^2 \frac{c^2}{n^2}\right\}^2 = \left(\frac{e^3 c^2 u_0}{8n}\right)^2 = o(n^{-1}).$$

To prove (b) we proceed as in the proof of Theorem 13: we estimate the probability of the existence of a triple (R_1, S, R_2), where $R_1 \cup S \cup R_2$ is the k-core, $|S| = k-1$, $|R_i| = r_i$, $2 \leq r_1 \leq r_2$, $u = n - r_1 - s - r_2 \leq 2u_0$ and G_p contains no $R_1 - R_2$ edge.

Setting $t = r_1 + k - 1$, we see that for $t \leq 2k$ the set $T = R_1 \cup S$ spans at least $t + k - 2$ edges. The probability of this is $O(n^{-3})$. Now if

$$2k < t \leq t_0 = \left\lfloor \left(\frac{8}{e^3 c^2}\right) n \right\rfloor + k - 1,$$

then T spans at least $2t$ edges. The probability of this is at most

$$\sum_{t=2k}^{t_0} \binom{n}{t}\binom{t^2/2}{2t} p^{2t} \leq \sum_{t=2k}^{t_0} \left\{\frac{en}{t}\left(\frac{et}{4}\right)^2 \frac{c^2}{n^2}\right\}^t = O(n^{-k}).$$

Finally, the probability that there is a triple (R_1, S, R_2) with $t_0 < t$ is at most

$$\sum_{u=0}^{2u_0} \sum_{r=t_0}^{\lfloor (n-u)/2 \rfloor} \binom{n}{k-1}\binom{n-k-1}{u}\binom{n-k-1-u}{r}(1-p)^{r(n-k+1-u-r)}$$

$$\leq n^{k-1} \sum_{u=0}^{2u_0} \sum_{r=t_0}^{\lfloor (n-u)/2 \rfloor} \left(\frac{en}{u}\right)^u \left(\frac{en}{r}\right)^r e^{-cr/3}$$

$$= n^{k-1} \sum_{u=0}^{2u_0} \sum_{r=t_0}^{\lfloor (n-u)/2 \rfloor} A(u, r)$$

$$\leq 4n^{k-1} \left(\frac{en}{2u_0}\right)^{2u_0} \left(\frac{en}{t_0}\right)^{t_0} e^{-ct_0/3} = o(n^{-1}).$$

Here the second inequality follows from the fact that in our range $A(u, r+1)/A(u, r) < \frac{1}{2}$ and $A(u+1, t_0)/A(u, t_0) > 2$. This completes our proof. □

5. A frivolous problem

As one has come to expect, in his closing address at the meeting Paul Erdös posed several open questions. One of these, which he had thought of at the banquet the night before, carried an award of \$10 and concerned random graphs. Is it true that a.e. graph of order n (i.e. a.e. $G_{1/2}$) is such that every spanned subgraph of it is connected or can be made connected by the addition of a single appropriate vertex of the graph? At the insistence of the present author the award was reduced to \$3. As it happens, even the latter amount is too generous. The following answer, obtained immediately after the talk, shows why.

We claim that a.e. $G_{1/2}$ contains $k = \lfloor \log_2 n \rfloor$ independent vertices which are not dominated by any vertex. Hence these k vertices form a disconnected subgraph of $G_{1/2}$ which cannot be made connected by the addition of a single vertex. To prove this, denote by X the number of such sets. (In order to make the formulae clearer, we revert to the usual notation of $p = P(\text{edge})$ and $q = 1 - p$, though in our case $p = q = \frac{1}{2}$.) Clearly

$$E(X) = \binom{n}{k} q^{-(k/2)} (1 - p^k)^{n-k} \to \infty.$$

Furthermore,

$$E(X^2) = \binom{n}{k} \sum_{r=0}^{k} \binom{k}{r} \binom{n-k}{r} (1 - 2p^k + p^{k+r})^{n-k-r},$$

since we have $\binom{n}{k}\binom{k}{r}\binom{n-k}{r}$ choices for an ordered pair of k-sets with $k - r$ vertices in common and the probability that a vertex not in these k-sets dominates neither of these k-sets is

$$1 - p^{k-r} + p^{k-r}(1 - p^r)^2 = 1 - 2p^k + p^{k+r}.$$

Straightforward calculations show that

$$E(X^2) \sim \binom{n}{k}\binom{n-k}{k}(1 - 2p^k + p^{-2k})^{n-2k} \leq E(X)^2.$$

Hence $\sigma^2(X) = o(E(X)^2)$, so $X > 0$ for a.e. G_p. (In fact,

$X = (1 + o(1))E(X)$ for a.e. G_p.) We hope to return to extensions of this problem later.

References

1. T. L. Austin, R. E. Fagen, W. F. Penney and J. Riordan, The number of components in random linear graphs. *Ann. Math. Statist.* **30** (1959), 747–754.
2. B. Bollobás, *Extremal Graph Theory*, Academic Press, London, 1978.
3. B. Bollobás, *Graph Theory – An Introductory Course*, Springer Verlag, New York, Heidelberg and Berlin, 1979.
4. B. Bollobás, Degree sequences of random graphs. *Discrete Math.* **33** (1981), 1–19.
5. B. Bollobás, The diameter of random graphs. *Trans. Amer. Math. Soc.* **267** (1981), 41–52.
6. B. Bollobás, Almost all regular graphs are Hamiltonian. *Europ. J. Combinat.* **4** (1983), 97–106.
7. B. Bollobás. The evolution of random graphs. *Trans. Amer. Math. Soc.* (in press).
8. P. Erdös, Problems and results in combinatorial analysis and combinatorial number theory. In *Proceeedings of the Ninth Southeastern Conference on Combinatorics, Graph Theory and Computing*, Congressus Numerantium XXI, Utilitas Mathematicae, Winnipeg, 1978, pp. 29–40.
9. P. Erdös and A. Rényi, On random graphs I. *Publ. Math. Debrecen* **6** (1959), 290–297.
10. P. Erdös and A. Rényi, On the evolution of random graphs. *Publ. Math. Inst. Hungar. Acad. Sci.* **5** (1960), 17–61.
11. P. Erdös and A. Rényi, On the evolution of random graphs. *Bull. Inst. Internat. Statist. Tokyo* **38** (1961), 343–347.
12. P. Erdös and A. Eényi, On the strength of connectedness of a random graph. *Acta. Math. Acad. Sci. Hungar.* **12** (1961), 261–267.
13. P. Erdös and A. Rényi, Asymmetric graphs. *Acta Math. Acad. Sci. Hungar.* **14** (1963), 295–315.
14. P. Erdös and A. Rényi, On the existence of a factor of degree one of a connected random graph. *Acta Math. Acad. Sci. Hungar.* **17** (1966), 359–368.
15. T. I. Fenner and A. M. Frieze, On the existence of Hamiltonian cycles in a class of random graphs. *Discrete Math.* **45** (1983), 301–305.
16. T. I. Fenner and A. M. Frieze, Hamiltonian cycles in random regular graphs. *J. Combinat. Theory B* (in press).
17. E. N. Gilbert, Random graphs. *Annals Math. Stat.* **30** (1959), 1141–1144.
18. G. I. Ivchenko, On the asymptotic behaviour of the degrees of vertices in a random graph. *Theory Prob. Appl.* **18** (1973), 188–195.
19. G. I. Ivchenko, The strength of connectivity of a random graph. *Theory Prob. Appl.* **18** (1973), 396–403.
20. G. I. Ivchenko and I. V. Medvedev, The probability of connectedness in a class of random graphs [in Russian]. *Vopr. Kibernetiki, Moscow* (1973), 60–66.
21. L. Katz, The probability of indecomposability of a random mapping function. *Ann. Math. Statist.* **26** (1955), 512–517.

22. A. K. Kelmans, On the connectedness of random graphs [in Russian]. *Automatika i Telemechanika* **28** (1967), 98–116.
23. V. L. Klee and D. Larman, Diameters of random graphs. *Canad. J. Math.* **33** (1981), 618–640.
24. V. L. Klee, D. G. Larman and E. M. Wright, The diameter of almost all bipartite graphs. *Studia Sci. Math. Hungar.* **15** (1980), 39–43.
25. J. Komlós and E. Szemerédi, Hamiltonian cycles in random graphs. In *Infinite and Finite Sets* (A. Hajnal. R. Rado and V. T. Sós, eds), Colloq. Math. Soc. J. Bolyai no. 10, North-Holland, Amsterdam, 1975, pp. 1003–1011.
26. W. Kordecki, Probability of connectedness of random graphs [in Russian]. *Prace Naukowe Inst. Mat. Fiz. Teor. Pol. Wroclaw* **9** (1973), 55–65.
27. A. D. Korshunov, On the diameter of random graphs. *Soviet Mat. Dokl.* **12** (1971), 302–305; MR 43 number 6124.
28. A. D. Korshunov, Solution of a problem of Erdös and Rényi on Hamilton cycles in non-oriented graphs. *Soviet Mat. Dokl.* **17** (1976), 760–764.
29. A. D. Korshunov, A solution of a problem of P. Erdös and A. Rényi about Hamilton cycles in non-oriented graphs [in Russian]. *Metody Diskr. Anal. v Teoriy Upr. Syst., Sbornik Trudov Novosibirsk* **31** (1977), 17–56.
30. I. N. Kovalenko, Theory of random graphs [in Russian]. *Kibernetika* **4** (1971), 1–4; English translation in *Cybernetics* **7**, 575–579.
31. J. W. Moon and L. Moser, On the distribution of 4-cycles in random bipartite tournaments. *Canad. Math. Bull.* **5** (1962), 5–12.
32. L. Pósa, Hamiltonian circuits in random graphs. *Discrete Math.* **14** (1976), 359–364.
33. A. Rényi, On connected graphs I. *Publ. Math. Inst. Hungar. Acad. Sci.* **4** (1959), 385–387.
34. V. E. Stepanov, Combinatorial algebra and random graphs. *Theory Prob. Appl.* **14** (1969), 373–399.
35. V. E. Stepanov, Phase transitions in random graphs. *Theory Prob. Appl.* **15** (1970), 187–203.
36. V. E. Stepanov, On the probability of connectedness of a random graph $G_m(t)$. *Theory Prob. Appl.* **15** (1970), 55–67.
37. E. M. Wright, The number of connected sparsely edged graphs. *J. Graph Theory* **1** (1977), 317–330.
38. E. M. Wright, Large cycles in large labelled graphs II. *Math. Proc. Camb. Phil. Soc.* **81** (1977), 1–2.
39. E. M. Wright, The number of connected sparsely edged graphs III. Asymptotic results. *J. Graph Theory* **4** (1980), 393–407.

LONG CYCLES IN SPARSE RANDOM GRAPHS

B. Bollobás, T. I. Fenner and A. M. Frieze

ABSTRACT   It is proved that if c is a sufficiently large constant then almost every graph of order n and size $\frac{1}{2}cn$ contains a cycle of length at least $(1 - c^6 e^{-c})n$.

This note is a continuation of [4]. As in [4], we shall study the maximal length of a path or cycle of a random graph $G_{c/n}$. As is customary, we write G_p for a random element of the probability space $\mathcal{G}(n, p(\text{edge}) = p)$ of all graphs with a fixed set of n labelled vertices, in which the edges are chosen independently and with probability p. Furthermore, we say that *almost every* (a.e.) G_p has a property Q if the probability that a graph $G \in \mathcal{G}(n, P(\text{edge}) = p)$ has Q tends to 1 as $n \to \infty$. For $c > 0$ set

$$1 - \beta(c) = \sup\{\beta \geq 0: \text{a.e. } G_{c/n} \text{ contains a cycle of length at least } \beta n\}.$$

It was proved by Ajtai, Komlós and Szemerédi [1] and by de la Vega [9] that a.e. $G_{c/n}$ contains long cycles and $\beta(c) \leq c_0/c$ for some absolute constant c_0. On the other hand, results of Erdös and Rényi [5] imply that $\beta(c) \geq (c + 1)e^{-c}$. In [4] it was shown that $\beta(c)$ decays exponentially: $\beta(c) < c^{24} e^{-c/2}$. As a consequence of our main result we find that, in fact, $-\log \beta(c) \sim c$ as $c \to \infty$.

THEOREM   *There is a polynomial P of degree at most 6 such that a.e. $G_{n,c/n}$ contains a cycle of length at least $(1 - P(c)e^{-c})n$. In particular,*

$$\beta(c) \leq P(c)e^{-c}.$$

PROOF   We shall show that if c is sufficiently large then $P(c) = c^6$ will do. Our proof is based on the method used by Fenner and Frieze [6] to solve a related problem and on the model of random graphs with a fixed degree sequence introduced by Bollobás [2]. In fact, the present proof is rather close to the proof of the theorem, due to Bollobás [3] and Fenner and Frieze [7], that if k is a sufficiently large constant then a.e. k-regular graph is Hamiltonian. Because of this similarity we shall not give all the details of the proof. We shall start with a lemma enabling us to locate an

GRAPH THEORY AND COMBINATORICS
ISBN 0-12-111760-X

appropriate large subgraph of a.e. $G_{c/n}$. Then we shall show that this subgraph is Hamiltonian for a.e. $G_{c/n}$.

As is customary, we take $V = \{1, 2, \ldots, n\}$. Consider a graph $G \in \mathcal{G}(n, P(\text{edge}) = c/n))$ and define

$$U_0 = \{x \in V : d(x) \leq 6 \text{ or } d(x) \geq 4c\}.$$

Suppose we have constructed a sequence of sets $U_0, U_1, \ldots, U_j$. Set

$$U'_{j+1} = \{x \in V - U_j : |\Gamma(x) \cap U_j| \geq 2\}.$$

If $U'_{j+1} = \varnothing$, stop the sequence. Otherwise let x_{j+1} be the minimal element of U'_{j+1} and put $U_{j+1} = U_j \cup \{x_{j+1}\}$. Suppose the sequence stops with $U_s \neq V$. Let H be the subgraph spanned by $V - U_s$ and write h for the order of H. Then every vertex of H has degree at least 6 and at most $4c$, since every vertex $x \in V - U_s$ has degree at least 7 in G and is joined to at most one vertex of U_s.

LEMMA 1   *Let $\varepsilon > 0$. If c is sufficiently large then a.e. $G_{c/n}$ is such that*

(i) *Any $t \leq n/(6c^3)$ vertices of $G_{c/n}$ span at most $3t/2$ edges.*
(ii) *Any $t \leq n/3$ vertices of $G_{c/n}$ span at most $ct/5$ edges.*
(iii) $n - h < c^6 e^{-c} n$.
(iv) *The set $W = \{x : (1 - \varepsilon)c < d_H(x) < (1 + \varepsilon)c\}$ has at least $(1 - c^{-4})h$ elements, and spans at least $(1 - \varepsilon)ch/2$ edges.*
(v) H *has at most $(1 + \varepsilon)ch/2$ edges.*

PROOF   The proofs of all the assertions are rather straightforward so we shall not give all the details. For $t \leq n/c$ the expected number of t-sets of vertices spanning at least $3t/2$ edges is at most

$$\sum_{u \geq 3t/2} \binom{n}{t} \binom{\binom{t}{2}}{u} \left(\frac{c}{n}\right)^u \left(1 - \frac{c}{n}\right)^{\binom{t}{2} - u} \leq c_i \sum_{u \geq 3t/2} \left(\frac{en}{t}\right)^t \left(\frac{ect^2}{2un}\right)^u e^{-ct^2/2n}$$

$$\leq c_2 \left(\frac{en}{t}\right)^t \left(\frac{ect^2}{3tn}\right)^{3t/2} e^{-ct^2/2n}$$

$$= c_2 \left(\left(\frac{c^3 e^5 t}{27n}\right) e^{-ct/n}\right)^{t/2}.$$

Since

$$\sum_{t=1}^{\lceil n/6c^3 \rceil} \left(\left(\frac{e^5 c^3 t}{27n}\right) e^{-ct/n}\right)^{t/2} = O(n^{-1/2}),$$

assertion (i) follows. The proof of (ii) is similar.

To prove (iii) note first that the expectation of $|U_0|$ is

$$n \sum_{k=0}^{6} \binom{n-1}{k} \left(\frac{c}{n}\right)^k \left(1 - \frac{c}{n}\right)^{n-1-k} + n \sum_{k \geq 4c} \binom{n-1}{k} \left(\frac{c}{n}\right)^k \left(1 - \frac{c}{n}\right)^{n-1-k}$$

$$< \left(\frac{c^6}{700}\right) e^{-c} n + 2 \left(\frac{enc}{4cn}\right)^{4c} e^{-c} n < \left(\frac{c^6}{600}\right) e^{-c} n,$$

provided that c is sufficiently large. It is easily checked that the variance of $|U_0|$ is $O(n)$, so by Chebyshev's inequality $|U_0| < (c^6/500)e^{-c}n$ for a.e. $G_{c/n}$. Now if this last inequality holds and the graph satisfies (i) then $n - h \geq c^6 e^{-c} n$ does not hold since otherwise for some j we have $|U_j| = \lfloor c^6 e^{-c} n \rfloor < n/6c^3$ and this set U_j spans at least $2j \geq \frac{3}{2}|U_j|$ edges. Hence (iii) follows.

Assertion (iv) follows from the fact that if c is sufficiently large and (iii) holds, then the degree sequence of H is close to the degree sequence of $G_{c/n}$: $d_H(X) \geq d(X) - 1$ for every $X \in V(H)$.

Assertion (v) is an immediate consequence of (iii). □

Let us assume that the graph $H = H(G_{c/n})$ in Lemma 1 has vertex set $V(H) = \{1, 2, \ldots, h\}$ and degree sequence $6 \leq d_1 \leq d_2 \leq \ldots \leq d_h \leq 4c$. Let $\mathcal{H} = \mathcal{H}(G_{c/n})$ be the set of all graphs with vertex set $\{1, 2, \ldots, h\}$ and degree sequence $(d_i)_1^h$. Turn $\mathcal{H}$ into a probability space by giving all members of $\mathcal{H}$ the same probability. Note that all members of $\mathcal{H}$ occur as $H = H(G_{c/n})$ with the same probability. Hence a.e. $G_{c/n}$ is such that a.e. element H of $\mathcal{H}(G_{c/n})$ satisfies the conclusions of Lemma 1 with $G_{c/n}$ replaced by H in (i) and (ii).

Consider a graph $G_{c/n}$ which satisfies these conditions. In view of the remarks above, our theorem will follow if we prove that for some $\varepsilon > 0$ and large enough c almost every graph in $\mathcal{H}$ is Hamiltonian.

The graphs in $\mathcal{H}$ are fairly close to being regular, so this assertion resembles the theorem, proved in [3] and [7], that a.e. regular graph is Hamiltonian. Hence it is no surprise that we can adapt the proofs in [3] and [7] to the present case.

In order to study $\mathcal{H}$, we consider the model defined in [2]. Let $D_1, D_2, \ldots, D_h$ be disjoint sets with $|D_i| = d_i$ and set

$$D = \bigcup_1^h D_i, \qquad 2m = |D| = \sum_1^h d_i.$$

A *configuration* C is a partition of D into m pairs, the *edges* of C. Let Φ be the set of all $N(m) = (2m - 1)!! = (2m)!2^{-m}/m!$ configurations. Turn Φ into a probability space by giving all members of Φ the same probability. For $C \in \Phi$ let $\phi(C)$ be the graph with vertex set $\{1, 2, \ldots, h\}$ in which i is joined to j $(i \neq j)$ if and only if C has an edge with one end-vertex in D_i and the other in D_j. Then clearly $\mathcal{H} \subset \phi(\Phi)$ and

$$|\phi^{-1}(H)| = \prod_{1}^{h} d_i!$$

for every $H \in \mathcal{H}$.

Let Q be a property of the graphs in $\mathcal{H}$ and let Q^* be a property of the configurations in Φ. Suppose that these properties are such that for $H \in \mathcal{H}$ and $C \in \phi^{-1}(H)$ the configuration C has Q^* if and only if H has Q. All we shall need from [3] is that in this case, if almost no C has Q^*, then almost no H has Q.

LEMMA 2   *A.e.* $H \in \mathcal{H}$ *is connected.*

PROOF   Let us say that $H \in \mathcal{H}$ has property Q if H is disconnected and satisfies the conclusions of Lemma 1. Let Q^* be such that $C \in \Phi$ has Q^* if and only if $\phi(C) \in \mathcal{H}$ and $\phi(C)$ has Q. We shall show that almost no $C \in \Phi$ has Q^*.

Note that if $C \in \Phi$ has Q^* then there is a set $U \subset \{1, 2, \ldots, h\}$, $1 \leq u = |U| \leq h/2$, such that C is the union of a partition of

$$X = \bigcup_{i \in U} D_i$$

and a partition of $Y = D - X$. If X is odd, this cannot happen. Suppose $|X| = 2x$ and $|Y| = 2y$. Then $6u \leq 2x \leq 4cu$ and $6(h - u) \leq 2y \leq 4c(h - u)$. The probability that C is the union of two such partitions is

$$\frac{(2x-1)!! \, (2y-1)!!}{(2x+2y-1)!!} \leq \left(\frac{(2x)!}{(2(x+y))_{2x}} \right)^{1/2} = \left(\frac{2(x+y)}{2x} \right)^{-1/2} \leq \left(\frac{6h}{6u} \right)^{-1/2}.$$

Hence the probability that C has Q^* is at most

$$\sum_{u=1}^{\lfloor h/2 \rfloor} \binom{h}{u} \binom{6h}{6u}^{-1/2} = O(h^{-2}) = o(1). \quad \square$$

REMARK   The simple proof above implies that if Δ is fixed and $6 \leq d_i = d_i(n) \leq \Delta$, $i = 1, 2, \ldots, n$, then a.e. graph with vertex set $\{1, 2, \ldots, n\}$ and degree sequence $(d_i)_1^n$ is connected.

Let us continue the proof of the theorem. Put $\mathcal{H}_0 = \{H \in \mathcal{H} : H$ satisfies the conclusions of Lemmas 1 and 2$\}$ and let Q be the property that $H \in \mathcal{H}_0$ and H is not Hamiltonian. Let Q^* be such that $C \in \Phi$ has Q^* if and only if $\phi(C)$ has Q and let $\Phi_0 = \{C : C$ has $Q^*\}$. To complete the proof of our theorem we shall show that almost no C has Q^*. This will be done by the colouring method introduced in [6] and used in [3] and [7].

Suppose that $C \in \Phi_0$. Let P_H be a longest path of $H = \phi(C)$, say of length l. Since H is connected and not Hamiltonian, it does not contain a cycle of length l. Consider all red–blue colourings of the edges of C in which there are exactly $3h$ red edges, the red edges join vertices of

$$E = \bigcup_{i \in W} D_i$$

and every edge mapped into an edge of P_H is blue. Colour each element of W with the colour of the edge incident with it. Denote by C^b the subconfiguration of C formed by the blue edges and denote by H^b the corresponding subgraph of H.

By making use of the properties guaranteed by Lemma 1, one can show that there are many colourings of C for which $B = H^b$ is such that

$$|U \cup \Gamma_B(U)| \geq 3|U|$$

whenever $U \subset \{1, 2, \ldots, h\}$ and $|U| \leq h/9$. Combining this with the lemma of Pósa [8], which is often used in the search for Hamilton cycles, one can prove the following crucial lemma. The proof, which is an exact analogue of the proofs in [3] and [7], is omitted.

LEMMA 3   *There is an absolute constant $c_1 > 0$ such that $c \in \Phi_0$ has at least*

$$(1 - 3\varepsilon)^{3h} \binom{m}{3h}$$

colourings with the following properties. Set $u = \lfloor c_1 h \rfloor$. There are distinct red elements $y_1, y_2, \ldots, y_u$ and not necessarily distinct sets $Y_1, Y_2, \ldots, Y_u$ such that each Y_i consists of red elements, $|Y_i| = u$ and the red edge incident with y_i does not join y_i to Y_i. □

Now we are ready to complete the proof of the theorem. Let Ψ_0 be the collection of all coloured configurations such that the configuration belongs to Φ_0 and the colouring satisfies the conclusions of Lemma 3. Then

$$|\Psi_0| \geq |\Phi_0| (1 - 3\varepsilon)^{3h} \binom{m}{3h}. \qquad (1)$$

In order to show that $|\Phi_0|/|\Phi|$ is small, we shall give a suitable upper bound for $|\Psi_0|$. Suppose that $B_0 = C_0^b$ for some $C_0 \in \Psi_0$. At most how many $C \in \Psi_0$ satisfy $C^b = B_0$? When extending B_0 to a configuration $C \in \Psi_0$, we have at most $6h - u$ choices for the red edge, incident with y_i. Hence we have at most $(6h - u)^{\lceil u/2 \rceil}$ choices for the first $w = \lceil u/2 \rceil$ edges incident with $y_1, y_2, \ldots, y_u$. The remaining $3h - w$ edges can be chosen in at most $N(3h - w)$ ways. Hence at most

$$(6h - u)^w N(3h - w) < N(3h) e^{-c_2 h}$$

configurations $C \in \Psi_0$ satisfy $C^b = B_0$, where c_2 is a positive absolute constant.

As clearly

$$|\{C^b: C \in \Psi_0\}| \leq N(m) \binom{m}{3h} \Big/ N(3h),$$

we have

$$|\Psi_0| \leq \left\{ N(m) \binom{m}{3h} \Big/ N(3h) \right\} N(3h) e^{-c_2 h}. \tag{2}$$

Inequalities (1) and (2) imply

$$|\Phi_0| / N(m) \leq |\Psi_0| \Big/ \left\{ N(m) \binom{m}{3h} (1 - 3\varepsilon)^{3h} \right\} = o(1),$$

provided that ε is small enough. This completes our proof. $\square$

With a little more work one can prove that a polynomial of degree 4 will do for P. However, as it is very likely that P can be chosen to be linear, the additional complications are hardly worthwhile.

References

1. M. Ajtai, J. Komlós and E. Szemerédi, The longest path in a random graph. *Combinatorica* **1** (1981), 1–12.
2. B. Bollobás, A probabilistic proof of an asymptotic formula for the number of labelled regular graphs. *Europ. J. Combinat.* **1** (1980), 311–316.
3. B. Bollobás, Almost all regular graphs are Hamiltonian. *Europ. J. Combinat.* **4** (1983), 97–106.
4. B. Bollobás, Long paths in sparse random graphs. *Combinatorica* **2** (1982), 223–228.
5. P. Erdös and A. Rényi, On the evolution of random graphs, *Publ. Math. Inst. Hungar. Acad. Sci.* **5** (1960), 17–61.
6. T. I. Fenner and A. M. Frieze, On the existence of Hamiltonian cycles in a class of random graphs. *Discrete Math.* **45** (1983), 301–305.
7. T. I. Fenner and A. M. Frieze, Hamiltonian cycles in random regular graphs. *J. Combinat. Theory B* (in press).
8. L. Pósa, Hamiltonian circuits in random graphs. *Discrete Math.* **14** (1976), 359–364.
9. W. F. de la Vega, Long paths in random graphs. *Studia Scient. Math. Hungar.* **14** (1979), 335–340.

6

ASPECTS OF THE RANDOM GRAPH

Peter J. Cameron

ABSTRACT Erdös and Rado proved that a certain graph R constructed by Rado is the essentially unique countable random graph. Among other things, in this paper it is shown that R is unchanged by small alterations, it admits a large number of cyclic automorphisms, and Aut(R) contains many countable groups as regular subgroups. Furthermore, several other probability measures on the space of countable graphs are discussed, including one in triangle-free random graphs. The paper contains a number of unsolved problems.

1. Introduction

It is well known that, if a large finite graph is chosen "at random", the probability that it admits any non-trivial automorphisms is very low. In a paper [3] mostly devoted to refining and quantifying this assertion, Erdös and Rényi showed that, for countable graphs, the situation is very different: a countable graph chosen at random will almost surely have infinitely many automorphisms. The reason for this fact is, at first sight, even more surprising than the fact itself. There is a particular graph (which I shall denote by R) which has the property that, if a countable graph is chosen at random, then it is almost surely isomorphic to R.

Erdös and Spencer, in an appendix (entitled "The Kitchen Sink") to their book [4] on probabilistic methods in combinatorics, claim that the theory of countable random graphs is demolished by this result. It is my contention that a new theory is created in its place. Both the graph R and the general construction method are of very great potential interest to permutation group theorists. I would like to take the opportunity of outlining some of the reasons for this. There is also much here to interest combinatorial theorists, including some Ramsey theory and enumeration questions. Connections with model theory also exist, though I shall not discuss these.

Section 2 proves the theorem of Erdös and Rényi, and Section 3 gives Rado's explicit construction of R. The next two sections adapt the Erdös–Rényi argument to show, first, that R admits many cyclic automorphisms, and then that a large number of countable groups can be embedded as regular subgroups of Aut(R). Section 5 shows that R is

GRAPH THEORY AND COMBINATORICS
ISBN 0-12-111760-X

unchanged by certain small alterations, and deduces the existence of further interesting permutation groups. The last three sections discuss some further situations in which similar results hold or might be discovered. The most interesting of these concerns triangle-free graphs, discussed in detail in Section 8. Section 7, on a Ramsey-type theorem, is not strictly necessary, but has been included in deference to another of Erdös' interests.

2. The Erdös–Rényi theorem

Because of its relevance to much that follows, I will take the liberty of repeating the beautiful argument given by Erdös and Renyi. I adopt the convention that subgraphs are always induced subgraphs, and embeddings of graphs are always as induced subgraphs.

Consider graphs Γ on a countable vertex set which have the following property:

(*) *Given any two finite disjoint sets U and V of vertices, there is a vertex z joined to every vertex in U and to none in V.*

If Γ is a graph satisfying (*), the following assertions hold for Γ:

(i) *Any finite graph Δ can be embedded in Γ.*

We construct an embedding ϕ one vertex at a time. If ϕ has been defined on vertices $y_0, \ldots, y_k$ of Γ, let U_0 (respectively V_0) be the set of vertices in $\{y_0, \ldots, y_k\}$ which are joined (respectively not joined) to y_{k+1}, and let $U = \phi(U_0)$, $V = \phi(V_0)$. If z is chosen according to property (*), then we can extend ϕ by setting $\phi(y_{k+1}) = z$.

(ii) *Any countable graph can be embedded in Γ.*

The proof is identical to that of (i).
(The Axiom of Choice is not needed here or subsequently – we are given an enumeration of the vertex set of Γ, and may choose z to be the first vertex in the enumeration which satisfies the appropriate conditions. Note the mechanical nature of the construction.)

(iii) *Any embedding of a finite subgraph Δ_0 of the finite or countable graph Δ into Γ can be extended to an embedding of Δ into Γ.*

Simply change the enumeration of Δ so that the vertices of Δ_0 come first, and begin the construction with the first vertex not in Δ_0.

Now let Γ_1 and Γ_2 be any two countable graphs having property (*).

(iv) *Γ_1 and Γ_2 are isomorphic.*

Modify the argument of (i) so that it proceeds "back and forth". (At even-numbered states, let y be the first vertex of Γ_1 on which ϕ is not yet defined, and extend ϕ to y; at odd-numbered stages, let z be the first vertex of Γ_2 not yet in the image of ϕ and extend ϕ^{-1} to z. This ensures that the embedding $\phi: \Gamma_1 \to \Gamma_2$ is onto.)

(v) *Any isomorphism between finite subgraphs of Γ_1 and Γ_2 can be extended to an isomorphism from Γ_1 to Γ_2.*

The same modification that led from (ii) to (iii) gives this.

At this point, we don't yet know whether any countable graphs with property (*) exist. This is established spectacularly, if non-constructively:

(vi) *If a countable graph is chosen "at random" by selecting edges independently with probability $\frac{1}{2}$, then it almost surely has property (*).*

REMARK   The proposed probability measure can be identified, up to a null set, with Lebesgue measure on the unit interval, by enumerating the pairs of vertices, and mapping the characteristic function of any graph to the real number having the corresponding dyadic expansion.

PROOF OF (vi)   For any given pair U, V of disjoint finite sets of vertices, the event that a given further vertex z is not "correctly joined" (i.e. does not satisfy the conclusions of (*)) is $1 - (1/2^n)$, where $n = |U \cup V|$. By independence, the probability that none of a given set of N further vertices is correctly joined is $(1 - (1/2^n))^N$. Since this can be made arbitrarily small by choice of N, the probability that no vertex is correctly joined is zero. Now the event that (*) fails is the union of countably many such null events, one for each choice of U and V, and hence is null.

(vii) *There exists a countable graph with property (*), unique up to isomorphism. It has infinitely many automorphisms.*

Immediate from (iv), (v) and (vi). Indeed, we see that the automorph-ism group of such a graph acts vertex-transitively, and even transitively on finite subgraphs of any isomorphism type. (This property is sometimes referred to as *ultrahomogeneity*.)

3. Rado's construction

At about the same time as the paper of Erdös and Rényi was published, Rado [5] gave the following construction of a countable graph, which I shall call R. The vertex set of R is the set of natural numbers (including zero); u and v are joined by an edge if and only if 2^u occurs in the dyadic expansion of v (the expression for v as a sum of distinct powers of u) or vice versa. (The edges are undirected.)

It is not immediately obvious that Rado's graph admits any non-trivial automorphisms at all. That it has infinitely many follows from the fact that it satisfies condition (*). This is easily proved. Let U and V be finite disjoint sets of vertices, and suppose (without loss) that $\max(U) > \max(V)$. Then the vertex $z = \sum_{u \in U} 2^u$ has the required properties.

We will use R as a "standard" copy of the countable random graph. Though the construction looks somewhat artificial, it can be motivated as follows. (For this argument I am indebted to Dr A. D. Sands.) A countable graph satisfying (*) can easily be constructed recursively by adding, at each step, new vertices joined to the existing ones in such a way that all instances of (*) for which U and V are made up of existing vertices hold. To simplify matters, we assume that one vertex is added for each pair for which a vertex was not added at an earlier stage, and that the obvious lexicographic order is followed. Then it can be checked that we obtain Rado's graph.

Of course, much more than the existence of infinitely many automorph-isms follows from property (*). For example, the paradox that R has no "obvious" automorphisms can be sharpened by observing that R admits a group of primitive recursive automorphisms acting ultrahomogeneously (in the sense of the last section). To see this, note that the construction of an automorphism extending a given finite isomorphism is entirely mechanical, given a subroutine for finding the first vertex z satisfying the conclusions of (*) for any given sets U and V. In general, such a vertex could be found by considering all vertices in order – success is guaranteed by (*) – but for R we can put an *a priori* upper bound on the length of such a search, viz.

$$\sum_{u \in U} 2^u + 2^{\max(V)+1}.$$

As an exercise, find an automorphism of R interchanging 0 and 1. (Using the above algorithm, 2 and 5 are interchanged, as are 2^2 and 2^5, 2^{2^2} and 2^{2^5}, etc. I know of no very simple description of this function.)

4. Cyclic automorphisms of R

How many automorphisms does R have? The answer has been known for some time. (L. Babai informs me that a more general problem was posed by M. Makkai in about 1969, and solved by a number of people, Babai among them.) The argument given here is designed to provide additional information. From now on, G will denote the automorphism group of R.

PROPOSITION 4.1 G contains $2^{\aleph_0}$ conjugacy classes, each of cardinality $2^{\aleph_0}$, of permutations acting as a single cycle on the vertex set of R.

PROOF   Let $g = (\ldots, x_{-1}, x_0, x_1, x_2, \ldots)$ denote a cyclic permutation of a countable set. If Γ is any G-invariant graph, and if x_0 and x_s are adjacent in Γ, then so are x_i and x_{i+s} for any $i \in \mathbb{Z}$, and in particular, x_{-s} is adjacent to x_0. So Γ is completely determined by the set

$$S = \{s \in \mathbb{Z} \mid s > 0, \ x_0 \text{ and } x_s \text{ are adjacent}\}.$$

Choose a g-invariant graph Γ at random by including natural numbers in S independently with probability $\frac{1}{2}$. I assert that almost surely Γ has property (*). For, given U and V, the vertices x_i for which the events that x_i is joined to vertices in U and V are *not* independent are those for which equations $i - j_1 = j_2 - i$ hold, where $x_{j_1}, x_{j_2} \in U \cup V$. Clearly there are only finitely many such. If they are excluded, then the probability that a further vertex z is not correctly joined is $1 - (1/2^n)$, as before.

Moreover, if $\{x_i \mid i \in I\}$ are such vertices, the events "x_i not correctly joined" (for $i \in I$) fail to be independent only if an equation $i_1 - j_1 = \pm(i_2 - j_2)$ holds, for $i_1, i_2 \in I$, $x_{j_1}, x_{j_2} \in U \cup V$. so, given N such vertices x_i, only finitely many vertices need be excluded for the next choice. The argument then proceeds as before.

This establishes that R admits cyclic automorphisms.

Now, for any cyclic automorphism g of R, given by $g = (\ldots, x_{-1}, x_0, x_1, \ldots)$ say, let $S_g = \{s > 0 \mid x_0 \text{ is adjacent to } x_s\}$. An easy argument shows that two cyclic automorphisms g and h are conjugate in G if and only if $S_g = S_h$. Since almost all subsets of $\mathbb{N} \setminus \{0\}$ occur as S_g for some g, there are $2^{\aleph_0}$ conjugacy classes. It follows that $|G| = 2^{\aleph_0}$. Moreover, the size of the conjugacy class of g is the index of the centralizer of g; since g commutes only with its powers, each such conjugacy class has size $2^{\aleph_0}$. This proves the proposition. $\quad\square$

PROBLEM 4.2   Determine all possible cycle structures for automorphisms of R, with the number and cardinalities of conjugacy classes for each type. (Note, for example, that G contains no finitary permutations.)

PROBLEM 4.3   Is G simple?

5. Cayley graphs and B-groups

The argument in Proposition 4.1 shows that the group G contains an infinite cyclic group acting regularly on the vertices of R (indeed, uncountably many such). In this form it can be generalized by replacing $\mathbb{Z}$ with many other countable groups.

Let X be a group, S a subset of X which is closed under taking inverses and does not contain the identity. The *Cayley graph* $\Gamma(X, S)$ of X with respect to S is the graph whose vertex set is X, in which x and y are adjacent if and only if $xy^{-1} \in S$. (We do not insist that S is a generating set for X; this holds if and only if $\Gamma(X, S)$ is connected.) Note that $\Gamma(X, S)$ admits X as a group of automorphisms, acting by right multiplication.

I shall call the subset S *special* if $\Gamma(X, S) \cong R$, or equivalently, if $\Gamma(X, S)$ has property (*). (It is implicit in the terminology that X is countable and that S is inverse-closed and does not contain 1.) Any special set defines an embedding of X as a regular subgroup of $G = \text{Aut}(R)$.

For elements x and y of a group X, let $T(x, y)$ be the subset $\{z \in X \mid xz^{-1} = zy^{-1}\}$ of X. A countable group X is said to have *property* (†) if the following is true: there do not exist finitely many elements $x_1, \ldots, x_n$, $y_1, \ldots, y_n$, $z_1, \ldots, z_m$ of X, with $x_i \neq y_i$ for $i = 1, \ldots, n$, for which

$$X = \bigcup_{i=1}^{n} T(x_i, y_i) \cup \{z_1, \ldots, z_m\}.$$

Note that $T(x, y) = \{z \in X \mid (xz^{-1})^2 = xy^{-1}\}$; so we see that a sufficient condition for (†) is that any non-identity element of X has only finitely many square roots. (However, this condition is not necessary. The multiplicative group of non-zero rational quaternions has property (†), but -1 has infinitely many square roots.)

PROPOSITION 5.1   *The following assertions about a countable group X are equivalent:*

   (i) X *is isomorphic to a regular subgroup of G.*

(ii)   *X has a special subset.*
(iii) *Almost every inverse-closed subset of $X \backslash \{1\}$ is special.*
(iv)   *X has property ($\dagger$).*

(*The probability measure in* (iii) *is defined by including inverse pairs of non-identity elements independently with probability $\frac{1}{2}$.*)

PROOF   We have already observed the equivalence of (i) and (ii). The proofs that (ii) $\Rightarrow$ (iv) $\Rightarrow$ (iii) follow those given for $X = \mathbb{Z}$ in the last section, while (iii) $\Rightarrow$ (ii) is trivial.   □

We remark that this theorem shows that, if X is a countable group with property ($\dagger$), then a random Cayley graph for X almost surely has automorphism group much larger than X. This too is in striking contrast with the finite case (see Babai and Godsil [1]).

The assertion that a special subset of X gives an embedding of X as a regular subgroup of G can be refined. It is straightforward to show that two regular subgroups of G isomorphic to X are conjugate if and only if the corresponding special subsets are equivalent under an automorphism of X. Hence:

COROLLARY 5.2   *If a countable group X has property ($\dagger$) and has fewer than $2^{\aleph_0}$ automorphisms, then it has $2^{\aleph_0}$ non-conjugate embeddings as a regular subgroup of G.*   □

PROBLEM 5.3   What if $|\mathrm{Aut}(X)| = 2^{\aleph_0}$? It seems unlikely that, even in this case, $\mathrm{Aut}(X)$ can have fewer than $2^{\aleph_0}$ orbits on special subsets of X, since almost all (inverse-closed) sets are special. I cannot prove this, but a very recent theorem of H. D. Macpherson (personal communication) is a hopeful pointer.

THEOREM 5.4   *Let H be a permutation group on a countable set Ω, and suppose that:*

(i) *H acts primitively on Ω (i.e. no non-trivial equivalence relation on Ω is invariant under H).*
(ii) *H has fewer than $2^{\aleph_0}$ orbits on the power set $\mathscr{P}(\Omega)$.*

Then H is k-fold transitive for every natural number k.   □

It should be noted that arbitrary groups satisfying (ii) have a fairly precise description in terms of primitive groups.

PROBLEM 5.5   Is there a permutation group on a countable set Ω with more than $\aleph_0$ but fewer than $2^{\aleph_0}$ orbits on $\mathcal{P}(\Omega)$? (Of course this is only meaningful if the Continuum Hypothesis fails.)

PROBLEM 5.6   What about analogues in higher cardinality?

In the terminology of Wielandt [7], a group X is called a B-*group* (or *Burnside group*) if any primitive permutation group containing a regular subgroup isomorphic to X is doubly transitive; that is, the adjunction of permutations to kill the non-trivial X-invariant equivalence relations must kill all non-trivial X-invariant binary relations. Many finite B-groups are known; the first examples, cyclic groups of composite order, were found by Burnside. Indeed, for almost all n, every group of order n is a B-group. However, no countable group is known to be a B-group at present. Since our group G is primitive but not double transitive, no countable group satisfying (†) can be a B-group. In particular, if non-identity elements of X have only finitely many square roots, then X is not a B-group. (This was first proved, by entirely different methods, by G. Higman (personal communication), in response to a question of K. W. Johnson.)

The group $X = \langle a, b \mid b^4 = 1, b^{-1}ab = a^{-1} \rangle$ does not satisfy (†), since $X = T(1, b^2) \cup T(b, b^3)$. I do not know whether or not X is a B-group. However, the direct product of two copies of X is an example of a group failing to satisfy (†) and yet not a B-group. (The Cayley graph $\Gamma(X \times X, S)$ has a primitive automorphism group, where $S = (X \times \{1\}) \cup (\{1\} \times X) \setminus \{(1, 1)\}$.)

PROBLEM 5.7   Is there a countable B-group? In particular, is the above group X a B-group?

6. Stability properties and further groups

A graph which is as ubiquitous as R might be expected to be impervious to small changes. This is indeed the case. The stability properties permit the construction of various interesting permutation groups related to R. I outline some of these.

First, if finitely many edges in R are deleted, and finitely many new edges are added, the resulting graph is still isomorphic to R. (It is easily seen that condition (*) implies the formally stronger version in which infinitely many vertices z with the required property are asserted to exist. This stronger condition clearly remains valid after a finite amount of tinkering.)

PROBLEM 6.1 Which countable graphs Γ have the property that, whenever finitely many adjacencies in Γ are altered (as above), the result is still isomorphic to Γ? (This property does not characterize R. For example, the graph whose connected components are all the finite connected graphs, each one occurring infinitely often, has the above property; so does its complement, and probably many others too.)

Now let G_1 be the group of permutations which are *almost automorphisms* of R, in the sense that a permutation g is in G_1 if and only if $g(R)$ is obtained from R by altering finitely many adjacencies. Then G_1 is k-fold transitive for every k. For let $(x_1, \ldots, x_k)$ and $(y_1, \ldots, y_k)$ be two k-tuples of distinct vertices. Let R_1 and R_2 be the graphs obtained from R by deleting all edges within $\{x_1, \ldots, x_k\}$ (respectively, within $\{y_1, \ldots, y_k\}$). Then R_1 and R_2 are isomorphic to R. The map $x_i \mapsto y_i$ ($i = 1, \ldots, k$) is an isomorphism between finite subgraphs of R_1 and R_2, and so extends to an isomorphism $g: R_1 \to R_2$; and g is almost an automorphism of R.

REMARK The graph mentioned after Problem 6.1 has a similar property; but we obtain no interesting group in this way, since any finitary permutation is almost an automorphism.

The operation of *switching* a graph with respect to a set Y of vertices, introduced by Seidel [6], consists of changing (as above) all adjacencies between Y and its complement, leaving all those wholly within or wholly outside Y unaltered. It is easy to show that the graph obtained from R by switching with respect to a finite subset is still isomorphic to R. It follows that the group G_2 of permutations which are automorphisms of R up to switching is doubly transitive. It is not triply transitive, since the parity of the number of edges within a 3-set is unaltered by switching. In fact, the stabilizer of a point in G_2 is isomorphic (as permutation group) to G, and G_2 also contains transitive subgroups isomorphic to G.

Finally, we note that R is isomorphic to its complement, since the complement also has property (*). The group G_3 of all automorphisms and anti-automorphisms of R is doubly transitive, and contains G as a normal subgroup of index 2. The group generated by G_2 and G_3 is triply transitive.

7. Hypergraphs

There is a wide generalization of the construction of R. The most general situation involves concepts from model theory, especially the amalgamation property. I do not intend to discuss this here. In this section I will simply give some properties of hypergraphs analogous to R.

The $(m+1)$-uniform hypergraph R_m $(m \geq 1)$ was defined by Rado [5] as follows. The vertex set is $\mathbb{N}$; the $(m+1)$-set $\{x_1, \ldots, x_{m+1}\}$ (with $x_1 < \ldots < x_{m+1}$) is an edge of R_m if and only if $2^{f(x_1, \ldots, x_m)}$ occurs in the dyadic expansion of x_{m+1}, where

$$f(x_1, \ldots, x_m) = \binom{x_1}{1} + \ldots + \binom{x_m}{m}.$$

(Note that f is a bijection from the set of m-tuples of natural numbers in increasing order and $\mathbb{N}$.) It has the following properties:

(i) *Any embedding of a finite sub-hypergraph Δ_0 of a finite or countable $(m+1)$-uniform hypergraph Δ into R_m can be extended to an embedding of Δ into R_m.*

(ii) *Any isomorphism between finite sub-hypergraphs of R_m can be extended to an automorphism of R_m.*

(iii) *If a countable $(m+1)$-uniform hypergraph is chosen at random by selecting edges independently with probability $\frac{1}{2}$, then it is almost surely isomorphic to R_m.*

(iv) $|\mathrm{Aut}(R_m)| = 2^{\aleph_0}$, *and* $\mathrm{Aut}(R_m)$ *is m-fold but not $(m+1)$-fold transitive.*

(The existence of permutation groups with any prescribed degree of transitivity was not widely known until fairly recently; it is interesting that operands for such groups were published twenty years ago.)

(v) *Any countable group can be embedded as a regular subgroup of* $\mathrm{Aut}(R_m)$ *if* $m > 1$.

(vi) *There is an embedding of the free group F of countable rank into* $\mathrm{Aut}(R_m)$ *in such a way that F is m-fold transitive.*

(A similar representation of F has been found, by entirely different methods, by A. Barlotti and K. Strambach (personal communication).)

8. Digression: a Ramsey-type theorem

For the next section, I need a lemma about colourings of subsets of a set. The result has a purely algebraic proof, but I will deduce it here from a Ramsey-type theorem.

Suppose that the k-element subsets of a set X are coloured with r colours $c_1, \ldots, c_r$ (all of which are used). The *colour scheme* of an l-set Y is the r-tuple $(a_1, \ldots, a_r)$, where a_i is the number of k-subsets of Y which have colour c_i. The *colour scheme matrix* of the colouring is the $r \times s$ matrix whose columns are all the (transposed) colour schemes which occur (s in number, say).

LEMMA 8.1 *If* $|X| \geq k + l$ *then the colour scheme matrix has rank* r; *in particular,* $s \geq r$. $\square$

This is proved in [2], and the number $k + l$ is easily seen to be best possible. An alternative proof for large X uses the following theorem, which was proved by R. Graham (personal communication), extending an earlier theorem of mine. It also includes the usual finite Ramsey theorem (which, however, is heavily used in the proof).

THEOREM 8.2 *If* X *is infinite or sufficiently large, then the colours and colour schemes can be ordered in such a way that the first* r *columns of the colour scheme matrix form an upper triangular matrix with non-zero diagonal.* $\square$

The conclusion of the theorem can be put as follows. With these orderings $(c_1, \ldots, c_r)$ and $(\sigma_1, \ldots, \sigma_s)$ of colours and colour schemes, we have that σ_1 only uses colour c_1 (this much is Ramsey's theorem); σ_2 uses c_2 and possibly c_1; σ_3 uses c_3 and possibly c_1 and c_2; and so on. Clearly Lemma 8.1 follows, at least for X sufficiently large.

PROBLEM 8.3 How large is "sufficiently large"? In other words, find estimates for the function $f(k, l, r)$ such that the theorem holds if and only if $|X| \geq f(k, l, r)$.

There are many other interesting questions related to Theorem 8.2, of which I shall mention one.

PROBLEM 8.4 Consider colourings of the k-subsets of X for which the number of colour schemes of l-sets is equal to the number of colours.

Show that there is a function $g(k, l)$ (independent of r) such that the conclusion of Theorem 8.2 holds of $|X| \geqslant g(k, l)$. (This is true for $k = 2$, $l = 3$, with $g(2, 3) = 9$; see [2].)

9. Other probability measures

It is possible to define many different probability measures on the space of all graphs on a given countable vertex set. Sometimes it happens that a random graph with respect to such a measure is almost surely isomorphic to another interesting graph, different from R. Keep in mind, as a prototype, the following situation. There is a countable graph T containing no triangles, and having the following property:

(**) *Whenever U and V are finite disjoint sets of vertices such that U contains no edges, there exists a vertex z joined to every vertex in U and to none in V. This property characterizes T (up to isomorphism) among countable triangle-free graphs. Furthermore, every embedding of a finite subgraph Δ_0 of a finite or countable triangle-free graph Δ into T extends to an embedding of Δ into T; and every isomorphism between finite subgraphs of T extends to an automorphism of T.*

(Note in passing that this gives another interesting permutation group. If x is a vertex of T and $(y_1, \ldots, y_k)$ and $(z_1, \ldots, z_k)$ are two k-tuples of distinct neighbours of x, then the induced subgraphs on $\{x, y_1, \ldots, y_k\}$ and $\{x, z_1, \ldots, z_k\}$ are both stars, and the map $x \mapsto x$, $y_i \mapsto z_i$ $(i = 1, \ldots, k)$ is an isomorphism between them. So the stabilizer of x in $\text{Aut}(T)$ is k-fold transitive, for all k, on the set of neighbours of x.)

How can we choose a triangle-free graph at random? Choosing edges independently with any fixed non-zero probability will almost surely create a triangle. If we consider the pairs of vertices in sequence, and join at random only if such joining would not create a triangle, we obtain a model which will depend very sensitively on the ordering of the pairs.

In general, the right approach is to prescribe the probability of occurrence of any given finite graph as the induced subgraph on a given set of vertices, this probability being the same anywhere in the graph. More formally, assign a number $f(\Gamma)$ to each labelled graph Γ on the vertex set $\{1, \ldots, n\}$ for every n, and decree that, if $x_1, \ldots, x_n$ are any vertices of the countable random graph, then the probability that the induced subgraph on $\{x_1, \ldots, x_n\}$ is isomorphic to Γ (under the map $x_i \mapsto i$, $i = 1, \ldots, n$) is $f(\Gamma)$. There are four obvious assertions that should hold for the function f:

P1. $f(\Gamma) \geq 0$ *for all* Γ.

P2. $f(\Gamma_0) = 1$, *where* Γ_0 *is the graph with no vertices*.

P3. *If* Γ *has vertex set* $\{1, \ldots, n\}$, *then* $f(\Gamma) = \sum f(\Gamma')$, *where the summation is over all graphs* Γ' *on* $\{1, \ldots, n+1\}$ *whose restriction to* $\{1, \ldots, n\}$ *is* Γ.

P4. f *is isomorphism-invariant*.

EXAMPLES   (1) The function $f(\Gamma) = 2^{-\binom{n}{2}}$ satisfies these conditions. It gives the measure described by the statement "choose edges independently with probability $\frac{1}{2}$" used earlier in this paper.
  (2) The function

$$f(\Gamma) = \begin{cases} 1 & \text{if } \Gamma \text{ has no edges,} \\ 0 & \text{otherwise,} \end{cases}$$

satisfies the conditions. Almost surely the random graph is null.
  (3) For the function

$$f(\Gamma) = \begin{cases} 1/3^{n-1} & \text{if } \Gamma \text{ is complete,} \\ 2/3^{n-1} & \text{if } \Gamma \text{ is the disjoint union of two or three complete graphs,} \\ 0 & \text{otherwise,} \end{cases}$$

the random graph is almost surely the disjoint union of three infinite complete graphs.
  (4) The function

$$f(\Gamma) = \frac{1}{2^n} \sum_B \frac{1}{2^{|B|(n-|B|)}}$$

(where the summation is over all bipartite blocks B of Γ) satisfies the conditions. Almost surely the random graph is bipartite. There is a "nice" highly symmetric bipartite graph which occurs with probability 1. This model can be described as follows: choose a random subset of the vertex set as a bipartite block, and then choose edges from that set to its complement independently.

  We can regard P3 and P4 as a system of equations for the values of F on $(n+1)$-vertex graphs, given its values on n-vertex graphs. The numbers of equations and unknowns are the numbers of n- and $(n+1)$-vertex graphs up to isomorphism. Now the matrix of this system of equations has rank equal to the number of equations. (This follows from Lemma 8.1. For consider the colouring of n-sets of vertices of R, where a colour is assigned to each isomorphism class of induced subgraphs. Now isomorphic $(n+1)$-vertex subgraphs have the same colour scheme. The converse is true if and only if the Reconstruction Conjecture holds. Thus, if the Reconstruction Conjecture is valid, then the matrix of the system of equations is just the colour scheme matrix; if not, then some columns may be repeated. But the assertion holds in any case.) We conclude that the equations have solutions, and indeed have many different solutions.

The same argument shows that, given any infinite graph X, we may add to P1–P4 the condition that $f(\Gamma) = 0$ if Γ is not an induced subgraph of X, and the equations will still have solutions. However, it seems difficult to guarantee the existence of solutions such that $f(\Gamma) > 0$ if Γ is an induced subgraph of X. If this could be done, say, for $X = T$ (the "nice" triangle-free graph), we would have a candidate for the required probability measure. Unfortunately, an arbitrary solution of these equations and inequalities has no strong claim to be "natural".

There is, however, another approach, based on the concept of probability as limiting frequency. If Γ is a triangle-free graph on the vertex set $\{1, \ldots, n\}$, and N is an integer greater than n, let $g_N(\Gamma)$ be the proportion of triangle-free graphs on $\{1, \ldots, N\}$ which induce Γ on $\{1, \ldots, n\}$.

PROBLEM 9.1   Does $\lim_{N \to \infty} g_N(\Gamma)$ exist?

If this limit, say $g(\Gamma)$, exists for all triangle-free graphs Γ, then it satisfies P1–P4 and so defines a probability measure with a good claim to naturalness. Of course, questions remain:

PROBLEM 9.2   If $g(\Gamma)$ exists for all Γ, is it positive for all Γ? If so, does the resulting probability measure have the property that a random triangle-free graph is almost surely isomorphic to T?

I conjecture an affirmative answer to all these questions. Consider the smallest interesting case, when Γ is a single edge. Then $g(\Gamma)$ is the limit (if it exists) of the edge-density of labelled triangle-free graphs on N vertices. One might expect bipartite graphs to be a reasonable sample of triangle-free graphs; on this basis, it seems plausible that $g(\Gamma)$ is about $\frac{1}{4}$, though I do not have a proof. For the record, the values of $g_N(\Gamma)$ for

$N \leq 9$ are tabulated:

N	g_N(edge)
2	0.5
3	0.428 57
4	0.390 24
5	0.363 40
6	0.343 24
7	0.327 19
8	0.313 97
9	0.302 79

These results were obtained partly by hand, and partly using a personal microcomputer – a tool which will probably be increasingly used by combinatorialists, being ideal for work of this kind.

References

1. L. Babai and C. D. Godsil, On the automorphism group of almost all Cayley graphs. *Europ. J. Combinat.* **3** (1982), 9–15.
2. P. J. Cameron, Colour schemes. *Ann. Discrete Math.* **15** (1982), 81–95.
3. P. Erdös and A. Rényi, Asymmetric graphs. *Acta Math. Acad. Sci. Hungar.* **14** (1963), 295–315.
4. P. Erdös and J. Spencer, *Probabilistic Methods in Combinatorics*, Academic Press, New York, 1974.
5. R. Rado, Universal graphs and universal functions. *Acta Arith.* **9** (1964), 331–340.
6. J. J. Seidel, Strongly regular graphs of L_2-type and of triangular type. *Indag. Math.* **29** (1967), 188–196.
7. H. Wielandt, *Finite Permutation Groups*, Academic Press, New York, 1964.

Notes added in proof

1.   Dr J. Truss (personal communication) has shown that Aut(R) is simple. All possible cycle structures for automorphisms have been determined. (See Problems 4.2, 4.3.)

2.   It follows from a theorem of Erdös, Kleitman and Rothschild that the probability measure described after Problem 9.1 exists, but that the random graph is almost surely the bipartite graph of Example (4). (I am grateful to Professor V. T. Sós for this information.) On the other hand, Ms J. Covington (personal communication) has found a "natural" probability measure on the space of triangle-free graphs for which the random graph is almost surely isomorphic to T.

PARTIAL EDGE-COLOURINGS OF COMPLETE GRAPHS OR OF GRAPHS WHICH ARE NEARLY COMPLETE

A. G. Chetwynd and A. J. W. Hilton

ABSTRACT Plantholt showed that if q edges are removed from K_{2n+1} then the resulting graph can be edge-coloured with $2n$ colours if and only if $q \geq n$. Andersen and Hilton showed that an edge-colouring of K_r with $2n-1$ colours can be extended to an edge-colouring of K_{2n} with the same set of colours if and only if there are at least $r-n$ edges of each colour in K_r. In this paper an analogue of this result is proved for edge-colourings of the graphs considered by Plantholt – that is graphs obtained from K_{2n+1} by removing n edges. Although any partial edge-colouring of K_n with $2n+1$ colours can be extended to a proper edge-colouring of K_{2n+1} with $2n+1$ colours, we show that the situation can be very different if we consider instead K_{2n+1} with n edges removed and the colouring uses $2n$ colours only. Plantholt's theorem verifies a special case of a conjecture of Vizing. A further special case of Vizing's conjecture is also proved.

1. Introduction

If G is a multigraph, an *edge-colouring* of G is a mapping $\phi : E(G) \to \mathscr{C}$, where $\mathscr{C}$ is a set of colours, such that, if e_1 and e_2 are edges with a common vertex, then $\phi(e_1) \neq \phi(e_2)$. The least number j for which there exists an edge-colouring of G with j colours is called the *chromatic index* of G and is denoted by $\chi'(G)$. Vizing [14] showed that if G is a simple graph then

$$\Delta(G) \leq \chi'(G) \leq \Delta(G) + 1,$$

where $\Delta(G)$ is the maximum degree of G; G is of *Class 1* if $\chi'(G) = \Delta(G)$, and is of *Class 2* otherwise. If $\chi'(G') < \chi'(G)$ for all proper subgraphs G' of G, and if G is Class 2, then G is said to be *critical*.

For an introduction to edge-colouring graphs, and, in particular, for an account in English of Vizing's original argument, the reader is referred to the book by Fiorini and Wilson [5].

In [2] Beineke and Wilson observed that any graph G obtained by removing q edges from K_{2n+1} was of Class 2 and thus needed $2n+1$ colours for an edge-colouring if $q < n$. In [7] Hilton conjectured that if $q = n$ then G would be Class 1, and thus could be edge-coloured with $2n$

GRAPH THEORY AND COMBINATORICS
ISBN 0-12-111760-X

colours. In other words, if $q = n - 1$, then G would be a critical graph. Hilton's conjecture was actually a special case of an earlier conjecture of Vizing [15] that if G is critical then

$$2 |E(G)| \geq |V(G)| (|\Delta(G)| - 1) + 3.$$

In [11] Plantholt proved Hilton's conjecture by an ingenious argument. In [1, Corollary 4.3.3], Andersen and Hilton showed that an edge-colouring of K_r with $2n - 1$ colours can be extended to an edge-colouring of K_{2n} with the same colours if and only if each colour is used on at least $r - n$ edges of K_r. It is natural to wonder whether an analogous result could be formulated and proved which could imply Plantholt's theorem. We show in Section 2 of this paper that there is such a result. A similar method is also used in Section 4 to prove the further case of Vizing's theorem to which attention was specifically drawn by Plantholt in [12]. Andersen and Hilton [1] also showed that any partial edge-colouring of K_n with $2n + 1$ colours can be extended to a proper edge-colouring of K_{2n+1} with $2n + 1$ colours. We show in Section 3 that there may be a very considerable change if n edges are removed from K_{2n+1} and the number of colours is reduced by 1 to $2n$.

We shall make use of the following results. An edge-colouring of a graph is *equalized* if $||E_i| - |E_j|| \leq 1$, whenever E_i and E_j are the sets of edges of G of two distinct colours. McDiarmid [10] and de Werra [16] showed that if G has an edge-colouring with a set of colours, then it has an equalized edge-colouring with the same set of colours. Hoffman and Kuhn [8] showed that a necessary and sufficient condition for a finite family $(A_i : i \in I)$ of subsets of a finite set to have a system $(x_i : i \in I)$ of distinct representatives such that $M \subseteq \bigcup_{i \in I} x_i$ is that both of the following sets of inequalities are satisfied:

I. $|\bigcup_{i \in I'} A_i| \geq |I'|$ $(\forall I' \subseteq I)$,
II. $|\{i : A_i \cap M' \neq \phi \text{ and } i \in I\}| \geq |M'|$ $(\forall M' \subseteq M)$.

We shall call condition I, *Hall's condition*, and condition II, the *marginal condition*.

If a graph is edge-coloured and a colour κ is present on one of the edges incident with some vertex v, we say that κ is *present* at v. If κ is not present on any of the edges incident with v, then we say that κ is *absent* at v. If V is a set of vertices, the set of neighbours in a graph G of vertices of V will be denoted by $N_G(V)$ (or possibly $N(V)$).

2. Step-by-step extensions of edge-colourings

Let G^* be a graph obtained by removing n edges from K_{2n+1}. Let the vertex set of G^* be $\{v_1, v_2, \ldots, v_{2n+1}\}$. If $d_{G^*}(v_1) \leq d_{G^*}(v_2) \leq \ldots \leq$

$d_{G^*}(v_{2n+1})$ then we say that G^* has a *standard vertex labelling*. With respect to a standard vertex labelling of G^*, for $1 \leq i \leq 2n+1$, let G_i^* be the restriction of G^* to $\{v_1, \ldots, v_i\}$, or, in other words, the maximal induced subgraph of G^* with vertex set $\{v_1, \ldots, v_i\}$; let $c_i = 2n - d_{G^*}(v_i)$. Given an edge-colouring with a set of $2n$ colours of G_r^* for some $r, 1 \leq r \leq 2n+1$, let $e_r(\sigma)$ be the number of edges of G_r^* coloured σ.

THEOREM 1   Let $1 \leq r \leq 2n+1$ and let G^* have a standard vertex labelling with respect to which G_r^* is defined.

A necessary and sufficient condition for an edge-colouring of G_r^* with a set $\mathscr{C}$ of $2n$ colours to be extendible to an edge-colouring of G^* with $\mathscr{C}$ is that there are pairwise disjoint sets $C_1, C_2, \ldots, C_r$ of colours of $\mathscr{C}$ such that

(i) $|C_i| = c_i$ $(1 \leq i \leq r)$.
(ii) No colour of C_i is used in G_r^* on an edge on the vertex v_i $(1 \leq i \leq r)$.
(iii)

$$e_r(\sigma) \geq \begin{cases} r - n - 1 & \text{if } \sigma \in C_1 \cup \ldots \cup C_r, \\ r - n & \text{otherwise.} \end{cases}$$

PROOF

(1) *Necessity*: Suppose G^* has an edge-colouring with a set $\mathscr{C}$ of $2n$ colours. For $1 \leq i \leq 2n+1$, the vertex v_i has a set C_i of $2n - d_{G^*}(v_i)$ colours missing from it. The number of edges of G^* is

$$\binom{2n+1}{2} - n = (2n)n,$$

so each colour is used on n edges exactly and is missing from exactly one vertex. Therefore each colour of C_i, the set of colours missing at v_i, is not missing in G^* at any other vertex. Therefore $C_1, \ldots, C_{2n+1}$ are pairwise disjoint, and so $C_1, \ldots, C_r$ are pairwise disjoint.

For $\sigma \in \mathscr{C}$, the number of edges of G^* coloured σ with at least one end on a vertex of $\{v_{r+1}, \ldots, v_{2n+1}\}$ is at most $2n+1-r$. Each colour is used in G^* on exactly n edges. Therefore the number of edges coloured σ in G^* is at least $n - (2n+1-r) = r - n - 1$. On the other hand, if $\sigma \in C_{r+1} \cup \ldots \cup C_{2n+1}$, then there is a vertex in $\{v_{r+1}, \ldots, v_{2n+1}\}$ at which σ does not appear, so the number of edges of G^* coloured σ with at least one end on a vertex of $\{v_{r+1}, \ldots, v_{2n+1}\}$ is at most $2n-r$. It follows that the number of edges coloured σ in G_r^* is at least $n - (2n-r) = r - n$.

(2) *Sufficiency*: Let $1 \leq r \leq 2n+1$, let $\mathscr{C} = \{\sigma_1, \ldots, \sigma_{2n}\}$ and let G_r^* satisfy conditions (i), (ii) and (iii). If $r = 2n+1$ then there is nothing to prove, so suppose $r < 2n+1$. Let $C_1 \cup \ldots \cup C_r = \{\sigma_1, \ldots, \sigma_{q_r}\}$; then $q_r \geq r$. We show that the given edge-colouring of G^* can be extended to an

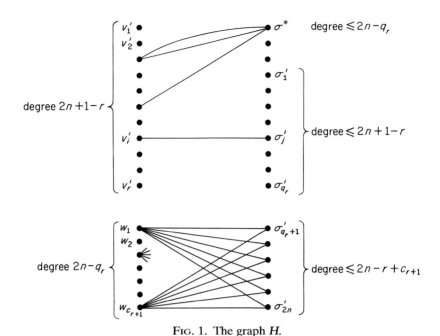

FIG. 1. The graph H.

edge-colouring of G_{r+1}^* satisfying (i), (ii) and (iii) (with $r+1$ replacing r, of course).

The description of the process we carry out is aided by the construction of the bipartite graph H we now describe. The vertex sets of H are $\{v_1', \ldots, v_r'\} \cup \{w_1, \ldots, w_{c_{r+1}}\}$ and $\{\sigma_1', \ldots, \sigma_{2n}', \sigma^*\}$. The edges of H are as follows. For $1 \leq i \leq r$, $1 \leq j \leq 2n$, v_i' is joined to σ_j' by an edge if σ_j is not used on an edge on v_i and is not in the set C_i. The vertex v_i' is joined to σ^* by

$$2n + 1 - r - |\{\sigma_j': \sigma_j' \text{ is joined to } v_i'\}|$$

edges. For $q_r + 1 \leq j \leq 2n$ and $1 \leq i \leq c_{r+1}$, σ_j' is joined to w_i. The bipartite graph H is illustrated in Fig. 1.

Let H' be $H \setminus \{\sigma^*\}$. Then

$$d_H(v_i') = 2n + 1 - r \qquad (1 \leq i \leq r),$$
$$d_H(w_i) = 2n - q_r \qquad (1 \leq i \leq c_{r+1}),$$
$$d_H(\sigma^*) \leq 2n - q_r$$
$$d_H(\sigma_j') \leq \begin{cases} 2n + 1 - r & (1 \leq j \leq q_r), \\ 2n - r + c_{r+1} & (q_r + 1 \leq j \leq 2n). \end{cases}$$

The first and second equalities follow immediately from the definitions of σ^* and of $w_1, \ldots, w_{c_{r+1}}$ respectively. To show the inequality for $d_H(\sigma^*)$,

$$d_H(\sigma^*) = \sum_{i=1}^{r} \{(2n+1-r) - d_{H'}(v_i')\}$$

$$= \sum_{i=1}^{r} \{(2n+1-r) - (d_{G^*}(v_i) - d_{G_r^*}(v_i))\},$$

and so is the number of edges of $\bar{G}$ which join vertices of $\{v_1, \ldots, v_r\}$ to vertices of $\{v_{r+1}, \ldots, v_{2n+1}\}$. This number is less than or equal to

$$2n - \sum_{i=1}^{r} c_i = 2n - q_r.$$

Therefore $d_H(\sigma^*) \leq 2n - q_r$. Finally to show the inequality for $d_H(\sigma_j')$: If $1 \leq j \leq q_r$ then, by assumption, $e_r(\sigma_j) \geq r - n - 1$. Therefore σ_j is not used on at most $r - 2(r-n-1) = 2n - r + 2$ vertices of G_r^*. However, σ_j is in $C_1 \cup \ldots \cup C_r$, so $d_H(\sigma_j') \leq 2n - r + 1$. The argument if $q_n + 1 \leq j \leq 2n$ is similar.

Let M be the set of those σ_j' such that

$$e_r(\sigma_j) = \begin{cases} r - n - 1 & \text{if} \quad 1 \leq j \leq q_r, \\ r - n & \text{if} \quad q_r + 1 \leq j \leq 2n. \end{cases}$$

These σ_j' will be called *marginal vertices* and the corresponding colours σ_j will be called *marginal colours*.

We would like to find a set J of independent edges of H' which covers each vertex of $\{v_1', \ldots, v_r'\} \cup \{w_1, \ldots, w_{c_{r+1}}\}$ except for those v_i' such that v_i is not joined in G_{r+1}^* to v_{r+1}, and which also covers each vertex of M. For if $v_i' \sigma_j' \in J$ then we colour $v_i v_{r+1}$ with σ_j, and if $w_i \sigma_j' \in J$ then we place σ_j in the set C_{r+1}. Then c_{r+1} colours will be placed in C_{r+1}. Furthermore, any marginal colour will *either* be used to colour an edge from v_{r+1} to one of $\{v_1, \ldots, v_r\}$, *or* will be placed in C_{r+1}. By relabelling if necessary, we may assume that $C_{r+1} = \{\sigma_{q_r+1}, \ldots, \sigma_{q_{r+1}}\}$. In the first case we will have

$$e_{r+1}(\sigma_j) = \begin{cases} (r+1) - n - 1 & \text{if} \quad 1 \leq j \leq q_r, \\ (r+1) - n & \text{if} \quad q_{r+1} + 1 \leq j \leq 2n. \end{cases}$$

In the second case we will have

$$e_{r+1}(\sigma_j) = r - n = (r+1) - n - 1 \quad \text{and} \quad q_r + 1 \leq j \leq q_{r+1}.$$

Thus marginal colours will satisfy (iii) with $r+1$ replacing r, but will be marginal in that case also. Clearly non-marginal elements will satisfy (iii) with $r+1$ replacing r.

Furthermore, $C_i \cap C_{r+1} = \phi (1 \leq i \leq r)$, and it is easy to check that the remaining parts of the conditions (i), (ii) and (iii) will be satisfied with $r+1$ replacing r. Thus it remains to demonstrate the existence of a suitable set J. We do this by verifying the Hoffman–Kuhn inequalities.

We show first that Hall's condition is satisfied. Let $W' \subseteq \{w_1, \ldots, w_{c_{r+1}}\}$ and $V' \subseteq \{v_1', \ldots, v_r'\}$ and let V' contain no vertex v_i' such that v_i is not joined in G to v_{r+1}. Clearly if $W' \neq \phi$ then $|N_H'(W')| = 2n - q_r \geq |W'|$. We consider various cases with $V' \neq \phi$.

Case H1 $c_{r+1} \geq 3$ or $(c_{r+1} = 2$ and $c_{r+2} \neq 0)$. Then

$$|N_{H'}(V' \cup W')| \geq |N_{H'}(V')|$$
$$\geq |N_{H'}(v_i')| \quad \text{for some } v_i',$$
$$\geq 2n - c_i - d_{G_r}(v_i).$$

If $2n - c_i - d_{G_r}(v_i) \geq r + c_{r+1}$ then we obtain

$$|N_{H'}(V' \cup W')| \geq |\{v_1, \ldots, v_r\} \cup \{w_1, \ldots, w_{c_r}\}| \geq |V'| + |W'|,$$

as required. If, however, $2n - c_i - d_{G_r}(v_i) < r + c_{r+1}$ then

$$\sum_{j=1}^{2n} c_j = 2n < 2r + c_i + c_{r+1} - 1 \leq 2r + c_1 + c_{r+1} - 1,$$

so

$$\sum_{j=2}^{r} c_j + \sum_{j=r+2}^{2n} c_j < 2r - 1.$$

Since

$$\left| \sum_{j=2}^{r} c_j \right| \geq 2(r-1)$$

it follows that

$$\sum_{j=r+2}^{2n} c_j = 0$$

and that $c_2 = \ldots = c_r = 2$. Consequently either $c_{r+1} \leq 1$, or $c_{r+1} = 2$ and $c_{r+2} = 0$, cases which are considered below.

Case H2 $c_{r+1} = 2$ and $c_{r+2} = \ldots = c_{2n+1} = 0$. Then

$$d_H(\sigma^*) \leq 2n - q_r = 2n - \sum_{i=1}^{r} c_i \leq 2n - (2n - 2) = 2.$$

Therefore

$$d_{H'}(v'_i) \geq (2n+1-r)-2 = 2n-r-1 \quad (1 \leq i \leq r).$$

If $2n-r-1 \geq r+2$ then, for some i, $1 \leq i \leq r$,

$$|N_{H'}(v'_i)| \geq 2n-r-1 \geq r+2 = r+c_{r+1} \geq |V'|+|W'|,$$

from which Hall's condition follows. On the other hand, if $2n-r-1 < r+2$ then $2n < 2r+3$, so $n-\frac{3}{2} < r$. But

$$2(r+1) \leq \sum_{i=1}^{r+1} c_i = 2n,$$

so $r \leq n-1$. Therefore, in this case, $r = n-1$ and $c_1 = \ldots = c_{r+1} = 2$. It follows that $d_{H'}(v'_i) = 2n-r$ or $2n-r+1$ $(1 \leq i \leq r)$. Therefore, for some i, $1 \leq i \leq r$,

$$|N_{H'}(v'_i)| \geq 2n-r = n+1 = r+2 \geq |V'|+|W'|,$$

from which Hall's condition again follows.

Case H3 $\quad c_{r+1} \leq 1$. Then $|W| \leq 1$. Consider the minimal subgraph H'' of H containing the set $V' \cup W'$ of vertices and all edges incident with these vertices. The number of edges in it is

$$|V'|(2n+1-r) \qquad \text{if} \quad |W'| = 0,$$
$$|V'|(2n+1-r)+(2n-q_r) \quad \text{if} \quad |W'| = 1.$$

Each vertex σ'_i has degree in H (and therefore in H'') at most $2n+1-r$, and σ^* has degree in H at most $2n-q_r$. If follows that if $d_{H''}(\sigma^*) < 2n-q_r$ then the number of σ-vertices in H'' is at least $|V'|+|W'|+1$, so

$$|N_{H'}(V')|+|N_{H'}(W')| \geq |V'|+|W'|. \tag{1}$$

If $d_{H''}(\sigma^*) = 2n-q_r = 0$, then (1) is similarly true. If $d_{H''}(\sigma^*) = 2n-q_r > 0$ then, for each v_i $(1 \leq i \leq r)$ such that v_i is joined in $\bar{G}$ to a vertex in $\{v_{r+1}, \ldots, v_{2n+1}\}$, the corresponding v'_i is in V'. Therefore, in this case, there is a vertex v_i joined in $\bar{G}$ to a vertex of $\{v_{r+1}, \ldots, v_{2n+1}\}$ and, for each such v_i, the corresponding vertex v'_i is in V'. (This follows from the fact that an edge $v'_i\sigma^*$ is in H if $|\{\sigma'_j: \sigma'_j \text{ is joined in } H \text{ to } v'_i\}| < 2n+1-r$.) In this case, therefore, $c_{r+1} = 1$ and there is a vertex v_i (with $v'_i \in V'$) joined in $\bar{G}$ to v_{r+1}; i.e. v_i is not joined in G to v_{r+1}. But this possibility is excluded from consideration by our initial choice of V'.

Thus, in all cases, Hall's condition is satisfied.

Next we show that the marginal condition is satisfied. First we note that if

$$e_r(\sigma) > \begin{cases} r-n-1 & \text{for } \sigma \in C_1 \cup \ldots \cup C_r, \\ r-n & \text{otherwise,} \end{cases}$$

then σ is not marginal. Therefore, if $n > r$ there are no marginal symbols, so that the marginal condition is satisfied vacuously.

Suppose therefore from now that $r \geq n$. Then it follows that $c_{r+1} \leq 1$. For if $c_{r+1} \geq 2$ we would obtain the following contradiction:

$$2n = \sum_{i=1}^{2n+1} c_i \geq \sum_{i=1}^{r+1} c_i \geq 2(r+1) > 2r \geq 2n.$$

Thus $c_{r+1} \leq 1$. Therefore there is at most one vertex, say $v^\dagger$, such that $v^\dagger \in \{v_1, \ldots, v_r\}$ and $v^\dagger$ is not joined in G to v_{r+1}.

Let M' be a set of marginal elements σ_i. We wish to show that

$$|N_{H'}(M')| \geq |M'| \qquad \text{if } v^\dagger \text{ does not exist,}$$

$$|N_{H'}(M') \backslash \{v\dagger'\}| \geq |M'| \qquad \text{otherwise.}$$

Consider the subgraph of H' consisting of the vertices of M' and the vertices of

$$N_{H'}(M') \qquad \text{if } v^\dagger \text{ does not exist,}$$

$$N_{H'}(M') \backslash v^{\dagger'} \qquad \text{otherwise,}$$

and all edges of H' between these vertices. The number of such edges

$$\begin{cases} = |M|(2n+1-r) & \text{if } v^\dagger \text{ does not exist,} \\ \geq |M|(2n+1-r)-(2n-r) = (|M|-1)(2n+1-r)+1 & \text{otherwise.} \end{cases}$$

The marginal condition now follows from the fact that the maximum degree in the subgraph is $2n+1-r$.

This proves the sufficiency. □

3. Extending partial edge-colourings

A *partial edge-colouring* of a graph G is an edge-colouring of some subgraph G' of G. It is shown in [1] that any partial edge-colouring of K_n with at most $2n-1$ colours can be extended to an edge-colouring of K_{2n} with $2n-1$ colours. We investigate the analogous problem for partial edge-colourings of G_r^*, and show that the situation is surprisingly different.

First, for $r \leq n+1$ we give necessary and sufficient conditions for extendibility.

THEOREM 2 *Let G^* have a standard vertex labelling. Let $1 \leq r \leq n+1$, let G_r^* be edge-coloured with $2n$ colours, and, for $1 \leq i \leq r$, let D_i be the set of colours not used on the vertex v_i. A necessary and sufficient condition for the edge-colouring of G_r^* with $2n$ colours to be extendible to an edge-colouring of G^* with $2n$ colours is that*

$$\left| \bigcup_{i \in I} D_i \right| \geq \sum_{i \in I} c_i \qquad (\forall I \subseteq \{1, \ldots, r\}). \tag{2}$$

PROOF In order to satisfy conditions (i) and (ii) of Theorem 1 with $r \leq n+1$, it is necessary and sufficient that the family

$$(D_{1,1}, \ldots, D_{1,c_1}, D_{2,1}, \ldots, D_{2c_2}, \ldots, D_{r,1}, \ldots, D_{r,c_r}),$$

where, for $1 \leq i \leq r$, $D_{i,1} = \ldots = D_{i,c_i} = D_i$, have a system of distinct representatives. Therefore, by Hall's theorem [4], it is necessary and sufficient that

$$\left| \bigcup_{i \in I} \bigcup_{j \in J_i} D_{ij} \right| \geq \sum_{i \in I} |J_i| \qquad (\forall I \subset \{1, \ldots, r\}, i \in I, J_i \subset \{1, \ldots, c_i\}),$$

or, in other words, that

$$\left| \bigcup_{i \in I} D_i \right| \geq \sum_{i \in I} c_i \qquad (\forall I \subset 1, \ldots, r\}).$$

This shows the necessity of (2).

To show the sufficiency, we have yet to show that, in addition, condition (iii) of Theorem 1 may be satisfied. If $r \leq n$ then condition (iii) is satisfied vacuously. If $r = n+1$ suppose that $c_1, \ldots, c_{n+1}$ have been chosen so that (i) and (ii) are satisfied, but (iii) is not satisfied. Then there is a colour σ which is not in $C_1 \cup \ldots \cup C_{n+1}$ and which has not been used on any edges of G_{n+1}. But

$$|C_1 \cup \ldots \cup C_{n+1}| \geq n+1,$$

and the number of colours used on edges of G_{n+1}^* is at least $n-1$, since there is a vertex of G_{n+1}^* of degree at least $n-1$ (and the edges on it will all be coloured differently). Since $(n-1)+(n+1) = 2n$, there is a colour used in the edge-colouring of G_{n+1}^* which is also in $C_1 \cup \ldots \cup C_{n+1}$. We can remove this colour from $C_1 \cup \ldots \cup C_{n+1}$ and substitute it with σ. This operation may be repeated if necessary. This proves Theorem 2. $\square$

COROLLARY 1   *If $2n - r + 1 \geq c_1 + \ldots + c_r$ then any partial edge-colouring of G_r^* with $2n$ colours can be extended to an edge-colouring of G^*.*

PROOF   Since the greatest degree in G_r^* of any vertex of G_r^* is less than or equal to $r - 1$, it follows that

$$|D_i| \geq 2n - r + 1 \qquad (1 \leq i \leq r).$$

Therefore

$$\left| \bigcup_{i \in I} D_i \right| \geq 2n - r + 1 \geq \sum_{i \in I} c_i \qquad (\forall I \subseteq \{1, \ldots, r\}),$$

so Corollary 1 follows from Theorem 2.   □

The next theorem shows a sharp contrast to the fact that a partial edge-colouring of K_n with $2n - 1$ colours can always be extended to an edge-colouring of K_{2n} with $2n - 1$ colours.

THEOREM 3   *If r is even and*

$$\binom{r/2}{2} \geq n \geq \frac{r}{2}$$

then there are examples of graphs G^* and partial edge-colourings of G_r^* with $2n$ colours such that the partial edge-colouring of G_r^* cannot be extended to an edge-colouring of G.

PROOF   Form a graph G^* by removing from K_{2n+1} n edges whose end-vertices all lie in a set $\{v_1, \ldots, v_{r/2}\}$ of $r/2$ vertices. This is possible since

$$2n \geq r \quad \text{and} \quad \binom{r/2}{2} \geq n.$$

Let F be a set of $r/2$ independent edges of G_r^*, each edge having exactly one end in the set $\{v_1, \ldots, v_{r/2}\}$, and colour G_r^* with ($\leq 2n$) colours in such a way that the edges of F all receive the same colour. Then

$$\left| \bigcup_{i \in \{1, \ldots, r\}} D_i \right| \leq 2n - 1$$

whereas

$$\sum_{i \in \{1, \ldots, r\}} c_i = 2n.$$

Therefore, by Theorem 2, the edge-colouring of G_r^* cannot be extended to an edge-colouring of G^* with $2n$ colours. This proves Theorem 3. $\square$

By contrast, we show in Theorem 4 that if $c_1 = n$, then any edge-colouring of G_{n+1}^* with $2n$ colours can be extended to an edge-colouring of G^* with $2n$ colours.

THEOREM 4   *If $c_1 = n$, then any edge-colouring of G_{n+1}^* with $2n$ colours can be extended to an edge-colouring of G^* with $2n$ colours.*

PROOF   Since $c_1 = n$ it follows that $c_2 = \ldots = c_{n+1} = 1$ and that, in G_r, v_1 is not joined to any other vertex; thus $|D_1| = 2n$. Since $|D_i| \geq n + 1 \ (2 \leq i \leq n + 1)$ it is easy to see that (2) is satisfied, so Theorem 4 follows from Theorem 2. $\square$

Given r even, let n_0 be the least integer such that if $n \geq n_0$ then, for any graph G^* formed from K_{2n+1} by removing n edges, any edge-colouring of G_r^* with $2n$ colours can be extended to an edge-colouring of G. Theorem 3 shows that

$$n_0 \geq \binom{r/2}{2} + 1$$

and focuses attention on what the value of n_0 might be.

CONJECTURE 1   *If r is even, then*

$$n_0 = \binom{r/2}{2} + 1.$$

4. Another proof of Plantholt's theorem and a proof of the "next case" of Vizing's conjecture

In this section we give another proof of Plantholt's theorem that if n edges are removed from K_{2n+1} then the resulting graph (our G^*) can be edge-coloured with $2n$ colours. We also prove the "next case" of Vizing's conjecture, namely that if $2n$ edges are removed from K_{2n+1}, giving a graph H, and if the maximum degree of H is $2n - 1$, then H is edge-colourable with $2n - 1$ colours.

First we need the following lemma:

LEMMA 1   *Let G be a multigraph with at most two vertices b (and possibly c) of highest degree, let all the non-simple edges be incident with b, and, if b and c are joined by more than one edge, let there be a vertex w such that w is joined to c but not to b. Let G not contain a subgraph on three vertices with $\Delta(G) + 1$ edges. Then*

$$\chi'(G) = \Delta(G).$$

PROOF   Let W be the set of vertices of G joined in G to b by non-simple edges. Let H and H^* be the simple subgraph and the sub-multigraph respectively of G induced by $W \cup \{b\}$. We show first that $\chi'(H^*) \leq \Delta(G)$.

If H is Class 1, then edge-colour H with $|V(H)| - 1$ colours, and extend this to an edge-colouring of H^* by colouring each of the extra edges on b with an extra colour. Then the number of colours used will be $d_{H^*}(b) \leq \Delta(G)$.

If H is Class 2, then edge-colour H with $|V(H)|$ colours. If the colour, say α, missing at b is also missing at some other vertex, say v^*, then colour one of the extra edges joining b to v^* with α, and colour the remaining extra edges on b with an extra colour. Then the number of colours used will be $d_{H^*}(b) \leq \Delta(G)$. On the other hand, if α is missing at no other vertex, then $|V(H)|$ is odd. Colour the extra edges on b with extra colours. Then the total number of colours used will be $d_{H^*}(b) + 1$. If $d_{H^*}(b) + 1 \leq \Delta(G)$ then this is the desired edge-colouring of H^*. Since $d_{H^*}(b) \leq \Delta(G)$, we have used at most $\Delta(G) + 1$ colours, or at most one colour too many. In that case we replace the colour α on all the $\frac{1}{2}(|V(H)| - 1)$ edges of H at which it now occurs by extra colours. Provided that $|V(H)| \geq 5$, there is such an extra colour which is not present in H^* on either of the vertices of an edge coloured α. Thus in this case, if $|V(H)| \geq 5$, then $\chi'(H^*) \leq \Delta(G)$. If $|V(H)| = 3$, since G does not contain a subgraph on three vertices with $\Delta(G) + 1$ edges, the number of edges in H^* is at most $\Delta(G)$, so clearly $\chi'(H^*) \leq \Delta(G)$.

We now show how to obtain from this an edge-colouring of G with $\Delta(G)$ colours. First colour $E(G \backslash H^*)$ with $\Delta(G)$ colours. This can be done by a result of Vizing [**14**], since $G \backslash H^*$ has at most one vertex of degree $\Delta(G)$. Then colour the edges of G joining H^* to $G \backslash H^*$ using Vizing's original argument [**14**], always having the pivot vertex of each of the fans in $V(G \backslash H^*)$, and choosing the last edge, or the last two edges, as follows. If b is the only vertex of maximum degree, choose an edge on b last. If b and c both exist, suppose first that $c \in V(G \backslash H^*)$. If c is joined to b, then colour the edge bc last. If c is not joined to b, colour an edge on b next to last, and an edge on c last. If $c \notin V(H^*)$ then choose the edge wc last and an edge on b next to last. Vizing's original argument will apply as, at any stage during the construction of the fans, there will always be a further colour available on the vertex at the other end from the pivot of the most recently adjoined edge; except at the final stage, such a further colour is then used to define the next edge of the fan. This yields the desired colouring of G with $\Delta(G)$ colours. $\square$

Theorem 1, of course, provides a direct proof of Plantholt's theorem. We next give another way of proving Plantholt's theorem, also using Theorem 1; the amount by which the proof depends on Theorem 1 varies (according to the choice of r).

PROOF OF PLANTHOLT'S THEOREM   Let $1 \leqslant r \leqslant 2n$. From G_r^* construct a multigraph G_r^{**} by adjoining a further vertex v^* and, for $1 \leqslant i \leqslant r$, c_i edges joining v^* to v_i. By Lemma 1,

$$\chi(G_r^{**}) \leqslant \sum_{1 \leqslant i \leqslant r} c_i \leqslant 2n.$$

An edge-colouring of G_r^{**} can be equalized; then the conditions of Theorem 1 are satisfied by G_r^*, so that G^* can be edge-coloured with $2n$ colours. $\square$

When $r = 2n$, Theorem 1 is so trivial that the above proof of Plantholt's theorem is really self-contained. The proof of the next theorem, the "next case" of Vizing's conjecture, is essentially an imitation of the "$r = 2n$ proof" above of Plantholt's theorem.

THEOREM 5   *Let H be a simple graph such that*

$$|V(H)| = 2n + 1, \quad |E(H)| = \binom{2n+1}{2} - 2n \quad and \quad \Delta(H) = 2n - 1.$$

Then $\chi'(H) = 2n - 1$.

PROOF   First suppose that $n \geqslant 5$. There is in H a vertex, say v_{2n+1}, of degree $2n - 1$. Let the other vertices be $v_1, \ldots, v_{2n}$. Let $H' = H \backslash v_{2n+1}$. Then H' has one vertex, say v_{2n}, of degree at most $2n - 1$, the remainder having degree at most $2n - 2$. For $1 \leqslant i \leqslant 2n + 1$, let $h_i = (2n-1) - d_H(v_i)$. Let H'^* be formed from H' by adjoining a further vertex v^* and inserting h_i edges between v^* and v_i, for $1 \leqslant i \leqslant 2n$. Then

$$d_{H'^*}(v^*) = \sum_{1 \leqslant i \leqslant 2n} h_i = 2n - 1.$$

Also $d_{H'^*}(v_{2n}) = 2n - 1$ and $d_{H'^*}(v_i) = 2n - 2$ ($1 \leqslant i \leqslant 2n - 1$). Except in the case when v_{2n} and v^* are joined by more than one edge, when there are no further multiple edges and when each other vertex is either joined to both, or to neither of v_{2n} and v^*, then, by Lemma 1, H'^* has an edge-colouring with $2n-1$ colours. In the exceptional case, remove an independent set F of n edges containing an edge on v_{2n} and an edge on v^* (but not an edge joining v_{2n} to v^*) and avoiding a vertex v' which was not joined to either of v_{2n} and v^*. Since $n \geq 5$, it is easy to see that there is such an F. Let e be an edge on v'. By Lemma 1, colour $H^* \backslash (F \cup e)$ with $2n-2$ colours. Then colour e also, using Vizing's argument with v' as the pivot vertex. Then again we obtain an edge-colouring of H'^* with $2n-1$ colours. In both cases, in the edge-colouring obtained, $v_1, \ldots, v_{2n-1}$ each have one colour missing, and, furthermore, each colour is missing from exactly one vertex. Therefore v_{2n+1} can be adjoined to H'^* and, for $1 \leq i \leq 2n-1$, the edge $v_i v_{2n+1}$ inserted with the

CHART 1

Number of edges	Class
$\binom{2n+1}{2}$	
$\vdots$	2
$\binom{2n+1}{2} - (n-1)$	
$\binom{2n+1}{2} - n$	1
$\binom{2n+1}{2} - (n+1)$	
$\vdots$	2 if $\Delta(G) = 2n-1$
$\binom{2n+1}{2} - (2n-1)$	1 if $\Delta(G) = 2n$
$\binom{2n+1}{2} - 2n$	1
$\binom{2n+1}{2} - (2n+1)$	
$\vdots$	2 if $\Delta(G) = 2n-2$
$\binom{2n+1}{2} - (3n-1)$	1 if $\Delta(G) \geq 2n-1$

colour on it being the colour missing at v_i in H'^*. Finally v^* can be deleted. This yields the graph H edge-coloured with $2n-1$ colours.

For $n \leq 4$, the theorem is easily deduced from results of Chetwynd and Yap [4] and Fiorini [5]. This proves Theorem 5. □

From Plantholt's theorem and Theorem 5 one may easily deduce that the edge-chromatic class which a graph on $2n+1$ vertices with at least $\binom{2n+1}{2} - (3n-1)$ edges belongs to is determined solely by the maximum degree and the number of edges, as indicated in Chart 1.

5. Some final remarks and conjectures

Plantholt's theorem and Theorem 5 verify the cases $r = 1$ and $r = 2$ of the following conjecture:

CONJECTURE 2 *Let* $1 \leq r \leq n$. *Let* G *be a simple graph with* $2n+1$ *vertices and maximum degree* $\Delta(G) = 2n+1-r$. *Then* G *is Class 2 if and only if, for some s such that* $0 \leq s \leq (r-1)/2$ *and for some set* $\{v_1, \ldots, v_{2s}\} \subset V(G)$,

$$|E(G) \setminus \{v_1, v_2, \ldots, v_{2s}\}| > \binom{2n+1-2s}{2} - (r-2s)(n-s).$$

For $1 \leq \sigma \leq (r-1)/2$, examples of graphs which do not satisfy one of these inequalities with $1 \leq s < \sigma$, but do satisfy the inequality with $s = \sigma$, may be constructed as follows. Let H be a graph obtained from a $K_{2(n-s)+1}$ by removing $(r-2s)(n-s)-1$ edges in such a way that the maximum degree $\Delta(H)$ of H is given by $\Delta(H) = 2(n-s)+1-(r-2s) = 2n-r+1$. The graph H^* consisting of H together with $2s$ isolated vertices is such an example. It is easy to see that H^* is Class 2 since

$$|E(H^*)| = \binom{2(n-s)+1}{2} - [(r-2s)(n-s)-1]$$

$$= [2(n-s)+1-(r-2s)](n-s)+1,$$

and no set of independent edges of H^* can consist of more than $(n-s)$ edges. Provided that the maximum degree is not increased, edges can be adjoined to this example to yield further examples.

When $r = 1$ or 2, Conjecture 2 is equivalent to the following conjecture:

CONJECTURE 3 *Let* $1 \leq r \leq n$. *Let* G *be a regular multigraph on* $2n+2$

vertices of degree $2n + 1 - r$ in which all non-simple edges are on the same vertex. Let G not contain a subgraph on three vertices with $2n + 2 - r$ edges. Then $\chi'(G) = 2n + 1 - r$.

In the case when G is a simple graph, we believe that Conjecture 3 has occurred to others, but, so far as we are aware, has not appeared in print. However, Dr R. Häggkvist has reported to us that he has proved Conjecture 3 (in the case when G is simple) for $1 \leqslant r \leqslant n/64$, but that, although some improvement may be possible, his method cannot be used as far as $r = n$. In the case when $r = 3$ and the complement of G is the union of three 1-factors, Conjecture 3 has been verified by Rosa and Wallis [13].

[*Late addition*: The authors have proved Conjecture 3 in the cases when $(2n + 1 - r) \geqslant 6(2n + 2)/7$ or $1 \leqslant r \leqslant 4$.]

Conjecture 2 implies the following conjecture.

CONJECTURE 4   *Let $1 \leqslant r \leqslant n$. Let G be a critical graph with $2n + 1$ vertices and maximum degree $2n + 1 - r$. Then*

$$|E(G)| = \binom{2n + 1}{2} - rn + 1.$$

For $2 < r \leqslant n$ and for graphs of maximum degree $2n + 1 - r$ with $2n + 1$ vertices, Conjecture 4 is stronger than the conjecture of Vizing referred to in the Introduction. Whereas in Conjecture 4 a critical graph has $2n^2 - (r - 1)n + 1$ edges, according to the Vizing conjecture a critical graph has at least $2n^2 - (r - 1)n - \frac{1}{2}(r - 3)$ edges. For $1 \leqslant r \leqslant 2$, the two conjectures coincide.

The restriction $r \leqslant n$ in Conjectures 2, 3 and 4 would be best possible, as the example of two disjoint K_{n+1}s, when n is even, shows that r cannot be increased to $n + 1$ in these conjectures.

A good first step towards proving the $r = 3$ case of Conjecture 2 might be to prove the following conjecture, which we feel is of some interest in its own right.

CONJECTURE 5   *The only connected Class 2 simple graphs with exactly three vertices of maximum degree are those obtained from K_{2n+1}, for some $n \geqslant 1$, by removing $(n - 1)$ independent edges.*

[*Late addition*: The authors have proved this conjecture.]

We also have the following conjecture.

CONJECTURE 6   *If a regular multigraph on $2n$ vertices of degree $2n - 1$ has* *no submultigraph consisting of three vertices and* $2n + 1$ *edges, then it can be edge-coloured with* $2n$ *colours.*

Of course, Conjecture 6 is true for simple graphs.

Finally we remark that in [3] we have obtained some results on the chromatic index of graphs of even order, generalizing a further earlier result of Plantholt [12].

References

1. L. D. Andersen and A. J. W. Hilton, Generalized Latin rectangles II: Embedding. *Discrete Math.* **31** (1980), 235–260.
2. L. W. Beineke and R. J. Wilson, On the edge-chromatic number of a graph. *Discrete Math.* **5** (1973), 15–20.
3. A. G. Chetwynd and A. J. W. Hilton, The chromatic index of graphs of even order with many edges. *J. Graph Theory* (to appear).
4. A. G. Chetwynd and H. P. Yap, Chromatic index critical graphs of order 9. *Discrete Math.* **47** (1983), 23–33.
5. S. Fiorini and R. J. Wilson, *Edge-colourings of Graphs*, Research Notes in Mathematics no. 16, Pitman, London, 1977.
6. P. Hall. On representatives of subsets. *J. London Math. Soc.* **10** (1935), 26–30.
7. A. J. W. Hilton, Definitions of criticality with respect to edge-colouring. *J. Graph Theory* **1** (1977), 61–68.
8. A. J. Hoffman and H. W. Kuhn, Systems of distinct representatives and linear programming. *Amer. Math. Monthly* **63** (1956), 455–460.
9. D. König, Über Graphen und ihre Anwendung auf Determinantentheorie und Mengenlehre. *Math. Ann.* **77** (1916), 453–465.
10. G. McDiarmid, The solution to a time-tabling problem, *J. Inst. Math. Applic.* **9** (1972), 23–34.
11. M. Plantholt, The chromatic index of graphs with a spanning star. *J. Graph Theory* **5** (1981), 5–13.
12. M. Plantholt, On the chromatic index of graphs with large maximum degree. *Discrete Math.* **47** (1983), 91–96.
13. A. Rosa and W. D. Wallis, Premature sets of 1-factors or how not to schedule round robin tournaments. *Discrete Appl. Math.* **4** (1982), 291–297.
14. V. G. Vizing, On an estimate of the chromatic class of a *p*-graph [in Russian]. *Diskret. Analiz.* **3** (1964), 25–30.
15. V. G. Vizing, Some unsolved problems in graph theory [in Russian]. *Uspekhi Mat. Nauk* **23** (1968), 117–134; English translation in *Russian Math. Surveys* **23** (1968), 125–142.
16. D. de Werra, Equitable colorations of graphs. *Rev. Fr. Rech. Oper.* **5** (1971), 3–8.

8

ANTICHAINS OF SUBSETS OF A FINITE SET

David E. Daykin

ABSTRACT Let $\mathcal{S}$ be an antichain in the set 2^n of subsets of $N = \{1, 2, \ldots, n\}$. The Lubell–Yamamoto–Meshalkin (LYM) inequality is

$$\sum_{X \in \mathcal{S}} 1 \Big/ \binom{n}{|X|} \leq 1.$$

We show that this holds for the "top half" or the "bottom half" of $\mathcal{S}$ in 2^{n-1}. We bound $\min\{|\mathcal{A}|, |\mathcal{B}|\}$, where $\mathcal{A} = \text{above } \mathcal{S}$ and $\mathcal{B} = \text{below } \mathcal{S}$, so that

$$\mathcal{A} = \{Y \subset N \colon \exists X \in \mathcal{S}, X \subset Y\}, \qquad \mathcal{B} = \{Y \colon \exists X \in \mathcal{S}, Y \subset X\}.$$

1. Introduction and statement of results

Let $\mathcal{S}$ be an *antichain* in the set 2^n of subsets of $N = \{1, 2, \ldots, n\}$. Thus $X, Y \in \mathcal{S}$ and $X \neq Y$ imply $X \not\subseteq Y$. For $0 \leq i \leq n$ let p_i be the number of sets X of cardinality $|X| = i$ in $\mathcal{S}$. The famous Lubell–Yamamoto–Meshalkin (LYM) inequality is

$$\sum_{0 \leq i \leq n} p_i \Big/ \binom{n}{i} \leq 1.$$

Define

$$\text{top}(\mathcal{S}) = \sum_{n/2 < i \leq n-1} p_i \Big/ \binom{n-1}{i},$$

$$\text{bot}(\mathcal{S}) = \sum_{1 \leq i < n/2} p_i \Big/ \binom{n-1}{i-1},$$

$$\text{mid}(\mathcal{S}) = \begin{cases} \frac{1}{2} p_m \Big/ \binom{n-1}{m} & \text{if } n = 2m, \\ 0 & \text{if } n = 2m+1, \end{cases}$$

$$\tau(\mathcal{S}) = \text{top}(\mathcal{S}) + \text{mid}(\mathcal{S}),$$

$$\zeta(\mathcal{S}) = \text{bot}(\mathcal{S}) + \text{mid}(\mathcal{S}).$$

Our first result shows that the LYM inequality holds for either the "top half" or the "bottom half" of $\mathcal{S}$ in 2^{n-1}; more precisely it is

THEOREM 1 $\min\{\tau(\mathcal{S}), \zeta(\mathcal{S})\} \leq 1$.

GRAPH THEORY AND COMBINATORICS
ISBN 0-12-111760-X

COROLLARY 1   *If* $n = 2m + 1$ *then*

$$\min\{|\mathcal{S}^*|, |\mathcal{S}_*|\} \leqslant \binom{2m}{m-1},$$

where

$$\mathcal{S}^* = \{X \in \mathcal{S}: m + 1 \leqslant |X|\}, \qquad \mathcal{S}_* = \{X \in \mathcal{S}: |X| \leqslant m\}.$$

COROLLARY 2   *If* $p_i = p_{n-i}$ *for* $0 \leqslant i \leqslant n$ *then* $\tau(\mathcal{S}) + \zeta(\mathcal{S}) \leqslant 2$ *(because* $\tau(\mathcal{S}) = \zeta(\mathcal{S}) \leqslant 1$).

Corollary 2 is due to Greene and Hilton [9] and it generalized an earlier result of Bollobás [1]. He had considered $\mathcal{S}$ such that $X \in \mathcal{S}$ implies $N \setminus X \in \mathcal{S}$. Clearly these have $p_i = p_{n-i}$. (Compare the theorems of Clements and Gronau [3].) The idea for Theorem 1 leads to many results, such as Theorem 6 below.

For the moment define

$$\text{above } \mathcal{S} = \{Y \subset N: \exists X \in \mathcal{S}, X \subset Y\},$$

$$\text{below } \mathcal{S} = \{Y: \exists X \in \mathcal{S}, Y \subset X\}.$$

THEOREM 2   *If* $\mathcal{A} = \text{above } \mathcal{S}$ *and* $\mathcal{B} = \text{below } \mathcal{S}$ *then*

$$\min\{|\mathcal{A}|, |\mathcal{B}|\} \leqslant \begin{cases} \dfrac{1}{2}\left(2^n + \dbinom{n}{m}\right) & \text{if} \quad n = 2m, \quad (1) \\[2ex] 2^{2m} + \dbinom{2m}{m-1} & \text{if} \quad n = 2m + 1. \quad (2) \end{cases}$$

Equality holds in (1) *if and only if* $\mathcal{S} = \{X \subset N, |X| = m\}$. *Equality holds in* (2) *if and only if* $\mathcal{S} = \mathcal{T} \cup \mathcal{U}$, *where* $\mathcal{T} = \{X \subset N: |X| = m + 1, d \notin X\}$ *for some* $d \in N$ *and* $\mathcal{U} = \{N \setminus X: X \in \mathcal{T}\}$.

This theorem is equivalent to the next one, which was conjectured by Daykin and Frankl [7]:

THEOREM 3   *Let* $\mathcal{A}, \mathcal{B} \subset 2^n$ *be such that* $X \in \mathcal{A}$, $Y \in \mathcal{B}$, $X \neq Y$ *implies* $X \not\subset Y$, *then the conclusions of Theorem 2 hold.*

For Theorems 2 and 3 we have a best possible inequality

$$|\mathcal{A}| + |\mathcal{B}| = |\mathcal{A} \cup \mathcal{B}| + |\mathcal{A} \cap \mathcal{B}| \leqslant 2^n + |\mathcal{A} \cap \mathcal{B}| \leqslant 2^n + \binom{n}{\lfloor \frac{1}{2}n \rfloor},$$

where the last inequality is due to Sperner and follows from the LYM inequality because $\mathcal{A} \cap \mathcal{B}$ is an antichain. The n even case of the two theorems follow immediately, but the odd case seems to be much harder, and it motivated this paper.

2. Cascades

Given a positive integer k, each positive integer s has a representation called a k-cascade:

$$s = \binom{b_k}{k} + \binom{b_{k-1}}{k-1} + \ldots + \binom{b_u}{u}, \tag{3}$$

where $b_k > b_{k-1} > \ldots > b_u \geq u \geq 0$. For $u \geq 1$ the cascade is *proper* and unique. When $u = 0$ it is *improper* and unique to within b_u. In this case we can make it proper by choosing $b_u = b_{u-1}$ and using the following identity a number of times on the right of the cascade:

$$\binom{x}{y} = \binom{x-1}{y} + \binom{x-1}{y-1} \quad \text{for} \quad 0 \leq y \leq x \text{ but not for } 0 = y = x. \tag{4}$$

If (3) is proper and $b_u > u$, by reversing this process we can make (3) improper.

For $0 \leq h < k$ define the h-cascade

$$\Delta(k \to h)s = \binom{b_k}{h} + \binom{b_{k-1}}{h-1} + \ldots + \binom{b_u}{u-k+h},$$

and note that this will have the same value whether (3) is proper or not.

3. Kruskal's theorem

For $0 \leq k \leq n$ let $\mathscr{L}_k$ denote the set of members X of 2^n with $|X| = k$. We order $\mathscr{L}_k$ by $X < Y$ if and only if $\max\{X \backslash Y\} < \max\{Y \backslash X\}$. Thus the ordering of $\mathscr{L}_3$ is 123, 124, 134, 234, 125, 135, ... because, for example, $\max\{124 \backslash 134\} = 2 < 3 = \max\{134 \backslash 124\}$. The first s members of $\mathscr{L}_k$ in this ordering is called an *initial section* (IS) and denoted by $\mathscr{F}_k(s)$. For $0 < k$ and $\mathscr{M} \subset \mathscr{L}_k$ define the *shadow* $\Delta\mathscr{M}$ of $\mathscr{M}$ by

$$\Delta\mathscr{M} = \{Y \in \mathscr{L}_{k-1} : \exists X \in \mathscr{M}, Y \subset X\}.$$

We now state a lemma which has a routine proof. It shows that the shadow of an IS is an IS.

Lemma 1 *If* $x = \Delta(k \to k-1)s$ *then* $\Delta\mathscr{F}_k(s) = \mathscr{F}_{k-1}(x)$.

THEOREM 4   *If $0 < k$ and $\mathcal{M} \subset \mathcal{L}_k$ and $s = |\mathcal{M}|$ then*

$$|\Delta \mathcal{F}_k(s)| = \Delta(k \to k - 1)s \leqslant |\Delta \mathcal{M}|.$$

This important theorem was discovered in 1963 by Kruskal [13] and in 1966 independently by Harper [10, 11] and Katona [12]. Simple proofs appear in [8, 4, 5].

4. Special antichains

Let $\mathcal{S}$ be a non-empty antichain in 2^n. Let k be the largest j with $p_j > 0$ and define $\mathcal{P}_k = \mathcal{R}_k = \mathcal{F}_k(p_k)$. For $0 \leqslant i < k$, given $\mathcal{P}_{i+1}$, we inductively define $\mathcal{Q}_i = \Delta \mathcal{P}_{i+1}$, $\mathcal{P}_i = \mathcal{F}_i(p_i + |\mathcal{Q}_i|)$ and $\mathcal{R}_i = \mathcal{P}_i \setminus \mathcal{Q}_i$, so $\mathcal{Q}_i$ is an IS contained in $\mathcal{P}_i$ and $|\mathcal{R}_i| = p_i$. Finally let $\mathcal{R} = \mathcal{R}_0 \cup \ldots \cup \mathcal{R}_k$.

THEOREM 5   *The above $\mathcal{R}$ is an antichain in 2^n.*

This was observed by Clements [2] and independently by Daykin, Godfrey and Hilton [8]. The proof is an easy application of Theorem 4. The important point is that the parameters $p_0, \ldots, p_n$ are the same for $\mathcal{R}$ and $\mathcal{S}$. We say that $\mathcal{S}$ is *special* if $\mathcal{R} = \mathcal{S}$.

5. Proof of Theorem 1

Consider first the case $n = 2m + 1$. Clearly we may assume that $\mathcal{S}$ is special and use the notation of Section 4. Suppose $\mathcal{P}_{m+1} \neq \varnothing$ for otherwise top$(\mathcal{S}) = 0$. Let Z be the last member of $\mathcal{P}_{m+1}$.

Case $n \in Z$ By induction n belongs to the last members of $\mathcal{Q}_i$, $\mathcal{P}_i$ for $i = m, \ldots, 1$ and hence to every member of $\mathcal{S}_* = \mathcal{R}_1 \cup \ldots \cup \mathcal{R}_m$. We delete n from every member of $\mathcal{S}_*$ to get an antichain in 2^{n-1} for which the LYM inequality says that $\zeta(\mathcal{S}) \leqslant 1$.

Case $n \notin Z$ Here n does not belong to any member of $\mathcal{S}^*$. So $\mathcal{S}^*$ is an antichain in 2^{n-1} for which the LYM inequality says that $\tau(\mathcal{S}) \leqslant 1$.

The foregoing proves the case $n = 2m + 1$. The case $n = 2m$ is proved similarly. One chooses Z as the "middle" set of $\mathcal{R}_m$. $\square$

THEOREM   6   *If $0 \leqslant \lambda, \mu, \nu \leqslant 1$ and $n - 2 \geqslant a > b > c \geqslant 2$ then*

$\min\{\alpha, \beta, \gamma, \delta\} \le 1$ *where* $w = n - 2$ *and*

$$\alpha = \left\{\lambda p_a \Big/ \binom{w}{a}\right\} + \sum_{a < i \le n-2} p_i \Big/ \binom{w}{i},$$

$$\beta = \left\{(1-\lambda)p_a \Big/ \binom{w}{a-1}\right\} + \mu p_b \Big/ \binom{w}{b-1} + \sum_{b < i < a} p_i \Big/ \binom{w}{i-1},$$

$$\gamma = (1-\mu)p_b \Big/ \binom{w}{b-1} + \left\{\nu p_c \Big/ \binom{w}{c-1}\right\} + \sum_{c < i < b} p_i \Big/ \binom{w}{i-1},$$

$$\delta = \left\{(1-\nu)p_c \Big/ \binom{w}{c-2}\right\} + \sum_{2 \le i < c} p_i \Big/ \binom{w}{i-2}.$$

This theorem is best possible in the sense that we can have $\alpha = \beta = \gamma = \delta = 1$; for example, make the four brackets $\{\ldots\}$ equal 1 and the p_i in all other terms 0. We omit details of the proof, it depends upon one of the following four alternatives holding. Every $X \in \mathscr{S}$ counted by α (β, γ, δ respectively) has $X \cap \{n-1, n\}$ equal to $\varnothing$ ($\{n-1\}$, $\{n\}$, $\{n-1, n\}$ respectively).

6. Proof of Theorem 3

Part 1   Suppose $X \subset N$ and $X \notin \mathscr{A} \cup \mathscr{B}$. If we cannot adjoin X to $\mathscr{A}$ there is a $Y \in \mathscr{B}$ with $X \subset Y$ and $X \ne Y$. If we also cannot adjoin X to $\mathscr{B}$ there is a $Z \in \mathscr{A}$ with $Z \subset X$. However, this gives the contradiction $Z \subset Y$. Hence we may assume that $\mathscr{A} \cup \mathscr{B} = 2^n$. By a similar argument we may assume that $\mathscr{A} = \text{above } \mathscr{S}$ and $\mathscr{B} = \text{below } \mathscr{S}$, where $\mathscr{S}$ is the antichain $\mathscr{A} \cap \mathscr{B}$. It is now clear that Theorems 2 and 3 are equivalent; also that if $X \subset Y \in \mathscr{B}$ then $X \in \mathscr{B}$.

Part 2   For $0 \le i \le n$ let q_i be the number of sets of cardinality i in $\mathscr{B}$ and put $\mathscr{D}_i = \mathscr{F}_i(q_i)$. Next put $\mathscr{C}_n = \{N\}$ and for $0 \le i < n$ put $\mathscr{C}_i = \mathscr{L}_i \setminus \Delta\mathscr{D}_{i+1}$, observing that $\Delta\mathscr{D}_{i+1} \subset \mathscr{D}_i$. By a routine application of Theorem 4 we know that $\mathscr{A}$ does not contain more sets of any cardinality i than $\mathscr{C}_i$. Let $\mathscr{C} = \mathscr{C}_0 \cup \ldots \cup \mathscr{C}_n$ and $\mathscr{D} = \mathscr{D}_0 \cup \ldots \cup \mathscr{D}_n$ and note that (i) $|\mathscr{A}| \le |\mathscr{C}|$, (ii) $|\mathscr{B}| = |\mathscr{D}|$, (iii) $\varnothing \ne \mathscr{S} = \mathscr{C} \cap \mathscr{D}$ is special, (iv) $\mathscr{C} = \text{above } \mathscr{S}$, (v) $\mathscr{D} = \text{below } \mathscr{S}$, (vi) $\mathscr{C} \cup \mathscr{D} = 2^n$ and (vii) $\mathscr{C}, \mathscr{D}$ satisfy the conditions of Theorem 3. From now on we will assume that $\mathscr{A} = \mathscr{C}$ and $\mathscr{B} = \mathscr{D}$. We let k be the greatest and $h - 1$ be the least integer i with $p_i \ne 0$.

REMARK   The shift operation of [4] is as follows. Given $J, K \subset N$ with

$J \cap K = \emptyset$ and $|J| = |K|$ the shift $(J \to K)X$ of $X \in \mathcal{A}$ is

$$(J \to K)X = \begin{cases} (X \backslash J) \cup K & \text{if } J \subset X, X \cap K = \emptyset, (X \backslash J) \cup K \notin \mathcal{A}, \\ X & \text{otherwise.} \end{cases}$$

Using this shift we can change $\mathcal{A}, \mathcal{B}$ as given by Part 1 step by step into $\mathcal{C}, \mathcal{D}$. We choose J, K with $J < K$ and $|J|$ minimal so that either $(J \to K)$ changes $\mathcal{A}$ or $(K \to J)$ changes $\mathcal{B}$. We use $(J \to K)$ on $\mathcal{A}$ and $(K \to J)$ on $\mathcal{B}$ and the hypothesis of Theorem 3 is preserved. Repeating this operation achieves the result.

Part 3 Only here do we assume $n = 2m + 1$.
Case $k \leq m$ Here $\mathcal{B}$ is contained in the lower half of 2^n so $|\mathcal{B}| \leq 2^{2m}$.
Case $m + 1 \leq h - 1$ As in the last case, $|\mathcal{A}| \leq 2^{2m}$.
Case $h - 1 = m$ and $k = m + 1$ Here $|\mathcal{A}| = 2^{2m} + p_m$ and $|\mathcal{B}| = 2^{2m} + p_{m+1}$ so we use Theorem 1. Either

$$\tau(\mathcal{S}) = p_{m+1} \Big/ \binom{2m}{m+1} \leq 1 \quad \text{or} \quad \zeta(\mathcal{S}) = p_m \Big/ \binom{2m}{m-1} \leq 1$$

so (2) holds with equality only as stated in Theorem 2.

For the remainder of the proof we assume only that $h < k$ and change $\mathcal{A}, \mathcal{B}$ until this part applies.

Part 4 We have $h < k$ and note that $p_k = q_k$ and $p_{h-1} = |\mathcal{L}_{h-1}| - |\Delta \mathcal{D}_h|$. Our objectives are as follows:

(i) To find an e in $1 \leq e \leq \min\{q_k, |\mathcal{L}_h| - q_h\}$.
(ii) To delete the last e sets from $\mathcal{D}_k$ so that $\mathcal{D}_k$ becomes $\mathcal{D}'_k = \mathcal{F}_k(q_k - e)$ and q_k becomes $q_k - e$.
(iii) To adjoin the next e sets to $\mathcal{D}_h$ so that $\mathcal{D}_h$ becomes $D'_h = \mathcal{F}_h(q_h + e)$ and q_h becomes $q_h + e$.
(iv) To adjoin $\mathcal{K} = (\Delta \mathcal{D}_k) \backslash (\Delta \mathcal{D}'_k)$ to $\mathcal{C}_{k-1}$, increasing $|\mathcal{C}_{k-1}|$ by $|\mathcal{K}|$.
(v) To delete $\mathcal{H} = (\Delta \mathcal{D}'_h) \backslash (\Delta \mathcal{D}_h)$ from $\mathcal{C}_{h-1}$, decreasing $|\mathcal{C}_{h-1}|$ by $|\mathcal{H}|$.

From (ii), (iii) we see that $|\mathcal{B}|$ does not change. Clearly the hypothesis of Theorem 3 will still hold. So all will be well if $|\mathcal{K}| \geq |\mathcal{H}|$, so $|\mathcal{A}|$ does not decrease. To find a suitable e we use the cascade algorithm of [5]. We use a sequence of such integers e until either $p_k = 0$, reducing the value of k, or $p_{h-1} = 0$, increasing the value of h. Repetition leads us into Part 3 as required.

Part 5 To find e, consider the following pair of cascades

$$s = |\mathcal{D}_k| = \binom{b_k}{k} + \ldots + \binom{b_u}{u} = p_k,$$

$$r = |\mathcal{D}_h| = \binom{a_h}{h} + \ldots + \binom{a_t}{t}.$$

Operation 1   (i) For $h \geq i \geq \max\{t, u\}$ if $a_i < b_i$ exchange a_i and b_i in r and s.

(ii) If $t > u$ subtract

$$\binom{b_{t+1}}{t+1} + \ldots + \binom{b_u}{u}$$

from s and add it to r. If this operation changes r, s, then it leaves them both as cascades, and we let e be the size of the change. When $e \geq 1$ both $\mathscr{A}, \mathscr{B}$ are changed as required for Part 4. However, neither $|\mathscr{A}|$ nor $|\mathscr{B}|$ change. To see this let Δ_w denote $\Delta(w \to w - 1)$ for $w = h, k$. Then notice that

$$\Delta_k s + \Delta_h r = \Delta_k (s - e) + \Delta_h (r + e)$$

because the same binomial coefficients occur on both sides of the equation.

Operation 2   If s is proper and $u < b_u$, make s improper.

Operation 3   Make r proper.

Operation 4   If r, s are proper and $u = b_u$ and $u \geq t$, delete $\binom{u}{u}$ from s

then add $\binom{t-1}{t-1}$ to r and put $e = 1$. Here

$$|\mathscr{H}| = \binom{u}{u-1} = u > t - 1 = \binom{t-1}{t-2} = |\mathscr{H}|$$

so $|\mathscr{A}|$ increases.

It is clear that Operations 2, 3 do not change $\mathscr{A}, \mathscr{B}$ but merely prepare the way for the other operations.

This ends the proof of Theorem 3. $\square$

Our manipulation of r, s in fact proved

LEMMA 2   *If* $1 \leq h < k$ *and* $0 \leq s$ *and* $\Delta(k \to h)s \leq r \leq v = \binom{n}{h}$ *then*

(i) $\binom{n}{h-1} - \Delta_h(r) \leq \Delta_k(s) - \Delta_k(s - v + r)$ *if* $v - r \leq s$.

(ii) $\Delta_h(r + s) - \Delta_h(r) \leq \Delta_k(s)$ *if* $s \leq v - r$.

References

1. B. Bollobás, Sperner systems consisting of pairs of complementary subsets. *J. Combinat. Theory A* **15** (1973), 363–366.

2. G. Clements, A minimization problem concerning subsets. *Discrete Math.* **4** (1973), 123–128.
3. G. F. Clements and H.-D. O. F. Gronau, On maximal antichains containing no set and its complement. *Discrete Math.* **33** (1981), 239–247.
4. D. E. Daykin, A simple proof of the Kruskal–Katona theorem. *J. Combinat. Theory A* **17** (1974), 252–253.
5. D. E. Daykin, An algorithm for cascades giving Katona-type inequalities. *Nanta Math.* **8** (1975), 78–83.
6. D. E. Daykin and P. Frankl, On Kruskal's cascades and counting containments in a set of subsets. *Mathematika* **30** (1983), 133–141.
7. D. E. Daykin and P. Frankl, Extremal sets of subsets satisfying conditions induced by a graph. In *Graph Theory and Combinatorics* (B. Bollabás, ed.), Academic Press, London, 1984, pp. 107–126.
8. D. E. Daykin, J. Godfrey and A. J. W. Hilton, Existence theorems for Sperner families. *J. Combinat. Theory A* **17** (1974), 245–251.
9. C. Greene and A. J. W. Hilton, Some results on Sperner families. *J. Combinat. Theory A* **26** (1979), 202–208.
10. L. H. Harper, Optimal assignments of numbers to vertices. *J. Soc. Indust. Appl. Math.* **12** (1964), 131–135.
11. L. H. Harper, Optimal numberings and isoperimetric problems on graphs. *J. Combinat. Theory* **1** (1966), 385–393.
12. G. Katona, A theorem for finite sets. In *Theory of Graphs* (P. Erdös and G. Katona, eds), Hungarian Academy of Sciences, Budapest, 1966, pp. 187–207.
13. J. B. Kruskal, The number of simplices in a complex. In *Mathematical Optimization Techniques* (R. Bellman, ed.), University of California Press, Berkeley, 1963, pp. 251–278.

9

EXTREMAL SETS OF SUBSETS SATISFYING CONDITIONS INDUCED BY A GRAPH

David E. Daykin and Peter Frankl

ABSTRACT Let $\mathscr{F}$ always denote a set of subsets of $N = \{1, 2, \ldots, n\}$. If $\mathscr{F}_1, \mathscr{F}_2$ are such that $X \in \mathscr{F}_1$, $Y \in \mathscr{F}_2$ imply $X \cap Y \neq \varnothing$ and $X \neq Y$, then $2^{-n} \min\{|\mathscr{F}_1|, |\mathscr{F}_2|\} \leq (3 - \sqrt{5})/2$, and this is essentially best possible. Now we introduce more general results.

Four conditions which $X, Y \subset N$ might satisfy are I: $X \cap Y \neq \varnothing$, U: $X \cup Y \neq N$, C: if $X \neq Y$ then $X \not\subset Y$, and $\neq$: $X \neq Y$. Let J be a subset of $\{I, U, C, \neq\}$. Let Γ be a finite loopless graph with vertex set $\{1, 2, \ldots, v\}$ and edge set E. Define

$$\lambda(n) = 2^{-n} \max\{|\mathscr{F}_1| + \ldots + |\mathscr{F}_v|\} \leq v,$$

$$\mu(n) = 2^{-n} \max\{\min\{|\mathscr{F}_1|, \ldots, |\mathscr{F}_v|\}\} \leq 1,$$

where the maxima are over all $\mathscr{F}_1, \ldots, \mathscr{F}_v$ such that $(i, j) \in E$, $X \in \mathscr{F}_i$, $Y \in F_j$ imply X, Y satisfy all the conditions in J.

We study the sequences $\lambda(n), \mu(n)$ and show they have limits λ, μ. They are non-decreasing except possibly if J is C, IC, UC, IUC. Clearly λ, μ depend only on Γ, J.

We determine λ in terms of fractional stability numbers of Γ. The case $J = C$ for λ generalizes to Kleitman–Lubell–Yamamoto–Meshalkin (or Lubell–Yamamoto–Meshalkin) posets.

When Γ is an edge, $\mu \in \{\frac{1}{4}, (3 - \sqrt{5})/2, \frac{1}{2}\}$. For Γ arbitrary $\mu(n) = \frac{1}{2}$ for $J = I, U$ and $\mu(n) = \frac{1}{4}$ for $J = IU$. When Γ is a directed circuit $\mu(n) = \frac{1}{4}$ for $J = C\neq$. When Γ is undirected and $J = \neq$ we determine μ in terms of the fractional chromatic number of Γ

The paper contains much more information.

1. Introduction

Let n be a positive integer, let N be the set $\{1, 2, \ldots, n\}$ and let 2^n be the set of subsets of N. We reserve the letters $\mathscr{F}, \mathscr{G}, \mathscr{H}$ for subsets of 2^n. Four conditions which $X, Y \subset N$ might satisfy are

I	(non-empty intersection)	$X \cap Y \neq \varnothing$;
U	(union not N)	$X \cup Y \neq N$;
C	(not properly contained)	if $X \neq Y$ then $X \not\subset Y$;
$\neq$	(not equal)	$X \neq Y$.

GRAPH THEORY AND COMBINATORICS
ISBN 0-12-111760-X

We reserve the letter J for a subset of $\{I, U, C, \neq\}$. For example if $J = C$ and $\mathscr{F}$ is such that all $X, Y \in \mathscr{F}$ satisfy J, this means that no member of $\mathscr{F}$ is a proper subset of another, in other words that $\mathscr{F}$ is an antichain. The origins of our work might be traced to a classical result of Sperner. In 1928 he showed that if $\mathscr{F}$ is an antichain then $|\mathscr{F}| \leq \mathrm{Sper}(n)$, where

$$\mathrm{Sper}(n) = \binom{n}{\lfloor \frac{1}{2}n \rfloor} = \max_{0 \leq i \leq n} \left\{ \binom{n}{i} \right\}.$$

This result was generalized independently by Yamamoto (in 1954), Meshalkin (in 1963) and Lubell (in 1966) to the well-known Lubell–Yamamoto–Meshalkin (LYM) inequality:

$$\sum_{X \in \mathscr{F}} 1 \Big/ \binom{n}{|X|} \leq 1 \quad \text{if} \quad \mathscr{F} \text{ is an antichain.}$$

In 1972 Brace and Daykin [3] studied $\max |\mathscr{F}|$ over all $\mathscr{F}$ such that all $X, Y \in \mathscr{F}$ satisfy a subset J of the conditions I, U, C. They conjectured that $\max |\mathscr{F}| = 2^n/4$ when $J = IU$. This was proved by Daykin, Hilton, Lovász, Schönheim, Seymour and others.

The above results all concern one set $\mathscr{F}$, and further details can be found in the recent survey by West [15]. In this paper we have two sets $\mathscr{F}, \mathscr{G}$ or more. Just one of our results is that $\max\{\min\{|\mathscr{F}|, |\mathscr{G}|\}\} = 2^n/4$ over all $\mathscr{F}, \mathscr{G}$ such that all $X \in \mathscr{F}, Y \in \mathscr{G}$ satisfy IU. This is clearly stronger than the above result about IU.

2. Statement of main results

Throughout this paper Γ will be a finite loopless graph without multiple edges having vertex set $\{1, 2, \ldots, v\}$ and edge set $E(\Gamma)$. We assume that $E(\Gamma) \neq \emptyset$ so $v \geq 2$. We think of sets $\mathscr{F}_1, \ldots, \mathscr{F}_v$ as sitting at the vertices of Γ.

Let J be a non-empty subset of the conditions $I, U, C, \neq$. Define

$$\lambda(n) = 2^{-n} \max\{|\mathscr{F}_1| + \ldots + |\mathscr{F}_v|\} \leq v,$$

$$\mu(n) = 2^{-n} \max\{\min\{|\mathscr{F}_1|, \ldots, |\mathscr{F}_v|\}\} \leq 1,$$

where these maxima are over all $\mathscr{F}_1, \ldots, \mathscr{F}_v$ such that $(i, j) \in E(\Gamma)$, $X \in \mathscr{F}_i, Y \in \mathscr{F}_j$ imply that X, Y satisfy all the conditions in J.

The motive for this paper was to study the sequences $\lambda(1), \lambda(2), \ldots$ and $\mu(1), \mu(2), \ldots$. They both depend on Γ and J and they are non-decreasing except possibly if J is C, IC, UC or IUC. We show in Section 3 that they always have limits λ, μ respectively. Clearly λ, μ depend only on Γ and J. Now Γ may be directed or undirected but if $C \notin J$ this makes no difference. So when Γ is directed we always assume $C \in J$.

Results on $\lambda(n)$ and λ

These are virtually complete. We remind the reader that a subset A of the vertex set $\{1, \ldots, v\}$ is independent if $i, j \in A$ implies that $(i, j) \notin E(\Gamma)$. Also the independence number $\alpha(\Gamma)$ of Γ is $\max\{|A|\}$ over independent sets A. Suppose that A is independent and $|A| = \alpha(\Gamma)$. For $1 \leq i \leq v$ put $\mathscr{F}_i = 2^n$ if $i \in A$ but $\mathscr{F}_i = \varnothing$ if $i \notin A$. This example is IUC$\neq$ and shows that $\alpha(\Gamma) \leq \lambda(n)$ always. In Section 4 we will define $\alpha^{(1)}(\Gamma)$ and $\alpha^{(2)}(\Gamma)$ for

THEOREM 2.1 *If Γ is undirected then*

$$\lambda(n) = \begin{cases} \alpha^{(1)}(\Gamma) & \text{for } J = \text{I, U,} \\ \alpha^{(2)}(\Gamma) & \text{if } J = \text{IU,} \\ \alpha(\Gamma) & \text{if } \neq \in J, \\ \alpha(\Gamma) & \text{for } n \geq n_0 = n_0(\Gamma) \text{ and } J = \text{C, IC, UC, IUC.} \end{cases}$$

When Γ is a triangle n_0 is $5, 2, 2, 1$ as J is C, IC, UC, IUC.

THEOREM 2.2 *If Γ is directed then $\lambda = \alpha(\Gamma)$ if $C \in J$ and $\lambda(n) = \alpha(\Gamma)$ if $C, \neq \in J$.*

THEOREM 2.3 *If Γ is a directed edge and $C \in J$ then $\alpha(\Gamma) = 1$ and*

$$\lambda(n) = \begin{cases} 1 + 2^{-n} \text{Sper}(n) & \text{if } J = C, \\ 1 & \text{otherwise.} \end{cases}$$

In view of these three theorems it only remains to evaluate $\lambda(n)$ for directed Γ in the cases C, IC, UC, IUC.

Results on $\mu(n)$ and μ

First we have a lower bound for $\mu(n)$.

LEMMA 2.1 *For $m = 2, 3, \ldots$ we have*

$$2^{1-2m} \leq \mu(n) \text{ for all } J, \Gamma \text{ with } v \leq \binom{2m-1}{m} \text{ and } 2m \leq n.$$

A pleasing discovery is

THEOREM 2.4

$$\mu(n) = \begin{cases} \frac{1}{2} & \text{for } J = \text{I}, \text{U}, \\ \frac{1}{4} & \text{if } J = \text{IU}. \end{cases}$$

We get an upper bound for $\mu(n)$ by looking at a single edge of Γ. When Γ is undirected and bipartite, doing so is in fact sufficient to evaluate $\mu(n)$. It turns out that when Γ is an edge the only possible values of μ are $\frac{1}{4}$, $(3-\sqrt{5})/2 = 0.381\,97$ and $\frac{1}{2}$, as shown in Table 2.1.

TABLE 2.1   Values of μ when Γ is an edge

Undirected	Directed	μ
I, U, $\neq$	C, C$\neq$	$\frac{1}{2}$
I$\neq$, U$\neq$	IC, UC, IC$\neq$, UC$\neq$	$(3-\sqrt{5})/2$
Otherwise	IUC, IUC$\neq$	$\frac{1}{4}$

When Γ is a directed circuit we prove that $\mu(n) = \frac{1}{4}$ if J is C$\neq$ in Theorem 11.1. When Γ is undirected and J is $\neq$ we determine μ in terms of the fractional chromatic number of Γ in Theorem 10.1. Clearly there is much remaining to be discovered about $\mu(n)$.

Results for Kleitman–LYM (KLYM) posets

Let P be a KLYM poset. The definition is given in Section 5, and 2^n is a KLYM poset. These seem to be the natural setting for dealing with the case $J = C$ of 2^n. For $1 \leq i \leq v$ let F_i be a subset of the elements of P. In this context the condition C takes the form

$$(i, j) \in E(\Gamma), \quad p \in F_i, \quad q \in F_j, \quad p \neq q \quad \text{imply} \quad p \not< q. \quad (2.1)$$

Let m be the maximum of the cardinalities of the ranks of P. We have three results in Section 5.

THEOREM 2.5   *If Γ is undirected and* (2.1) *holds and*

$$m(v + 1 - \alpha(\Gamma))\alpha(\Gamma) \leq |P|, \quad (2.2)$$

then

$$|P|^{-1} \max\left\{ \sum |F_i| \right\} = \alpha(\Gamma).$$

THEOREM 2.6   *If Γ is directed and (2.1) holds then*

$$\alpha(\Gamma) \leqslant |P|^{-1} \max\{\textstyle\sum |F_i|\} \leqslant \alpha(\Gamma) + |P|^{-1} m(v-1),$$

and these bounds are best possible.

THEOREM 2.7   *If Γ is directed and (2.1) holds and condition $\neq$ holds then*

$$|P|^{-1} \max\{\textstyle\sum |F_i|\} = \alpha(\Gamma).$$

3. The convergence of $\lambda(n)$ and $\mu(n)$

Let $\mathcal{F}_1, \ldots, \mathcal{F}_v$ be an example for $n = m \geqslant 2$ with properties J. For $1 \leqslant i \leqslant v$ put

$$\mathcal{G}_i = \{X \setminus \{m\} : m \in X \in F_i\} \quad \text{and} \quad \mathcal{H}_i = \{X : m \notin X \in \mathcal{F}_i\}.$$

Then $\mathcal{G}_1, \ldots, \mathcal{G}_v$ has properties $J \setminus I$ and $\mathcal{H}_1, \ldots, \mathcal{H}_v$ has properties $J \setminus U$ for $n = m - 1$. This proves that

$$\max\{\lambda_{J \setminus I}(n), \lambda_{J \setminus U}(n)\} \geqslant \lambda_J(n+1) \quad \text{for} \quad n = 1, 2, \ldots$$

and in particular establishes

LEMMA 3.1   *We have $\lambda(1) \geqslant \lambda(2) \geqslant \ldots \geqslant 1$ for $J = C, \neq, C\neq$.* $\square$

Suppose we have an example $\mathcal{F}_1, \ldots, \mathcal{F}_v$ for $n = m - 1 \geqslant 1$. We form a second example for $n = m$ by replacing each member X of each $\mathcal{F}_i$ by the two sets X and $X \cup \{m\}$. We call this operation *doubling*. The properties $I, U, \neq, C\neq$ clearly carry over under doubling, and this fact yields

LEMMA 3.2   *We have $\lambda(1) \leqslant \lambda(2) \leqslant \ldots \leqslant v$ and $\mu(1) \leqslant \mu(2) \leqslant \ldots \leqslant 1$ except possibly if J is C, IC, UC or IUC.* $\square$

Next suppose that $\mathcal{F}_1, \ldots, \mathcal{F}_v$ is an example with properties J and $C \in J$. For $1 \leqslant i \leqslant v$ put

$$\mathcal{G}_i = \mathcal{F}_i \setminus \mathcal{H}_i \quad \text{where} \quad \mathcal{H}_i = \mathcal{F}_i \cap \left(\bigcup_{(i,j) \in E(\Gamma)} \mathcal{F}_j \right).$$

Then $\mathcal{H}_i$ is an antichain, and so

$$|\mathcal{F}_i| - \text{Sper}(n) \leqslant |\mathcal{F}_i| - |\mathcal{H}_i| \leqslant |\mathcal{G}_i|.$$

Also $\mathcal{G}_1, \ldots, \mathcal{G}_v$ has properties J and $\neq$. For any $k > n$ let us double $\mathcal{G}_1, \ldots, \mathcal{G}_v$ a total of $k - n$ times. The resulting example has properties J. This shows that

$$\lambda(n) - v 2^{-n} \text{Sper}(n) \leqslant \lambda(k) \quad \text{for} \quad n \leqslant k \text{ if } C \in J,$$

and similarly for $\mu(n)$, except that the factor v is deleted. It is now easy to see that for all J the sequences $\lambda(n)$ and $\mu(n)$ converge with limits λ and μ respectively because $2^{-n} \operatorname{Sper}(n) \to 0$ as $n \to \infty$.

4. Independence numbers of Γ

A subset A of $V(\Gamma)$ is independent if $i, j \in A$ implies $(i, j) \notin E(\Gamma)$. The independence number $\alpha(\Gamma)$ of Γ is $\max\{|A|\}$ over independent sets A. If $r \geq 1$ the rth power fractional independence number $\alpha^{(r)}(\Gamma)$ of Γ is

$$\max\{x_1^r + \ldots + x_v^r\}$$

over all choices of real numbers $x_1, \ldots, x_v$ such that

$$0 \leq x_i \leq 1 \quad \text{for} \quad 1 \leq i \leq v, \tag{4.1}$$

$$x_i + x_j \leq 1 \quad \text{for all} \quad (i, j) \in E(\Gamma). \tag{4.2}$$

Clearly $\alpha^{(r)}(\Gamma)$ decreases as r increases and is not less than $\alpha(\Gamma)$. We need

LEMMA 4.1   *We can achieve $\alpha^{(r)}(\Gamma)$ with all $x_i \in \{0, \frac{1}{2}, 1\}$, and when $r > 1$ we cannot achieve it in any other way.*

PROOF   Suppose some $x_i \notin \{0, \frac{1}{2}, 1\}$. Put $\alpha = \sum x_k^r$,

$$a = \min\{x_i : 0 < x_i < 1\}, \qquad A = \{i : 1 \leq i \leq v, x_i = a\},$$

$$b = \max\{x_i : 0 < x_i < 1\}, \qquad B = \{i : 1 \leq i \leq v, x_i = b\},$$

so $A \neq \emptyset$, $B \neq \emptyset$ and $a \leq b$.

Case $a + b < 1$ Here if $i \in A$ and $(i, j) \in E(\Gamma)$ then $x_i + x_j < 1$ so we can increase x_i and hence α.

Case $a + b > 1$ Here if $j \in B$ and $(i, j) \in E(\Gamma)$ then $x_i = 0$ so we can increase x_j.

Case $a + b = 1$ Here we eliminate $b = 1 - a$ and consider α as a function of a, keeping all x_i with $i \notin A \cup B$ constant. The result follows by the convexity of α.   □

5. Theorems on KLYM posets

Let P be a ranked poset. That means there is a map rank: $P \to \{0, 1, \ldots, h\}$ such that for all $p, q \in P$ with $p < q$, firstly $\operatorname{rank}(p) < \operatorname{rank}(q)$, and secondly if $2 + \operatorname{rank}(p) \leq \operatorname{rank}(q)$ then $p < p' < q$ for some $p' \in P$. We assume that for $0 \leq i \leq h$ rank i of P, which is the set $\{p \in P : \operatorname{rank}(p) = i\}$, has cardinality $\nu(i) \geqslant 1$. We put $m = \max\{\nu(0), \ldots, \nu(h)\}$ and clearly $\nu(0) + \ldots + \nu(h) = |P|$.

Suppose further that P is a KLYM poset. By this we mean that there is non-empty list $\Lambda_1, \ldots, \Lambda_c$ of maximal chains Λ_i of P such that for every $p \in P$ the number of chains which contain p in the list is $c/\nu(\text{rank}(p))$. There is a growing literature about KLYM posets, cf [**1, 2, 4, 6–8, 10, 13, 15**]. Other writers have called them LYM posets, but our K is in honour of Kleitman [**13**]. The set 2^n of subsets of N is perhaps the most important example of a KLYM poset.

For $1 \leqslant i \leqslant v$ let F_i be a subset of the elements of P. We are here interested in

$$|P|^{-1} \max\{|F_1| + \ldots + |F_v|\}$$

over all $F_1, \ldots, F_v$ satisfying conditions like I, U, C, $\neq$. However, conditions I, U do not carry over to KLYM posets. Further, the maximum is trivially $|P|^{-1} \alpha(\Gamma)$ if $\neq$ is among the conditions. Hence we are only left with condition C, which here takes the form (2.1). Our results are Theorems 2.5, 2.6 and 2.7.

PROOF OF THEOREM 2.5   Let $F_1, \ldots, F_v$ satisfying (2.1) be given. For any maximal chain Λ of P consider a bipartite graph $\Delta = \Delta(\Lambda)$. If Λ is $p_0 < p_1 < \ldots < p_h$ say, then the vertices in one part of Δ are $p_0, p_1, \ldots, p_h$ and the vertices in the other part are $F_1, F_2, \ldots, F_v$. We have (p_i, F_j) as an edge of Δ if and only if $p_i \in F_j$.

Suppose that Δ had $\alpha + 1$ independent edges $(P_{i_0}, F_{j_0}), \ldots, (P_{i_\alpha}, F_{j_\alpha})$. Then by definition of α there are $j_r \neq j_s$ such that $(j_r, j_s) \in E(\Gamma)$. Since $p_{i_r} \in F_{j_r}$ and $p_{i_s} \in F_{j_s}$ and either $p_{i_r} < p_{i_s}$ or $p_{i_s} < p_{i_r}$ this contradicts (2.1). Hence Δ does not have $\alpha + 1$ independent edges.

We now distinguish two types of maximal chain Λ.

Type 1 $\deg(p \text{ in } \Delta(\Lambda)) \leqslant \alpha$ for all $p \in \Lambda$.

Type 2   Not type 1. For this type of Λ there is some $p \in \Lambda$ with $\deg(p$ in $\Delta(\Lambda)) \geqslant \alpha + 1$. Put

$$\theta(\Lambda) = \{p \in \Lambda, \deg(p \text{ in } \Delta(\Lambda)) \geqslant \alpha\}.$$

Since Δ does not have $\alpha + 1$ independent edges, by Hall's theorem on systems of distinct representatives, we know that $|\theta(\Lambda)| \leqslant \alpha$.

Next we claim that for any Λ we have

$$\sum_{p \in \Lambda} \nu(\text{rank}(p)) \deg(p \text{ in } \Delta(\Lambda)) \leqslant \alpha |P|. \tag{5.1}$$

For chains of Type 1 this is trivial. For chains of Type 2 the left side of (5.1) is not greater than

$$(\alpha - 1) \sum_{\substack{p \in \Lambda \\ \deg p \leqslant \alpha - 1}} \nu(\text{rank}(p)) + v \sum_{\substack{p \in \Lambda \\ \deg p \geqslant \alpha}} \nu(\text{rank}(p))$$

$$= (\alpha - 1) \sum_{p \in \Lambda} \nu(\text{rank}(p)) + (v - \alpha + 1) \sum_{\substack{p \in \Lambda \\ \deg p \geqslant \alpha}} \nu(\text{rank}(p))$$

$$\leqslant (\alpha - 1)|P| + (v - \alpha + 1) \sum_{\substack{p \in \Lambda \\ \deg p \geqslant \alpha}} m$$

$$\leqslant (\alpha - 1)|P| + (v - \alpha + 1)\alpha m$$

$$\leqslant \alpha |P|,$$

by (2.2). So (5.1) holds for any Λ.

Next let χ be the characteristic function defined by $\chi(w, W)$ is 1 if $w \in W$ but 0 otherwise. Then clearly we have

$$\deg(p_i \text{ in } \Delta(\Lambda)) = \sum_{1 \leqslant j \leqslant v} \chi(p_i, F_j).$$

Also for any $F \subset P$ we have

$$c |F| = \sum_{\Lambda \in \text{list}} \sum_{p \in \Lambda} \nu(\text{rank}(p))\chi(p, F).$$

It now follows that

$$c \sum_{1 \leqslant j \leqslant v} |F_j| = \sum_{\Lambda \in \text{list}} \sum_{p \in \Lambda} \nu(\text{rank}(p)) \sum_{1 \leqslant j \leqslant v} \chi(p, F_j)$$

$$= \sum_{\Lambda \in \text{list}} \sum_{p \in \Lambda} \nu(\text{rank}(p)) \deg(p \text{ in } \Delta(\Lambda))$$

$$\leqslant \sum_{\Lambda \in \text{list}} \alpha |P| = c\alpha |P|, \qquad (5.2)$$

and Theorem 2.5 is proved. □

PROOF OF THEOREM 2.6   Let $F_1, \ldots, F_v$ satisfying (2.1) be given. For any maximal chain Λ let $\Delta(\Lambda)$ be as defined in the last proof. Since Γ is now directed we modify $\Delta(\Lambda)$ to get a bipartite graph $\Delta'(\Lambda)$. For $1 \leqslant j \leqslant v$, if there is an edge (p_i, F_j) in Δ, we remove the one with i as small as possible from Δ. Thus if R is the set of removed edges then $|R| \leqslant v$. Finally, if $R \neq \emptyset$ we choose exactly one edge $(p_i, F_j) \in R$ with i as small as possible and replace it. The result is Δ'. Notice that Δ' is a subgraph of Δ obtained by deleting at most $v - 1$ edges.

We claim that $\deg(p_i$ in $\Delta'(\Lambda))\leq\alpha$ for all Λ and $0\leq i\leq h$. If true we can bound the sum in (5.1) as follows:

$$\sum_{p\in\Lambda}\nu(\text{rank}(p))\deg(p\text{ in }\Delta)\leq m(v-1)+\sum_{p\in\Lambda}\nu(\text{rank}(p))\deg(p\text{ in }\Delta')$$

$$\leq m(v-1)+\alpha\sum_{p\in\Lambda}\nu(\text{rank}(p))$$

$$=m(v-1)+\alpha\,|P|. \tag{5.3}$$

The term $m(v-1)$ above arises because an edge (p, F) in $\Delta\setminus\Delta'$ contributes $\nu(\text{rank}(p))\leq m$ to the sum, and there are not more than $v-1$ such edges. Using (5.3) in (5.2) instead of (5.1) we get the right-hand side inequality of Theorem 2.6. The left-hand side is trivial, so it remains to establish our claim that $\deg(p$ in $\Delta')\leq\alpha$.

Suppose therefore that there is a k with $\deg(p_k$ in $\Delta')\geq\alpha+1$. Then by definition of α there are r, s such that (p_k, F_r), $(p_k, F_s)\in\Delta'$ and $(r, s)\in E(\Gamma)$. In turn, by definition of R, there exist (p_i, F_r), $(p_j, F_s)\in R$. Recall that we replaced one edge, e say, of R. If $e=(p_k, F_r)$ then $k=i$ and $e\neq(p_k, F_s)$ so $j<k$, contradicting the definition of e. Hence $e\neq(p_k, F_r)$ and $i<k$. We now have $(r, s)\in E(\Gamma)$, $p_i\in F_r$, $p_k\in F_s$ and $p_i<p_k$. This contradicts (2.1), establishes our claim, and ends the proof of the inequality of Theorem 2.6.

Finally we give an example to show that the upper bound can be attained. Let Γ be the complete directed transitive graph so $(i, j)\in E(\Gamma)$ if and only if $1\leq i<j\leq v$. Let P be a KLYM poset. Choose i with $\nu(i)=m$. Put $F_1=\{p\in P:\text{rank}(p)\geq i\}$, $F_2=\ldots=F_{v-1}=\text{rank }i$ of P, and $F_v=\{p\in P:\text{rank}(p)\leq i\}$. Then $\sum|F_i|=|P|+m(v-1)$ as desired. $\square$

PROOF OF THEOREM 2.2   Put $P=2^n$ in Theorems 2.6 and 2.7. $\square$

6. To find $\lambda(n)$ when Γ is undirected

PROOF OF THEOREM 2.1   This proof is constructed for the various cases as follows:

Case $J=\text{I}$ Let $\mathscr{F}_1,\ldots,\mathscr{F}_v$ satisfy $J=\text{I}$. This means that if $(i, j)\in E(\Gamma)$ and $X\in\mathscr{F}_i$ and $Y\in\mathscr{F}_j$ then $X\cap Y\neq\varnothing$. Put $x_i=2^{-n}|\mathscr{F}_i|$ for $1\leq i\leq v$. Then clearly (4.1) holds. Also (4.2) holds, for if $x_i+x_j>1$ there is an $X\subset N$ such that $X\in\mathscr{F}_i$ and $N\setminus X\in\mathscr{F}_j$. Hence $\sum x_i\leq\alpha^{(1)}(\Gamma)$ or in other words $2^{-n}(|\mathscr{F}_1|+\ldots+|\mathscr{F}_v|)\leq\alpha^{(1)}(\Gamma)$.

To show that this is best possible we must give an example. For use here and elsewhere put

$$\mathcal{G}_{i,j} = \{X \subset N : i \in X, j \notin X\} \quad \text{for} \quad 1 \leq i, j \leq n;$$

thus

$$\mathcal{G}_{1,0} = \{X \subset N : 1 \in X\} \quad \text{and} \quad \mathcal{G}_{0,1} = \{X \subset N : 1 \notin X\}.$$

As shown in Lemma 4.1 we can choose $x_i \in \{0, \frac{1}{2}, 1\}$ with $\sum x_i = \alpha^{(1)}(\Gamma)$. Then for $1 \leq i \leq v$ we let $\mathcal{F}_i$ be $\varnothing$, $\mathcal{G}_{1,0}$, 2^n accordingly as x_i is $0, \frac{1}{2}, 1$. This example completes the proof of case $J = I$.

Case $J = U$ This follows from case $J = I$ by replacing each set X by its complement $N \setminus X$.

We will need the fundamental result

LEMMA 6.1 (Seymour [14]) $\sqrt{|\mathcal{F}|} + \sqrt{|\text{incomp } \mathcal{F}|} \leq \sqrt{(2^n)}$.

Here and elsewhere

$$\text{incomp } \mathcal{F} = \{Y \subset N : \text{both } X \not\subset Y \text{ and } Y \not\subset X \text{ for all } X \in \mathcal{F}\}.$$

Note that we use $\subset$ in the sense that $X \subset Y$ allows $X = Y$.

Case $J = IU$ Let $\mathcal{F}_1, \ldots, \mathcal{F}_v$ satisfy $J = IU$. If $(i, j) \in E(\Gamma)$ then $\mathcal{F}_j^c = \{N \setminus X : X \in \mathcal{F}_i\} \subset \text{incomp } \mathcal{F}_i$ so $\sqrt{|\mathcal{F}_i|} + \sqrt{|\mathcal{F}_j^c|} = \sqrt{|\mathcal{F}_i|} + \sqrt{|\mathcal{F}_j|} \leq \sqrt{2^n}$. We put $x_i = \sqrt{(2^{-n} |\mathcal{F}_i|)}$ for $1 \leq i \leq v$. Then (4.1), (4.2) hold so $\sum x_i^2 \leq \alpha^{(2)}(\Lambda)$ or $2^{-n} \sum |\mathcal{F}| \leq \alpha^{(2)}(\Gamma)$.

For our example let $\mathcal{F} = \mathcal{G}_{1,2}$, so $\mathcal{F}$ is clearly IU. We choose $x_i \in \{0, \frac{1}{2}, 1\}$ with $\sum x_i^2 = \alpha^{(2)}(\Gamma)$. Then for $1 \leq i \leq v$ we let $\mathcal{F}_i$ be $\varnothing$, $\mathcal{F}$, 2^n according as x_i is $0, \frac{1}{2}, 1$. This example completes the proof of the case $J = IU$.

Case $\neq \in J$ We always have $\alpha(\Gamma) \leq \lambda(n)$. Given $X \subset N$ let A be the set of vertices i such that $X \in \mathcal{F}_i$. Because $\neq \in J$ this set A is independent so $|A| \leq \alpha(\Gamma)$. It follows immediately that $2^{-n} \sum |\mathcal{F}| \leq \alpha(\Gamma)$. Notice that we did not use the structure of 2^n, so if 2^n was replaced by any set B we would have $\max \sum |\mathcal{F}| = |B| \, \alpha(\Gamma)$.

Cases $J = C, IC, UC, IUC$ These follow from Theorem 2.5 with $P = 2^n$ because then (2.2) holds for n sufficiently large. $\square$

7. Two examples

Let a be a real number in $0 < a \leq \frac{1}{2}$. Let $N = \{1, \ldots, n\}$ be partitioned into $N = M_1 \cup \ldots \cup M_k$. For $1 \leq i \leq k$ let $m_i = |M_i|$ and do the following. Choose any ordering $<_i$ of the set of all subsets of M_i, such that Z, $Z' \subset M_i$ and $|Z| < |Z'|$ imply $Z <_i Z'$. Let $\mathcal{A}_i$ be the first $\lfloor a2^{m_i} \rfloor$ subsets of M_i in this ordering. Let $\mathcal{C}_i$ be complementary to $\mathcal{A}_i$ so $\mathcal{C}_i = \{M_i \setminus Z : Z \in \mathcal{A}_i\}$. Let $\mathcal{B}_i$ be the remaining subsets of M_i so $\mathcal{B}_i =$

$\{Z \subset M_i: Z \notin \mathcal{A}_i \cup \mathcal{C}_i\}$ and $|\mathcal{B}_i| \sim (1-2a)2^{m_i}$. Notice carefully that $\mathcal{A}_i, \mathcal{B}_i, \mathcal{C}_i$ are pairwise disjoint. Moreover if $X \in \mathcal{A}_i$, $Y \in \mathcal{B}_i$ and $Z \in \mathcal{C}_i$ then $Y \cap Z \neq \varnothing$, $X \cup Y \neq N$, $Y \not\subset X$, $Z \not\subset Y$ and $X \neq Y \neq Z$.

Next we define

$$\mathcal{F}^{(1)} = \{X \subset N: X \cap M_i \in \mathcal{C}_i \text{ for at least one } i\},$$

$$\mathcal{F}^{(2)} = \{Z \subset N: Z \cap M_i \in \mathcal{A}_i \cup \mathcal{B}_i \text{ for all } i\},$$

$$\mathcal{G} = \{Y \subset N; Y \cap M_i \in \mathcal{B}_i \text{ for all } i\}.$$

Clearly if $X \in \mathcal{F}^{(1)}$ and $Y \in \mathcal{G}$ then $X \cap Y \neq \varnothing$, $X \not\subset Y$ and $X \neq Y$. Using the fact that $m_1 + \ldots + m_k = n$ we obtain

$$|\mathcal{G}| = \prod_{1 \leq i \leq k} |\mathcal{B}_i| \sim 2^n (1-2a)^k = 2^n (1-a)^{2k} b,$$

where

$$b = ((1-2a)/(1-a)^2)^k < 1,$$

and

$$|\mathcal{F}^{(2)}| \sim 2^n (1-a)^k,$$

so

$$|\mathcal{F}^{(1)}| = 2^n - |\mathcal{F}^{(2)}| \sim 2^n (1-(1-a)^k).$$

We choose a so that $(1-a)^{2k} = 1-(1-a)^k = c$ say. Then $c = (3-\sqrt{5})/2$ because $(1-a)^k$ is the golden section $(-1+\sqrt{5})/2$. We will use the fact that as $k, m_1, \ldots, m_k$ all go to ∞ we have $a \to 0$ and $b \to 1$.

With a new value of a and $\mathcal{G}$ as above we put

$$\mathcal{F}^{(3)} = \{X \subset N: \text{both } X \cap M_i \in \mathcal{A}_i \text{ for at least one } i$$
$$\text{and } X \cap M_i \in \mathcal{C}_i \text{ for at least one } i\}.$$

Clearly if $X \in \mathcal{F}^{(3)}$ and $Y \in \mathcal{G}$ then $X \cap Y \neq \varnothing$, $X \cup Y \neq N$, $X \not\subset Y$, $Y \not\subset X$ and $X \neq Y$. Now if

$$\mathcal{F}^{(4)} = \{Z \subset N: Z \in \mathcal{B}_i \cup \mathcal{C}_i \text{ for all } i\},$$

then $|\mathcal{F}^{(2)}| = |\mathcal{F}^{(4)}|$ so

$$|\mathcal{F}^{(3)}| = 2^n - |\mathcal{F}^{(2)}| - |\mathcal{F}^{(4)}| + |\mathcal{G}| \sim 2^n \{1 - 2(1-a)^k + (1-2a)^k\}.$$

We now choose a so that $(1-a)^k = \frac{1}{2}$ and then $|\mathcal{F}^{(3)}| \to 2^n/4$ and $|\mathcal{G}| \to 2^n/4$ as $k, m_1, \ldots, m_k$ all $\to \infty$.

EXAMPLE 7.1 Let $v = 2$, $\mathcal{F}_1 = \mathcal{F}^{(1)}$, $\mathcal{F}_2 = \mathcal{G}$ and Γ be the edge $(1, 2)$. As $k, m_1, \ldots, m_k$ all go to ∞ we have $2^{-n} \min\{|\mathcal{F}_1|, |\mathcal{F}_2|\} \to (3-\sqrt{5})/2$. If Γ is undirected then I$\neq$ hold but if Γ is directed then IC$\neq$ hold.

EXAMPLE 7.2 Let $v = 2$, $\mathcal{F}_1 = \mathcal{F}^{(3)}$, $\mathcal{F}_2 = \mathcal{G}$ and Γ be the edge $(1, 2)$.

Whether Γ is directed or undirected IUC$\neq$ hold and $2^{-n} \min\{|\mathcal{F}_1|, |\mathcal{F}_2|\} \to \frac{1}{4}$.

8. To find $\lambda(n)$, $\mu(n)$ when Γ is an undirected edge

We have $v = 2$, $E(\Gamma) = \{(1, 2)\}$ and write $\mathcal{F}$ for $\mathcal{F}_1$ and $\mathcal{G}$ for $\mathcal{F}_2$.

THEOREM 8.1 *If Γ is an undirected edge then $\lambda(n) = 1$ for all J and $n \geqslant 1$.*

PROOF We have $\alpha = \alpha^{(1)} = \alpha^{(2)} = 1$, and we apply Theorems 2.1 and 2.5. $\square$

THEOREM 8.2 *When Γ is an undirected edge,*

$$
\begin{array}{lll}
\mu(n) = \tfrac{1}{2} & \text{for} & J = I, U, \neq, \\
\mu(n) \leqslant \mu = (3 - \sqrt{5})/2 & \text{for} & J = I\neq, U\neq, \\
\mu(n) \geqslant \mu = \tfrac{1}{4} & \text{if} & J = C, \\
\mu(n) = \tfrac{1}{4} & \text{for} & J = IU, C\neq, \\
\mu(n) \leqslant \mu = \tfrac{1}{4} & \text{if} & IU \subset J \text{ or } C \neq \subset J, \\
\mu = \tfrac{1}{4} & \text{for} & J = IC, UC.
\end{array}
$$

REMARK 8.1 The authors conjectured in [6] that $\mu(n) = \max\{\tfrac{1}{4}, 2^{-n}\,\text{Sper}(n)\}$ for $J = C$. There is an earlier weaker conjecture of Gronau [9].

PROOF OF THEOREM 8.2 This proof is again presented in terms of the various cases:

Cases $J = I, U, \neq$ By Theorem 8.1 we have $|\mathcal{F}| + |\mathcal{G}| \leqslant 2^n$ so $2^{-n} \min\{|\mathcal{F}|, |\mathcal{G}|\} \leqslant \tfrac{1}{2}$ so $\mu(n) \leqslant \tfrac{1}{2}$. To see that we have $\mu(n) = \tfrac{1}{2}$ consider examples where $\mathcal{F}$ and $\mathcal{G}$ are $\mathcal{G}_{0,1}$ or $\mathcal{G}_{1,0}$.

Case $J = C$ The example $\mathcal{F} = \mathcal{G}_{1,2}$ and $\mathcal{G} = \mathcal{G}_{2,1}$ shows that $\tfrac{1}{4} \leqslant \mu(n)$. Then the example $\mathcal{F} = \mathcal{G} = \{X \subset N : |X| = \lfloor \tfrac{1}{2}n \rfloor\}$ shows that $2^{-n}\,\text{Sper}(n) \leqslant \mu(n)$. Thus we have the left side of our result

$$\max\{\tfrac{1}{4}, 2^{-n}\,\text{Sper}(n)\} \leqslant \mu(n) \leqslant \tfrac{1}{4} + 2^{-n}\,\text{Sper}(n).$$

For the right-hand inequality let $\mathcal{A} = \mathcal{F} \cap \mathcal{G}$. Then $\mathcal{A}$ is an antichain so $|\mathcal{A}| \leqslant \text{Sper}(n)$. Put $\mathcal{F}' = \mathcal{F} \setminus \mathcal{A}$ and $\mathcal{G}' = \mathcal{G} \setminus \mathcal{A}$, then $\mathcal{G}' \subset \text{incomp}\,\mathcal{F}'$, so by Lemma 6.1 we have $\sqrt{|\mathcal{F}'|} + \sqrt{|\mathcal{G}'|} \leqslant \sqrt{2^n}$. Hence $2^{-n} \min\{|\mathcal{F}'|, |\mathcal{G}'|\} \leqslant \tfrac{1}{4}$ and the inequality follows.

Case $J = C\neq$ The example $\mathcal{F} = \mathcal{G}_{1,2}$ and $\mathcal{G} = \mathcal{G}_{2,1}$ shows that $\tfrac{1}{4} \leqslant \mu(n)$. The reverse inequality follows from Lemma 6.1 as in the case $J = C$ because here $\mathcal{F} \subset \text{incomp}\,\mathcal{G}$.

Case $J = IU$ We use the example $\mathcal{F} = \mathcal{G} = \mathcal{G}_{1,2}$. Then we note that $\mathcal{F} \subset \text{incomp}\{N \setminus X : X \in \mathcal{G}\}$ and apply Lemma 6.1.

Case $IU \subset J$ By the case $J = IU$ we have $\mu(n) \le \frac{1}{4}$. Then Example 7.2 shows that $\frac{1}{4} \le \mu$.

Case $C \ne \subset J$ Here, by Lemma 3.2, we have $\mu(1) \le \mu(2) \le \ldots \le \mu$. By the case $J = C$ we have $\mu \le \frac{1}{4}$. Then Example 7.2 shows that $\frac{1}{4} \le \mu$.

Cases $J = IC, UC$ We have $\mu \le \frac{1}{4}$ by the case $J = C$ and $\frac{1}{4} \le \mu$ by Example 7.2.

Cases $J = I\ne, U\ne$ We will only deal with the case $J = I\ne$ because the case $J = U\ne$ will then follow by taking complements in N. In view of Example 7.1 it is sufficient to show that $\mu(n) \le (3 - \sqrt{5})/2$ for $J = I\ne$.

So suppose $\mathcal{F}, \mathcal{G}$ are $I\ne$. Put

$$\mathcal{F}^{(1)} = \{X \in \mathcal{F} : \exists Y \in \mathcal{G}, \, Y \subset X\}, \quad \mathcal{F}^{(2)} = \mathcal{F} \setminus \mathcal{F}^{(1)},$$

$$\mathcal{G}^{(1)} = \{Y \in \mathcal{G} : \exists X \in \mathcal{F}, \, X \subset Y\}, \quad \mathcal{G}^{(2)} = \mathcal{G} \setminus \mathcal{G}^{(1)}.$$

Since condition $\ne$ holds we have $\mathcal{F} \cap \mathcal{G} = \varnothing$ and

$$|\mathcal{F} \cup \mathcal{G}| = |\mathcal{F}^{(1)}| + |\mathcal{F}^{(2)}| + |\mathcal{G}^{(1)}| + |\mathcal{G}^{(2)}|.$$

It follows from the definitions that $\mathcal{F}^{(2)} \subset \text{incomp}(\mathcal{G}^{(2)})$ so

$$\sqrt{|\mathcal{F}^{(2)}|} + \sqrt{|\mathcal{G}^{(2)}|} \le \sqrt{2^n}.$$

Next let $\mathcal{H} = \{N \setminus X : X \in \mathcal{F}^{(1)} \cup \mathcal{G}^{(1)}\}$ so

$$|\mathcal{H}| = |\mathcal{F}^{(1)}| + |\mathcal{G}^{(1)}|.$$

Assume that $Z \in \mathcal{F} \cap \mathcal{H}$. Then $Z = N \setminus X$ with either $X \in \mathcal{F}^{(1)}$ or $X \in \mathcal{G}^{(1)}$. In the first case there is a $Y \in \mathcal{G}$ with $Y \subset X$ and hence $Z \cap Y = \varnothing$, contradicting I. In the second case $Z \cap X = \varnothing$, again contradicting I. This shows that $\mathcal{F} \cap \mathcal{H} = \varnothing$ and by symmetry $(\mathcal{F} \cup \mathcal{G}) \cap \mathcal{H} = \varnothing$. Let us define real numbers

$$a_1 = |\mathcal{F}^{(1)}|/2^n, \quad a_2 = |\mathcal{F}^{(2)}|/2^n, \quad b_1 = |\mathcal{G}^{(1)}|/2^n, \quad b_2 = |\mathcal{G}^{(2)}|/2^n.$$

Then since $|\mathcal{F} \cup \mathcal{G} \cup \mathcal{H}| \le 2^n$ the inequality $\mu(n) \le (3 - \sqrt{5})/2$ comes from

LEMMA 8.1   *Over reals* $a_1, a_2, b_1, b_2 \ge 0$ *satisfying*

$$\sqrt{a_2} + \sqrt{b_2} \le 1, \tag{8.1}$$

and

$$2a_1 + a_2 + 2b_1 + b_2 \le 1, \tag{8.2}$$

we have

$$\max\{\min\{a_1 + a_2, \, b_1 + b_2\}\} = (3 - \sqrt{5})/2,$$

with equality if and only if, possibly by exchanging a and b,

$$0 = b_1 < a_2 = (7 - 3\sqrt{5})/2 < a_1 = -2 + \sqrt{5} < b_2 = a_1 + a_2 = (3 - \sqrt{5})/2.$$

PROOF   Let a_1, a_2, b_1, b_2 be chosen to maximize the minimum. Without loss of generality we can assume $a_1 + a_2 = b_1 + b_2$. Let $x = \min\{a_2, b_1\}$. Then by squaring twice we can verify that

$$\sqrt{(a_2 - x)} + \sqrt{(b_2 + x)} \leq \sqrt{a_2} + \sqrt{b_2}.$$

We change variables by adding x to a_1, b_2 and subtracting x from a_2, b_1. Clearly (8.1), (8.2) still hold and two cases arise.

Case $a_2 = 0$ Here (8.2) implies (8.1). So we want $\max\{b_1 + b_2\}$ subject to $4b_1 + 3b_2 \leq 1$. This max is clearly $\frac{1}{3}$ and $\frac{1}{3} < (3 - \sqrt{5})/2$.

Case $b_1 = 0$ Here we want $\max\{a_1 + a_2\}$ subject to $\sqrt{a_2} + \sqrt{(a_1 + a_2)} \leq 1$ and $3a_1 + 2a_2 \leq 1$. Put $y = \min\{1 - 3a_1 - 2a_2, a_2\}$ so $y \geq 0$. We change variables by adding y to a_1 and subtracting y from a_2. Then we assume $3a_1 + 2a_2 = 1$, for otherwise we are back in the case $a_2 = 0$. Eliminating a_2 we now want $\max\{(1 - a_1)/2\}$ subject to

$$\sqrt{\{(1 - 3a_1)/2\}} + \sqrt{\{(1 - a_1)/2\}} \leq 1.$$

Squaring this inequality twice and solving a quadratic equation shows that the best value of a_1 is $-2 + \sqrt{5}$ and Lemma 8.1 follows.

9. To find $\lambda(n)$, $\mu(n)$ when Γ is a directed edge

We continue using the notation of Section 8 but now the edge $(1, 2)$ is directed. We assume that $C \in J$ because the other cases are covered by our work on undirected Γ.

THEOREM 9.1   *When Γ is a directed edge*

$$\lambda(n) = \begin{cases} 1 + 2^{-n} \operatorname{Sper}(n) & \text{if } J = C, \\ 1 & \text{otherwise.} \end{cases}$$

PROOF   Two cases must be considered:

Case $J = C$ Since $\mathscr{F} \cap \mathscr{G}$ is an antichain we have $|\mathscr{F}| + |\mathscr{G}| = |\mathscr{F} \cup \mathscr{G}| + |\mathscr{F} \cap \mathscr{G}| \leq 2^n + \operatorname{Sper}(n)$. There is equality if $\mathscr{F} = \{X \subset N: \lfloor \frac{1}{2}n \rfloor \leq |X|\}$ and $\mathscr{G} = \{X \subset N: |X| \leq \lfloor \frac{1}{2}n \rfloor\}$.

Case $J \neq C$ We have $1 \leq \lambda(n)$ by $\mathscr{F} = 2^n$ and $\mathscr{G} = \varnothing$. Ignoring the condition C we see by Theorem 8.1 that $\lambda(n) \leq 1$. $\square$

THEOREM 9.2 *When Γ is a directed edge*

$$\mu(n) = \tfrac{1}{2}\{1 + 2^{-n}\,\mathrm{Sper}(n)\} \quad \text{if} \quad n = 2m \text{ and } J = C,$$

$$\mu(n) = \tfrac{1}{2}\left\{1 + 2^{-2m}\binom{2m}{m-1}\right\} \quad \text{if} \quad n = 2m+1 \text{ and } J = C,$$

$$\mu(n) = \tfrac{1}{2} \quad \text{if} \quad J = C\neq,$$

$$\mu = (3 - \sqrt{5})/2 \quad \text{for} \quad J = IC, UC,$$

$$\mu(n) \leqslant \mu = (3 - \sqrt{5})/2 \quad \text{for} \quad J = IC\neq, UC\neq,$$

$$\mu(n) \leqslant \mu = \tfrac{1}{4} \quad \text{for} \quad J = IUC, IUC\neq.$$

PROOF Again a number of cases must be considered:

Case $J = C$ When n is even the result follows from Theorem 9.1. The authors conjectured the result for n odd but it seems to be much deeper than n even. It is the main result in a paper of Daykin [4], who does not celebrate birthdays but wishes his friend Paul everlasting life in happiness.

Case $J = IC$ If $\mathcal{H} = \mathcal{F} \cap \mathcal{G}$ then $\mathcal{H}$ is an antichain. Let $\mathcal{F}' = \mathcal{F}\backslash\mathcal{H}$ and $\mathcal{G}' = \mathcal{G}\backslash\mathcal{H}$. Then $\mathcal{F}', \mathcal{G}'$ have properties I$\neq$ so $\min\{|\mathcal{F}'|, |\mathcal{G}'|\} \leqslant 2^n(3 - \sqrt{5})/2$ by Theorem 8.2. It follows that

$$\mu(n) \leqslant (3 - \sqrt{5})/2 + 2^{-n}\,\mathrm{Sper}(n).$$

The result now follows by Example 7.1. The example $\mathcal{F} = \{X \subset N: 7 \leqslant |X|\}$ and $\mathcal{G} = \{X \subset N: 6 \leqslant |X| \leqslant 7\}$ with $n = 12$ shows that $\mu(12) \geqslant 0.3872 > 0.381\,97 = (3 - \sqrt{5})/2$.

Remaining cases These follow from Theorem 8.2 by deleting condition C.

10. To find $\mu(n)$ when $v > 2$ and Γ is undirected

Let $\mathcal{H}$ be a set $\{H_1, \ldots, H_m\}$ of distinct finite sets H_i so $\mathcal{H}$ is a hypergraph. We recall that the fractional chromatic number $\chi^*(\mathcal{H})$ is defined as

$$\chi^*(\mathcal{H}) = \min\left\{\sum_{1 \leqslant i \leqslant m} x(H_i)\right\} \tag{10.1}$$

evaluated over all real numbers $x(H_1), \ldots, x(H_m) \geqslant 0$ such that

$$1 \leqslant \sum_{\substack{1 \leqslant i \leqslant m \\ p \in H_i}} x(H_i) \quad \text{for all} \quad p \in H_1 \cup \ldots \cup H_m. \tag{10.2}$$

We need χ^* for

THEOREM 10.1 *Let $\mathcal{H}$ be the set of all distinct, maximal by inclusion,* *independent subsets of* $V(\Gamma)$. *If* Γ *is undirected and* J *is* $\neq$ *then*

$$\mu = 1/\chi^*(\mathcal{H})$$

and

$$\mu - 2^{-n}\operatorname{Sper}(v) \leq \mu(n) \leq \mu.$$

PROOF (i) Let $\mathcal{F}_1, \ldots, \mathcal{F}_v$ satisfy $J = \neq$. In this part we show that if $\delta = \min\{|\mathcal{F}_i|\}$ then $\delta \leq 1/\chi^*$. If $\delta = 0$ we have nothing to do so we assume $\delta > 0$. Let $M_1, \ldots, M_{2^n}$ be the distinct subsets of N. Put

$$A_j = \{k: 1 \leq k \leq v, M_j \in \mathcal{F}_k\} \quad \text{for} \quad 1 \leq j \leq 2^n,$$

so A_j is an independent subset of $V(\Gamma)$. Let $H_1, \ldots, H_m$ be the distinct members of $\mathcal{H}$. Choose any map $\omega : \{1, \ldots, 2^n\} \to \{1, \ldots, m\}$ such that $A_j \subset H_{\omega(j)}$ for all j. For $1 \leq i \leq m$ let ν_i be the number of times that H_i occurs in the list $H_{\omega(1)}, \ldots, H_{\omega(2^n)}$, and give to H_i the weight $x(H_i) = \nu_i/\delta \geq 0$.

Now $p \in H_1 \cup \ldots \cup H_m = V(\Gamma)$ means that $1 \leq p \leq v$, and for each such p we have

$$1 \leq \delta^{-1}|\mathcal{F}_p| = \delta^{-1} \sum_{\substack{1 \leq j \leq 2^n \\ p \in A_j}} 1$$

$$\leq \delta^{-1} \sum_{\substack{1 \leq j \leq 2^n \\ p \in H_{\omega(j)}}} 1$$

$$= \delta^{-1} \sum_{\substack{1 \leq i \leq m \\ p \in H_i}} \left(\sum_{\substack{1 \leq j \leq 2^n \\ H_i = H_{\omega(j)}}} 1 \right)$$

$$= \delta^{-1} \sum_{\substack{1 \leq i \leq m \\ p \in H_i}} \nu_i = \sum_{\substack{1 \leq i \leq m \\ p \in H_i}} x(H_i).$$

So these weights satisfy the conditions (10.2). Hence

$$\chi^* \leq \sum_{1 \leq i \leq m} x(H_i) = \delta^{-1} \sum_{1 \leq i \leq m} \nu_i = \delta^{-1} 2^n,$$

giving $\mu(n) \leq 1/\chi^*$ as required.

(ii) We here show that $(1/\chi^*) - 2^{-n}\operatorname{Sper}(v) \leq \mu(n)$. Now $\mathcal{H} = \{H_1, \ldots, H_m\}$ is an antichain on $V(\Gamma)$, so $m \leq \operatorname{Sper}(v)$. Choose any solution x of the $\chi^*(\mathcal{H})$ problem, this means that

$$\chi^*(\mathcal{H}) = \sum_{1 \leq i \leq m} x(H_i)$$

and (10.2) holds. For $1 \leq i \leq m$ put

$$d_i = \lfloor 2^n x(H_i)/\chi^* \rfloor,$$

so

$$\sum d_i \leq \sum 2^n x(H_i)/\chi^* = 2^n.$$

Let $\mathcal{G}_1, \ldots, \mathcal{G}_m$ be pairwise disjoint sets of subsets of N with $d_i = |\mathcal{G}_i|$ for $1 \leq i \leq m$. Then put

$$\mathcal{F}_p = \bigcup_{\substack{1 \leq i \leq m \\ p \in H_i}} \mathcal{G}_i \quad \text{for} \quad 1 \leq p \leq v.$$

Then for each p, by using (10.2), we have

$$|\mathcal{F}_p| = \sum_{\substack{1 \leq i \leq m \\ p \in H_i}} d_i \geq -m + \sum_{\substack{1 \leq i \leq m \\ p \in H_i}} 2^n x(H_i)/\chi^* \geq -m + 2^n/\chi^*,$$

and the inequality for $\mu(n)$ follows. As $n \to \infty$ we have $2^{-n} \text{Sper}(v) \to 0$ so $\mu = 1/\chi^*$.

Finally we must verify that $\mathcal{F}_1, \ldots, \mathcal{F}_v$ satisfy the condition $\neq$. So suppose that (i, j) is an edge and $X \in \mathcal{F}_i \cap \mathcal{F}_j$. Since $\mathcal{G}_1, \ldots, \mathcal{G}_m$ are disjoint, there is a unique k with $X \in \mathcal{G}_k$. Hence $\mathcal{G}_k \subset \mathcal{F}_i \cap \mathcal{F}_j$ and this implies that $i, j \in H_k$, contradicting the fact that H_k is an independent set. This completes the proof of Theorem 10.1. $\square$

REMARK 10.1   An obvious candidate for an undirected Γ is the complete graph K_v. Even for this graph and $J = C$ we do not know $\mu(n)$. When $v = \text{Sper}(m)$ for some m an obvious example shows that $1/2^m \leq \mu(n)$ for $m \leq n$. The best example we could find for $v = 4$ has $\mu(n) = \frac{5}{64}$. For this example let Q_1, Q_2, Q_3, Q_4 be a partition of the 20 subsets of $\{1, \ldots, 6\}$ of cardinality 3 with each Q_i having five of the subsets. Then put $\mathcal{F}_i = \{X \subset N: X \cap \{1, \ldots, 6\} \in Q_i\}$ for $1 \leq i \leq 4$.

PROOF OF LEMMA 2.1   We must construct an example. Let Γ be the complete undirected graph on

$$v = \binom{2m-1}{m} \geq 3$$

vertices. Let $Y_1, \ldots, Y_v$ be the v members of $\{Y \subset \{2, 3, \ldots, 2m\}: |Y| = m\}$. For $1 \leq i \leq v$ let $Y_i^c = \{1, 2, \ldots, 2m\} \setminus Y_i$. Finally for $n \geq 2m$ and $1 \leq i \leq v$ put

$$\mathcal{F}_i = \{X \subset N: X \cap \{1, 2, \ldots, 2m\} \text{ is } Y_i \text{ or } Y_i^c\},$$

so $|\mathcal{F}_i| = 2^{n-2m+1}$. This example is IUC$\neq$ and the lemma follows. $\square$

11. When Γ is directed

Here and elsewhere

$$\text{above } \mathcal{F} = \{X \subset N : \exists\, Y \in \mathcal{F},\ Y \subset X\},$$
$$\text{below } \mathcal{F} = \{X \subset N : \exists\, Y \in \mathcal{F},\ X \subset Y\}.$$

We need this result of great importance:

LEMMA 11.1 (Kleitman [12]) $|\mathcal{F}|\, 2^n \leqslant |\text{above } \mathcal{F}|\, |\text{below } \mathcal{F}|$.

THEOREM 11.1   *If $v \geqslant 3$ and Γ is a directed circuit and $J = C \neq$ then* $\sum \sqrt{|\mathcal{F}_i|} \leqslant (v/2)\sqrt{2^n}$ *and* $\mu(n) = \tfrac{1}{4}$.

PROOF   The conditions are that if $X \in \mathcal{F}_i$ and $Y \in \mathcal{F}_{i+1}$ then $X \not\subset Y$ and $X \neq Y$. Subscripts are taken mod v. For $1 \leqslant i \leqslant v$ let $a_i = 2^{-n}\, |\text{above } \mathcal{F}_i|$, $b_i = 2^{-n}\, |\text{below } \mathcal{F}_i|$. Since $J = C \neq$ we have $(\text{above } \mathcal{F}_i) \cap (\text{below } \mathcal{F}_{i+1}) = \varnothing$ so $a_i + b_{i+1} \leqslant 1$. Consequently, summing over i, we get $\sum (a_i + b_i) \leqslant v$. Now Lemma 11.1 says that

$$2^{-n}\, |\mathcal{F}_i| \leqslant a_i b_i \leqslant \{(a_i + b_i)/2\}^2.$$

We take the square root and sum to get our first result. For the second result, $\mu(n) = \tfrac{1}{4}$, notice that since $\sum (a_i + b_i) \leqslant v$ there is an i with $a_i + b_i \leqslant 1$. This implies that $a_i b_i \leqslant \tfrac{1}{4}$ so $2^{-n}\, |\mathcal{F}_i| \leqslant \tfrac{1}{4}$. Then examples to show the theorem is best possible are, for n even $\mathcal{G}_{12}, \mathcal{G}_{21}, \mathcal{G}_{12}, \mathcal{G}_{21}, \ldots$, and for n odd $\mathcal{G}_{12}, \mathcal{G}_{31}, \mathcal{G}_{23}, \mathcal{G}_{12}, \mathcal{G}_{21}, \ldots$.   □

REMARK 11.1   We think that Theorem 11.1 holds for $J = C$ if n is large.

THEOREM 11.2   *If Γ is the directed path $(1, 2), (2, 3)$ and $J = C \neq$ then*

$$\sqrt{|\mathcal{F}_2|} + \sqrt{(|\mathcal{F}_1| + |\mathcal{F}_3|)} \leqslant (\tfrac{3}{2})\sqrt{2^n}.$$

PROOF   Without loss of generality we may assume $\mathcal{F}_1 = \text{above } \mathcal{F}_1$, $\mathcal{F}_3 = \text{below } \mathcal{F}_3$ and $\mathcal{F}_2 \subset (2^n \backslash \mathcal{F}_1) \cap (2^n \backslash \mathcal{F}_3)$. Put $a = 2^{-n}\, |\mathcal{F}_1|$, $b = 2^{-n}\, |\mathcal{F}_2|$, $c = 2^{-n}\, |\mathcal{F}_3|$ then Lemma 11.1 says that $b \leqslant (1 - a)(1 - c)$. Hence if $d = (a + c)/2$ then

$$\sqrt{b} + \sqrt{(a + c)} \leqslant \sqrt{(2d)} + \sqrt{\{(1 - a)(1 - c)\}} \leqslant \sqrt{(2d)} + (1 - d) \leqslant \tfrac{3}{2}.$$

The example $\mathcal{F}_1 = \mathcal{G}_{1,0}$, $\mathcal{F}_2 = \mathcal{G}_{2,1}$, $\mathcal{F}_3 = \mathcal{G}_{0,2}$ shows that the theorem is best possible.   □

REMARK 11.2   We now mention a result of a different kind. Let

$\mathcal{F}_1, \mathcal{F}_2, \mathcal{F}_3, \mathcal{F}_4$ be such that if $1 \leq i \leq 4$ and $X \in \mathcal{F}_i$, $Y \in \mathcal{F}_{i+1}$, $Z \in \mathcal{F}_{i+2}$ then we do not have either $X \subset Y \subset Z$ or $Y \subset X$, where suffices are taken mod 4. Then Hilton [11] has proved that $\sum \sqrt{|\mathcal{F}_i|} \leq 2(\sqrt{2^n})$.

12. Further problems

The KLYM poset P is log convex if $\nu(i-1)\nu(i+1) \leq \nu(i)\nu(i)$ for $0 < i < h$. Harper [10] proved that the direct product of such KLYM posets forms a KLYM poset. It would be interesting to rework this paper with 2^n replaced by P^n for such a P. If P was also a distributive lattice then so too would be P^n. Lemmas 11.1 and 6.1 of Kleitman and Seymour which we used hold in distributive lattices [1, 2].

The number $(3-\sqrt{5})/2$ does not appear to play a role in distributive lattices. This can be seen as follows. First note that the proofs of the upper bounds for the cases $J = \neq, C, C\neq$ of Theorem 8.2 are valid for distributive lattices. Then secondly consider the two examples below in the (distributive) lattice of divisors of the integer $2^r 3^r$.

Example 12.1 Let $F = \{2^s 3^t : 0 < s < t < r\}$ and $G = \{2^s 3^t : 0 < t < s < r\}$. Then we get properties corresponding to $IU\neq$ and $\frac{1}{2} \leq \mu(n)$.

Example 12.2 Let $F = \{2^s 3^t : 0 < s < \frac{1}{2}r < t < r\}$ and $G = \{2^s 3^t : 0 < t < \frac{1}{2}r < s < r\}$. Then we get properties like $IUC \neq$ and $\frac{1}{4} \leq \mu(n)$.

References

1. R. Ahlswede and D. E. Daykin, An inequality for the weights of two families of sets, their unions and intersections. *Z. Wahrscheinlichkeitstheorie Verw. Gebiete* **43** (1978), 183–185.
2. R. Ahlswede and D. E. Daykin, Inequalities for a pair of maps $S \times S \rightarrow S$ with S a finite set. *Math. Zeit.* **165** (1979), 267–289.
3. A. Brace and D. E. Daykin, Sperner type theorems for finite sets. In *Combinatorics, Proceedings of a Conference, Oxford, 1972*, International Mathematical Association, Oxford, 1972, pp. 18–37.
4. D. E. Daykin, Antichains of subsets of a finite set. In *Graph Theory and Combinatorics* (B. Bollabás, ed.), Academic Press, London, 1984, pp. 99–106.
5. D. E. Daykin and P. Frankl, Sets of finite sets satisfying union conditions. *Mathematika* **29** (1982), 128–134.
6. D. E. Daykin and P. Frankl, Inequalities for subsets of a set and KLYM posets. *SIAM J. Alg. Disc. Meth.* **4** (1983), 67–69.
7. D. E. Daykin and P. Frankl, On Kruskal's cascades and counting containments in a set of subsets. *Mathematika* **30** (1983), 133–141.

8. D. E. Daykin, L. Harper and D. B. West, Some remarks on normalized matching. *J. Combinat. Theory A* (in press).
9. H. D. O. F. Gronau, On maximal families of subsets of a finite set. *Discrete Math.* **34** (1981), 119–130.
10. L. Harper, The morphology of partially ordered sets. *J. Combinat. Theory A* **17** (1974), 44–58.
11. A. J. W. Hilton, private communication.
12. D. J. Kleitman, Families of non-disjoint subsets. *J. Combinat. Theory* **1** (1966), 153–155.
13. D. J. Kleitman, On an extremal property of antichains, etc. In *Combinatorics, Proceedings of a Conference, Breukelen, 1974* (M. Hall and J. H. van Lint, eds), Math. Centrum Tract 55, Amsterdam, 1974, pp. 77–90.
14. P. D. Seymour, On incomparable families of sets. *Mathematika*, **20** (1973), 208–209. ·
15. D. B. West, Extremal problems in partially ordered sets. In *Ordered Sets, Proceedings of a Conference, Banff, 1981* (I. Rival, ed.), D. Reidel, Amsterdam, 1982, pp. 473–521.

10

COMPLETION OF SPARSE PARTIAL LATIN SQUARES

David E. Daykin and Roland Häggkvist

ABSTRACT   Let P be a partial $n \times n$ latin square with $0, 1, \ldots, n-1$ as symbols. (1) If $n = 16k$ and each row, column and symbol is used at most $(\sqrt{k})/32$ times then P can be completed. (2) Form a partial $mn \times mn$ latin square Q from P as follows. Replace each cell c of P by an $m \times m$ array $A(c)$. If c is empty then $A(c)$ is empty. If c has symbol x then exactly one cell of $A(c)$ is filled from among $mx, mx + 1, \ldots, mx + m - 1$ in any way. We conjecture that Q can be completed for $m \geq 2$ and prove it for $m \equiv 0 \pmod{16}$.

1. Introduction

A partial $n \times n$ latin square P is an $n \times n$ array where some cells are filled with one of the symbols $0, 1, \ldots, n-1$ in such a way that no symbol occurs twice in a row or column. If every cell is filled then P is a latin square. If every empty cell can be filled so that the result is a latin square then we say that P can be completed. Our main result, which we think to be the first of its kind, is in the style of Evan's problems.

PROPOSITION 1   *If P is a partial $16k \times 16k$ latin square where each row, column and symbol is used at most $(\sqrt{k})/32$ times then P can be completed.*

The word sparse is in our title because $(\sqrt{k})/32$ is small. We think that the proposition would still be true if $16k$ and $(\sqrt{k})/32$ were replaced by k and uk respectively, where u is some constant, maybe $u = \frac{1}{4}$. A famous result of Ryser is

PROPOSITION 2   *If P is a partial $n \times n$ latin square whose filled cells consist of all the cells in an $r \times s$ rectangle then P can be completed if and only if each symbol occurs at least $r + s - n$ times in P.*

This result trivially implies that if $n = 2d - 1$ (respectively $n = 2d$) and P has all its filled cells in the first $d - 1$ (respectively d) rows and the first d columns then P can be completed. From this observation we easily got

PROPOSITION 3   *Any partial $n \times n$ latin square P can be partitioned into*

GRAPH THEORY AND COMBINATORICS
ISBN 0-12-111760-X

four partial $n \times n$ *latin squares each of which can be completed independently.*

We think that this proposition would still be true if four was replaced by two.

Given integers $m \ge t \ge 1$ and a partial $n \times n$ latin square P one can form a partial $mn \times mn$ latin square Q as follows. Replace each cell c of P by an $m \times m$ array $A(c)$. If c is empty then $A(c)$ is empty. If c is filled with some symbol x then exactly one cell of $A(c)$ is filled with one of the symbols $tx, tx + 1, \ldots, tx + t - 1$ in any way. Thus P and Q have the same number of filled cells. Let $P(m, t)$ be the set of all such Q. We conjecture that for $m \ge 2$ every member of $P(m, m)$ can be completed, but we can only prove

PROPOSITION 4   *If P is any partial latin square and $m \equiv 0 \pmod{16}$ and $Q \in P(m, m)$ then Q can be completed.*

This result is used to prove Proposition 1, and its proof depends on Proposition 3. Thus improvement of either Proposition 3 or Proposition 4 will yield a strengthening of Proposition 1.

2. Proof of Proposition 4

For this section we must slightly generalize our definitions to allow a symbol to occur several times in a row or column. We write $P = (n, \sigma, \rho)$, read P is a partial latin (n, σ, ρ) square, to mean that P is an $n \times n$ array where some cells are filled with one of the symbols $0, 1, \ldots, \sigma - 1$ in such a way that no symbol occurs $\rho + 1$ times in a row or column. If $n \ge \sigma \ge \rho = 1$ then P is in fact a partial $n \times n$ latin square. If every cell of P is filled and $n = \sigma\rho$ then P is called a latin $(\sigma\rho, \sigma, \rho)$ square. We reserve script letters $\mathcal{P}, \mathcal{Q}, \mathcal{R}, \mathcal{S}$ for these squares. Thus, for example, $\mathcal{P} = (n, n, 1)$ means that $\mathcal{P}$ is an $n \times n$ latin square.

If m, t are positive integers and $P = (n, \sigma, \rho)$ we form the set $P(m, t)$ exactly as in the introduction, but now we do not require $m \ge t$. Note that if $Q \in P(m, t)$ then $Q = (mn, t\sigma, \rho)$. Further if m', t' are also positive integers then $P(mm', tt')$ is the union of the sets $Q(m', t')$ with $Q \in P(m, t)$.

Our proof of Proposition 4 is illustrated in Figure 1. To start it we let P be any partial $n \times n$ latin square so $P = (n, n, 1)$. Using Proposition 3 we partition P into $P_0, \ldots, P_3$, which can be completed into four latin

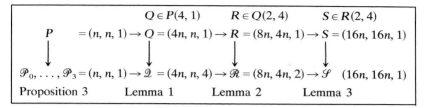

FIG. 1. Diagrammatic representation of the proof of Proposition 4.

squares $\mathscr{P}_0, \ldots, \mathscr{P}_3$ respectively. Then we choose any $Q \in P(4, 1)$ and fix on it. So Q is a partial $4n \times 4n$ latin square and in fact $Q = (4n, n, 1)$.

LEMMA 1   *We can embed Q in some $\mathscr{Q} = (4n, n, 4)$.*

PROOF   We must show that the empty cells of Q can be filled from among the symbols $0, 1, \ldots, n-1$ so that in the result $\mathscr{Q}$ each symbol occurs exactly four times in each row and column.

Let c be an arbitrary cell of P. Let x_g be the symbol in the corresponding cell of $\mathscr{P}_g$ for $0 \le g \le 3$. Note that $x_0, \ldots, x_3$ may not be distinct.

Case (i): c is empty   Then the 4×4 array $A(c)$ is empty in Q. We fill $A(c)$ with the square $(y_{g,h})$ where $y_{g,h} = x_{g+h(\mathrm{mod}\,4)}$ for $0 \le g, h \le 3$.

Case (ii): c contains symbol x, say   Then exactly one of $P_0, \ldots, P_3$ has x in the corresponding cell, and hence at least one of $x_0, \ldots, x_3$ is x. Further exactly one cell of $A(c)$ is non-empty and it contains x. Let the coordinates of this x be g', h' where $0 \le g', h' \le 3$. We choose an integer e so that $x = x_{e+g'+h'(\mathrm{mod}\,4)}$. Then we fill $A(c)$ with the square $(y_{g,h})$ where $y_{g,h} = x_{e+g+h(\mathrm{mod}\,4)}$ for $0 \le g, h \le 3$.

We do the above for each cell c and the lemma is proved.   □

Next we choose any $R \in Q(2, 4)$ and fix it.

LEMMA 2   *We can embed R in some $\mathscr{R} = (8n, 4n, 2)$.*

PROOF   For $0 \le x \le n-1$ we will say where to put the four symbols $4x, \ldots, 4x+3$ in empty cells of R, ensuring that each of these symbols then occurs exactly twice in each row and column. We deal with each x independently in the same way, and this will prove the lemma. So let x be fixed arbitrarily. Of course by cell (i, j) we mean the cell in row i and column j.

We recall that Q is embedded in $\mathscr{Q}$ and use $\mathscr{Q}$ to define a bipartite graph B. There are $8n$ vertices in both parts of B, namely $r_0, \ldots, r_{8n-1}$

and $c_0, \ldots, c_{8n-1}$. For $0 \le i, j \le 8n-1$ there is an edge (r_i, c_j) in B if and only if the cell $(\lfloor i/2 \rfloor, \lfloor j/2 \rfloor)$ of $\mathcal{Q}$ contains the symbol x. Consider the subgraph G of B induced by the vertices $r_0, r_2, \ldots, r_{8n-2}$ and $c_0, c_2, \ldots, c_{8n-2}$. We will use the notation

$$\alpha_{ij} = (r_{2i}, c_{2j}), \qquad \beta_{ij} = (r_{2i+1}, c_{2j+1}),$$

$$\gamma_{ij} = (r_{2i}, c_{2j+1}), \qquad \delta_{ij} = (r_{2i+1}, c_{2j}).$$

Clearly all four of $\alpha_{ij}, \beta_{ij}, \gamma_{ij}, \delta_{ij}$ are edges of B if and only if any one of the four is an edge of B. By definition G is 4-regular so B is 8-regular. Because G is 4-regular it has four 1-factors and hence two 2-factors, say C, D. Using these we define four 2-factors $F_0, \ldots, F_3$ of B by

$$F_0 = \{\alpha_{ij} ; \alpha_{ij} \in C\} \cup \{\beta_{ij} ; \alpha_{ij} \in D\},$$

$$F_1 = \{\beta_{ij} ; \alpha_{ij} \in C\} \cup \{\alpha_{ij} ; \alpha_{ij} \in D\},$$

$$F_2 = \{\gamma_{ij} ; \alpha_{ij} \in C\} \cup \{\delta_{ij} ; \alpha_{ij} \in D\},$$

$$F_3 = \{\delta_{ij} ; \alpha_{ij} \in C\} \cup \{\gamma_{ij} ; \alpha_{ij} \in D\}.$$

Our idea is to use these four 2-factors to fill cells of R as follows. For $0 \le g \le 3$ and $0 \le i, j \le 8n-1$ if (r_i, c_j) is an edge of F_g then we want to fill the cell (i, j) of R with the symbol $4x + g$. However, we might change existing entries of R, so to examine these we put

$$E = \{(r_i, c_j); (r_i, c_j) \text{ is an edge of } B \text{ and cell } (i, j) \text{ of } R \text{ is occupied}\}.$$

Since $R \in Q(2, 4)$ and $Q = (4n, n, 1)$ the edges of E are all independent, and further, for $0 \le i, j \le 4n-1$ at most one of the vertices r_{2i}, r_{2i+1} or of the vertices c_{2j}, c_{2j+1} is incident with an edge of E.

Now suppose that $F_0, \ldots, F_3$ would put the wrong symbol $4x + e$ in a cell of Q where they should put $4x + f$, say. Then there are i, j such that the offending $4x + e$ comes from one of $\alpha_{ij}, \beta_{ij}, \gamma_{ij}, \delta_{ij}$. Consider a permutation π of the vertices of E which exchanges r_{2i}, r_{2i+1} or exchanges c_{2j}, c_{2j+1} but leaves all other vertices fixed. There are three such π and clearly one of them, when applied as a graph automorphism to $F_0, \ldots, F_3$, corrects the offence. We choose a π for each offence and the lemma is proved. $\square$

Finally we choose any $S \in R(2, 4)$ and fix it.

LEMMA 3   *We can embed S in some $\mathcal{S} = (16n, 16n, 1)$.*

PROOF   This is virtually the same as the last one. The graph corresponding to B has $16n$ vertices in each part instead of $8n$ and is 4-regular. The subgraph corresponding to G is 2-regular and so C, D become 1-factors.

The definition of $F_0, \ldots, F_3$ can be left unchanged but now they are 1-factors. There are no other important differences. In the last proof 2-factors caused each symbol to occur twice in a row of $\mathcal{R}$. Because we now have 1-factors $\mathcal{S}$ is a latin square. This ends the proof of Lemma 3 and Proposition 4. $\square$

3. Proof of Proposition 1

We will show that there is a partial $k \times k$ latin square M such that $P \in M(16, 16)$. The result then follows by Proposition 4. If P is empty it is easy to find the M and we will use induction on the number of filled cells of P. So for the induction step we will assume that some symbol occurs in P and that removing this symbol leaves a partial latin square $Q \in N(16, 16)$ for some partial $k \times k$ latin square N.

For both P and Q let us label the rows $a_1, \ldots, a_{16k}$ and the columns $a_{16k+1}, \ldots, a_{32k}$ and the symbols $a_{32k+1}, \ldots, a_{48k}$. We may assume that the removed symbol was a_{48k} from row a_{16k} and column a_{32k} of P. Consider the tripartite graph T with vertex sets $a_1, \ldots, a_{16k}$ and $a_{16k+1}, \ldots, a_{32k}$ and $a_{32k+1}, \ldots, a_{48k}$, where $(a_i, a_j), (a_j, a_h), (a_h, a_i)$ are edges if and only if a_h occurs in the cell of Q in row a_i and column a_j. That $Q \in N(16, 16)$ means the following. We can make three partitions

$$\{a_{16d+1}, \ldots, a_{16(d+k)}\} = A_{d+1} \cup \ldots \cup A_{d+k} \quad \text{for} \quad d = 0, k, 2k, \quad (1)$$

such that $|A_p| = 16$ and there is at most one edge between the sets A_p, A_q of vertices of T for $1 \leq p, q \leq 3k$.

Let U be the corresponding graph for P. Then U is T together with the three edges (a_{16k}, a_{32k}), (a_{32k}, a_{48k}), (a_{48k}, a_{16k}). If there is at most one edge between A_p, A_q in U the proposition is proved. So we suppose otherwise and modify the partitions (1) until this is the case.

So assume that there are two edges between A_p and A_q. Clearly there can not be more than two. Then there is a vertex a_i in A_p or A_q, say in the former, with only one edge from a_i to A_q. To simplify the notation we suppose $p = 1$. Now by hypothesis no row, column or symbol is used more than $(\sqrt{k})/32$ times in P. Hence for any A_e the number of sets A_f in one of the two parts of U not containing A_e with an edge between A_e and A_f is at most $|A_e|(\sqrt{k})/32 = (\sqrt{k})/2$. If a vertex is distance 2 from A_1 it must be in $\{a_1, \ldots, a_{16k}\}$, and hence the number of sets A_e containing such a vertex is at most $(\sqrt{k})(((\sqrt{k})/2) - 1) < k - 1$. Hence there is an A_f not of this form with $2 \leq f \leq k$. Let a_j be any vertex in A_f. We now replace (1) by the partition $A'_1, \ldots, A'_{3k}$ defined by

$$A'_1 = (A_1 \setminus a_i) \cup \{a_j\} \quad \text{and} \quad A'_f = (A_f \setminus a_j) \cup \{a_i\}$$

and $A'_g = A_g$ otherwise. The proposition follows by at most three repetitions of this process. $\square$

Bibliography

1. J. Denes and A. D. Keedwell, *Latin Squares and Their Applications*, English Universities Press, London, 1974.
2. T. Evans, Embedding incomplete latin squares. *Amer. Math. Monthly* **67** (1960), 959–961.
3. C. C. Lindner, *A Survey of Finite Embedding Theorems for Partial Latin Squares and Quasigroups*, Lecture Notes in Mathematics no. 406, Springer Verlag, Berlin, 1974.

11

THE ANTIVOTER PROBLEM: RANDOM 2-COLOURINGS OF GRAPHS

Peter Donnelly and Dominic Welsh

ABSTRACT The antivoter model is a stochastic process of 2-colourings of the vertices of a finite graph. The main aim of the paper is to relate the behaviour of these colourings to the underlying graph. In the proofs of our results, we exploit the connection between the antivoter problem and annihilating random walks, first studied by Erdös and Ney.

1. Introduction

Consider a random sequence of 2-colourings of a finite graph G which evolves as follows. At time $t = 0$ a subset A of the vertex set V is black and $V \backslash A$ is white. At random intervals of time a vertex v chooses a neighbour u at random and if u and v are similarly coloured v changes colour. If we let η_t^A denote the set of black vertices at time t, then η_t^A is the process known in the study of particle systems as the *antivoter model*.

Because of the "repulsive element" in its definition η_t^A should evolve towards a collection of 2-colourings of G which are optimum in some sense. Certainly if G is bipartite, then η_t^A converges as $t \to \infty$ to one of the subsets of V induced by any proper 2-colouring.

An extensive duality theory has been developed for particle systems. The details of this, together with a good list of references, may be found in Griffeath [3]. In our case it means that η_t^A is intimately related to a system of processes $\hat{\eta}_t^B$ consisting of *annihilating random walks* on G. In this model, first introduced by Erdös and Ney [2], particles initially at the vertices of B perform independent random walks on V, the vertex set of G, with the added proviso that if a particle jumps to a vertex already occupied by another particle both particles are annihilated. We will exploit the relationship between these two systems heavily in what follows.

Our interest in this paper will be in relating the behaviour of the process η_t, particularly when t is large, to properties of the underlying graph G.

GRAPH THEORY AND COMBINATORICS
ISBN 0-12-111760-X

2. Basic properties and the duality relation

Throughout this paper we assume that G is a finite connected graph with vertex set V, edge set E.

To define the antivoter process η_t^A formally we associate with each vertex of G a "random clock" which rings, independently for each vertex, at the instances of a Poisson process of rate 1, in other words the intervals between a given clock ringing are independent exponential random variables with mean 1 and these are independent for each clock. When the clock at a vertex v rings the vertex chooses a neighbour u at random and adopts the colour opposite to that of u. Thus if u and v are different colours no change occurs. At time $t = 0$ the vertices of A are coloured black. The set of vertices which are black at time t is denoted by η_t^A.

Clearly η_t^A is a Markov process with state space 2^V and we have the following obvious observation:

When G is bipartite η_t^A has two absorbing states induced by the essentially unique proper 2-colouring of G. $\qquad\qquad$ (1)

We will often wish to consider the more general process η_t^μ in which the initial black set A is chosen according to a distribution μ on 2^V.

It is also convenient to introduce the function $\eta_t^A(\cdot)$ on V defined by

$$\eta_t^A(x) = \begin{cases} 1 & \text{if } x \in \eta_t^A, \text{ i.e. if } x \text{ is black at time } t, \\ 0 & \text{otherwise.} \end{cases}$$

The annihilating random walk process $\hat{\eta}_t^B$ is defined formally in the same way by associating a similar set of random exponential clocks with the vertices. When the clock at vertex v rings, if there is a particle at v it jumps to a randomly chosen neighbour u. If there is a particle at u both particles are annihilated. At time $t = 0$ the particles are located exactly at the members of B. The set of vertices occupied by particles at time t is denoted by $\hat{\eta}_t^B$.

The *duality relationship* between these two systems is expressed in the following theorem.

THEOREM 1 *Let A and B be arbitrary subsets of V, then*

$$P(|\eta_t^A \cap B| \text{ even}) = P(|\hat{\eta}_t^B \cap A| + \varepsilon_t^B \text{ even}) \qquad (2)$$

where ε_t^B is the number of jumps made by the particles of the process $\hat{\eta}_t^B$ in the interval $(0, t)$.

The proof of this theorem depends upon a rather complicated construction of both processes by means of a random mechanism known as a percolation substructure. The details can be found in Griffeath [3, p. 68].

We will use two special cases of this duality relationship each of which has a more intuitive interpretation.

Take B as a singleton set $\{x\}$ in (2). Then we have (writing x for $\{x\}$ where necessary)

$$P(|\eta_t^A \cap \{x\}| \text{ even}) = P(|\hat{\eta}_t^x \cap A| + \varepsilon_t^x \text{ even}). \tag{3}$$

For an intuitive interpretation of this we go back in time and consider the following sequence of "vertex–time" pairs

$$(u_0, t), \quad (u_1, t_1), \quad (u_2, t_2), \quad \ldots, \quad (u_k, t_k),$$

where $t > t_1 > t_2 \ldots > t_k > 0$ and where $u_0 = x$ and $u_i \in V$ are such that the clock at x "most recently" (relative to t) rang at time t_1 when x looked at u_1; the clock at u_1 most recently (relative to t_1) rang at t_2 and u_1 looked at u_2 and so on until we arrive at a vertex u_k whose clock did not ring before time t_k. Note that the colour of x at t will be opposite to that of u_1 at t_1 which in turn will be opposite to u_2 at t_2 and so on. Thus at time t, x will be opposite in colour to the colour of u_k at time $t = 0$ if k is odd and the same colour as u_k if k is even.

By constructing a random walk on V by initially placing a particle at x ($=u_0$) and making it jump from u_i to u_{i+1} at time $t - t_{i+1}$ we obtain a realization of the (vacuously) annihilating random walk $\hat{\eta}_t^x$. The event that x is white at time t is the disjoint union of the event that k is odd (that is that $\hat{\eta}_t^x$ has made an odd number of jumps) and u_k ($=\hat{\eta}_t^x$) is initially black together with the event that k is even (that is $\hat{\eta}_t^x$ has made an even number of jumps) and $u_k (=\hat{\eta}_t^x)$ is initially white. The relationship (3) is now evident.

A similar but more complicated argument which we omit can be used to justify (2) when B is a 2-point set.

As a first application of duality we have

THEOREM 2   *Suppose that the initial distribution μ of black points has the property that the probability that any particular vertex is black is the constant p, then for any vertex x and time $t \geq 0$, the probability that x is black at time t is given for any graph G by*

$$P(x \in \eta_t^\mu) = \tfrac{1}{2}(1 + (2p - 1)e^{-2t}). \tag{4}$$

PROOF   Note that $P(x \in \eta_t^\mu) = P(|\eta_t^\mu \cap \{x\}| \text{ odd})$. Now an argument similar to the one used to derive (3) shows that

$$P(|\eta_t^\mu \cap \{x\}| \text{ odd}) = P(\{\hat{\eta}_t^x \text{ is coloured black initially}\} \cap \varepsilon_t^x \text{ even})$$
$$+ P(\{\hat{\eta}_t^x \text{ is coloured white initially}\} \cap \varepsilon_t^x \text{ odd})$$
$$= pP(\varepsilon_t^x \text{ even}) + (1 - p)P(\varepsilon_t^x \text{ odd})$$

since the initial colouring of V is independent of the behaviour of the random walk which determines ε_t^x. But

$$P(\varepsilon_t^x \text{ even}) = e^{-t} \cosh(t),$$
$$P(\varepsilon_t^x \text{ odd}) = e^{-t} \sinh(t),$$

and (4) follows. $\square$

Immediate consequences of Theorem 2 are

COROLLARY 1 *For any graph G and μ as above,*

$$\lim_{t \to \infty} P(x \text{ is black at time } t) = \tfrac{1}{2}. \quad \square$$

COROLLARY 2 *When G is bipartite and μ has constant one-dimensional marginals, each of the two absorbing states is equally likely.* $\square$

3. Equilibrium

Since the state space of the Markov process η_t^A is finite it must have an equilibrium distribution ν^A as $t \to \infty$, where

$$\nu^A(X) = \lim_{t \to \infty} P(\eta_t^A = X) \qquad (X \subseteq V).$$

There are three different cases to consider.

G is bipartite

If G is bipartite with edges joining V_1 to V_2 then

$$\nu^A(X) = 0, \qquad X \neq V_1 \text{ or } V_2,$$

and from Corollary 2 above,

$$\nu^\mu(V_1) = \nu^\mu(V_2) = \tfrac{1}{2}$$

for any μ with constant one-dimensional marginals.

G is an odd circuit

Suppose that G is a circuit C_{2m+1} with $2m+1$ vertices. Then it is easy to see that η_t^A has a closed set of recurrent states. These are the $4m+2$ states in which exactly one pair of adjacent vertices is the same colour and every other pair of adjacent vertices has opposite colours.

The theory of finite Markov chains shows that the equilibrium distribution is independent of the initial set and by symmetry gives equal weight to each of these configurations.

Notice that the equilibrium states are as close as possible to proper 2-colourings of C_{2m+1}; they represent "frustrated disorder".

G is non-bipartite and not an odd circuit

In contrast to the above two cases we have the following result:

THEOREM 3   *Apart from the two obvious transient states V and ϕ, if G is not bipartite nor an odd circuit then every other state is recurrent and the Markov chain η_t^A is irreducible for any A.*

As an immediate consequence, in this case, ν^A is independent of A.

The proof of this is surprisingly intricate, we will show that any two of the non-trivial states communicate. We proceed by a sequence of lemmas to show how we can change from a configuration or colouring ω of G to any other configuration ω' on G, provided that ω' is not trivial (V or ϕ). We write $\omega(V)$ to denote the colour of v in the colouring ω.

LEMMA 1   *If $\sigma = \{v_1, v_2, v_3, \ldots, v_k\}$ is a path with any colouring then we can change the colouring of σ to an alternating black/white colouring leaving v_1 unchanged.*

PROOF   Obvious.   □

LEMMA 2   *If $C = \{v_1, \ldots, v_{2m+1}\}$ is an odd circuit and ω is any colouring of C then we can change ω to ω' such that $\omega'(v_i) = \omega(v_{i+k})$ with addition modulo $2m+1$. (In other words we can permute any colouring of an odd circuit.)*

PROOF   It is obviously sufficient to prove this lemma for $k = 1$. Divide the 2-colouring of C into successive monochromatic blocks and alternating paths. Choose a block B_1, say, and change the colour of a vertex v at one end. Then change the path between the blocks B_1 and B_2 so that v and this path form a larger alternating path. But this will move the boundary of B_2 towards B_1. Repeating this argument for the successive blocks completes the proof.   □

LEMMA 3   *It is sufficient to consider the case where G consists of a spanning tree T together with an additional edge forming an odd circuit in T.*

PROOF   Any non-bipartite graph G has an odd circuit C. By assumption it has at least one other edge. Delete an edge e from C and augment $C \backslash e$ to a spanning tree of G. Replace the deleted edge. Changing the configurations on this subgraph of G allows us to change configurations on the whole of G.   □

PROOF OF THEOREM 3   By Lemma 3 it is sufficient to prove the result when G consists of an odd circuit $C = \{v_1, v_2, \ldots, v_{2m+1}\}$ together with a path $v_1, u_1, u_2, \ldots, u_k$ with $u_i \notin C$.

Case 1: ω' restricted to C contains two distinct colours   We can assume that the configuration now on C contains two colours. By rotating this configuration on C we can colour v_1 black or white as we need. By successively applying Lemma 1 to the path $v_1, u_1, \ldots, u_i$ with $i = k$, $k-1, \ldots, 1$ we can turn $\omega(u_1, \ldots, u_k)$ to $\omega'(u_1, \ldots, u_k)$.

Again by rotating the configuration on C we can colour u_1 as desired and rotate C back to its original configuration. Now using this procedure together with Lemma 1 at most $2m+1$ times we can colour in turn the vertices $v_{2m+1}, v_{2m}, \ldots$ as desired. Now if necessary rotate the circuit to get u_1 to its desired colour and rotate circuit C back.

Case 2: ω' restricted to C is monochromatic, say black   Let j be the minimum integer such that u_j is white in ω'. Notice that such a j must exist.

Use the argument of Case 1 to colour $u_{j+1}, \ldots, u_k$ correctly and to colour $u_1, u_2, \ldots, u_j$ white.

Now use the argument of Case 1 to make C all black, changing only u_1. Note that after doing this u_1 is white.

Now change $u_1, u_2, \ldots, u_{j-1}$ successively to black as required by ω' and we are done.   □

NOTE   Although in theory the above theorem says that all non-trivial states are "possible", we should emphasize that in the Monto Carlo simulations which we have done for this problem most of these states seem to have very low probability.

4. Correlation at equilibrium

Henceforth we only consider graphs which are neither bipartite nor odd circuits.

Because of Theorem 3 we know there exists an equilibrium process η with density ν so that for any subset Y of V

$$P(\eta = Y) = \nu(Y),$$

and η is the limit in distribution of our antivoter process.

In general it seems difficult to calculate ν explicitly, however we note that:

In equilibrium any vertex is equally likely to be black or white.   (5)

PROOF   From Corollary 1 we know that for suitable μ

$$\lim_{t \to \infty} P(x \in \eta_t^\mu) = \tfrac{1}{2}.$$

But this limit does not depend on μ and the result follows. $\square$

We calculate the two-dimensional distributions by means of the *similarity function* $s(x, y)$ defined by

$$s(x, y) = P(\eta(x) = \eta(y)),$$

which represents the probability that x and y are the same colour at equilibrium.

Note that since by symmetry arguments

$$\nu(V \backslash X) = \nu(X), \qquad X \subseteq V.$$

$s(x, y)$ determines the distribution of the colouring of the pair $\{x, y\}$.

THEOREM 4   *Let G be a graph which is neither bipartite nor an odd circuit. Further suppose that G has a transitive abelian group of automorphisms and let its adjacency matrix be A. Then for any pair of vertices x, y,*

$$s(x, y) = \frac{1}{2} \left[1 + \frac{(I + \alpha A)_{xy}^{-1}}{(I + \alpha A)_{yy}^{-1}} \right],$$

where α^{-1} is the common vertex degree ρ.

PROOF

$$s(x, y) = P(|\eta \cap \{x, y\}| \text{ even})$$

$$= \lim_{t \to \infty} P(|\eta_t^A \cap \{x, y\}| \text{ even}) \qquad (A \subseteq V),$$

$$= \lim_{t \to \infty} P(|\hat{\eta}_t^{\{x,y\}} \cap A| + \varepsilon_t^{\{x,y\}} \text{ even})$$

$$= P(\varepsilon_\infty^{\{x,y\}} \text{ even}),$$

where $\varepsilon_\infty^{\{x,y\}}$ is the number of steps taken by the particles initially at x and y before their (certain) annihilation.

We now introduce a simple random walk ξ^x on the vertex set of G which starts at x and which has the property that

$$P(\varepsilon_\infty^{\{x,y\}} \text{ even}) = P(E_{xy}),$$ (6)

where E_{xy} is the event that ξ^x first reaches y after an even number of steps. We delay the proof of (6) until the end of the theorem.

Define

$$f_n^{xy} = P(\xi^x \text{ first reaches } y \text{ after } n \text{ steps}),$$
$$u_n^{xy} = P(\xi_n^x = y).$$

Then if $U^{xy}(z)$ and $F^{xy}(z)$ denote the generating functions of the u_n^{xy} and f_n^{xy} sequences we have

$$F^{xy}(z) = U^{xy}(z)/(1 + U^{yy}(z)), \tag{7}$$

by the standard arguments.

But $P(E_{xy}) = [1 + F^{xy}(-1)]/2$ and since αA is the transition matrix of the random walk ξ^x we have

$$U^{xy}(-1) = (-\alpha A + \alpha^2 A^2 - \alpha^3 A^3 + \ldots)_{xy}$$
$$= ((I + \alpha A)^{-1} - I)_{xy}, \tag{8}$$

where we know that $I + \alpha A$ is non-singular since by taking G to be non-bipartite we ensure that $-\rho$ is not an eigenvalue of A (see [5]). The theorem follows from (7) and (8) once we have proved (6). To do this we associate with the process $\hat{\eta}^{xy}$ a simple random walk process ξ on the same graph.

Denote the transitive abelian group of automorphisms by $\mathcal{A}$. It is easy to check that the stabilizer of any point is the identity (that is that the group $\mathcal{A}$ is regular) and that for any pair of vertices x_1, x_2 of G there is a unique $\tau \in \mathcal{A}$ such that $x_1\tau = x_2$.

We will construct a realization of ξ^x with the following property:

Suppose that after the nth jump $(n \leq \varepsilon_\infty^{\{x,y\}})$ of $\hat{\eta}^{\{x,y\}}$ the particle which was originally at x is at u, and the other particle is at v, and that $\xi_n^x = z$; then there is an automorphism $\pi_n \in \mathcal{A}$ with the property that $u\pi_n = z$ and $v\pi_n = y$.

Note that this implies that the distance from u to v is the same as that from ξ^x to y and so in particular ξ^x will first reach y exactly when the particles in $\hat{\eta}^{\{x,y\}}$ annihilate.

The proof is by induction. Choose $\pi_0 = 1$. Suppose π_n is given. The next move of $\hat{\eta}^{\{x,y\}}$ will be one of two types:

(1) The particle at u moves to (a randomly chosen neighbour) u'. Put $\xi_{n+1}^x = u'\pi_n$. Since there is a one-to-one correspondence (under π_n) between the neighbours of u and those of ξ_n^x, this is equivalent to making ξ^x jump to a randomly chosen neighbour. It is easy to see that the choice $\pi_{n+1} = \pi_n$ now gives an automorphism with the desired property.

(2) The particle at v moves (at random) to a neighbour v'. Now choose (the unique) $\sigma \in \mathcal{A}$ with $v'\sigma = v$. Define $z' = z\sigma = \xi_n^x \sigma$. We claim that z' is a neighbour of z. To see this, choose $\tau \in A$ such that $z\tau = v'$. Then $z'\tau = z\sigma\tau = z\tau\sigma = v'\sigma\tau$, so τ: $(z, z') \rightarrow (v', v'\sigma)$ and since τ preserves edges and non-edges, z' is a neighbour of z. Once more this gives a one-to-one correspondence between the neighbours of z and those of v, so if we put $\xi_{n+1}^x = z' = \xi_n^x \sigma$, this results in ξ_n^x jumping to a random neighbour. Choosing $\pi_{n+1} = \sigma\pi_n$ now gives the desired automorphism since $x\pi_{n+1} = x\sigma\pi_n = x\pi_n\sigma = \xi_n^x \sigma = \xi_{x+1}^x$.
This completes the proof of Theorem 4. $\square$

Intuitively, because of the repellent nature of the model, we would expect some form of "negative correlation" in our equilibrium distribution ν. Certainly we might hope that adjacent vertices are more likely than not to be opposite colours. The following example shows that some care is needed.

EXAMPLE   Let G have vertex set $V_1 \cup V_2$ where $V_1 = \{v_1, v_2, \ldots, v_m\}$ and $V_2 = \{u_1, u_2, \ldots, u_m\}$ and edges (v_i, v_{i+1}), (v_i, v_{i-1}), (u_i, u_{i+1}), (u_i, u_{i-1}), $i = 1, 2, \ldots, m$ (with addition modulo m), and (v_i, u_j), $i = 1, 2, \ldots, m$, $j = 1, 2, \ldots, m$. Thus V_1 and V_2 are both circuits and we add all possible edges between V_1 and V_2. It is routine to prove that, as $m \rightarrow \infty$,

$$s(v_i, v_j) = s(u_i, u_j) \rightarrow 1, \qquad i, j = 1, 2, \ldots, m,$$

so that in particular, for example, $s(v_1, v_2) \rightarrow 1$. Thus when m is large v_1 and v_2 are highly likely to be the same colour at equilibrium. This result seems reasonable heuristically, as we would expect V_1 and V_2 to be different colours at equilibrium.

The above graph is vertex-transitive. If we insist on slightly more symmetry and impose the condition of edge transitivity we can prove

THEOREM 5   *If G is edge-transitive and satisfies the conditions of Theorem 4 then if x, y are neighbouring vertices*

$$s(x, y) \leqslant \frac{n-2}{2n-3} < \tfrac{1}{2}. \tag{9}$$

PROOF   By edge-transitivity, for x and y neighbours in G,

$$u_n^{yy} = P(\xi_n^y = y) = \sum \rho^{-1} p(\xi_{n-1}^y = z) = u_{n-1}^{xy},$$

where the sum is over all neighbours z of y. Thus

$$U^{xy}(-1) = -U^{yy}(-1)$$

and the right-hand side is independent of y because G is vertex-transitive. We write

$$d = U^{yy}(-1) + 1 = (I + \alpha A)_{yy}^{-1}$$

for the diagonal element of $(I + \alpha A)^{-1}$. Noting that for x, y neighbours

$$s(x, y) = \tfrac{1}{2}[1 + F^{xy}(-1)]$$

and

$$
\begin{aligned}
F^{xy}(-1) &= U^{xy}(-1)/[1 + U^{yy}(-1)] \\
&= -U^{yy}(-1)/[1 + U^{yy}(-1)] \\
&= (1 - d)/d,
\end{aligned}
$$

we see that

$$s(x, y) = (2d)^{-1}.$$

But noting that

$$nd = \text{trace } (I + \alpha A)^{-1} = \sum_{i=1}^{n} (1 + \alpha \lambda_i)^{-1}, \tag{10}$$

where $n = |V|$ and the λ_i are the eigenvalues of A, the theorem follows since we have

LEMMA 4 *If G is regular of degree ρ and not bipartite, with eigenvalues λ_i, $i = 1, 2, \ldots, n$, then*

$$\sum_{i=1}^{n} \frac{\rho}{\rho + \lambda_i} \ge n + \frac{n}{2(n-2)}.$$

PROOF By the Perron–Frobenius theorem, $\max \lambda_i = \rho = \lambda_1$, say, and $|\lambda_i| \le \rho$, $i = 1, 2, \ldots, n$. Moreover, since G is not bipartite, $\lambda_i \ne -\rho$ for any i. Thus $\rho + \lambda_i$ is positive, and hence, by the Cauchy–Schwartz inequality,

$$\sum_{i=2}^{n} \left(\frac{\rho + \lambda_i}{\rho}\right) \sum_{i=2}^{n} \left(\frac{\rho}{\rho + \lambda_i}\right) \ge (n-1)^2. \tag{11}$$

But

$$2 + \sum_{i=2}^{n} \frac{\rho + \lambda_i}{\rho} = \sum_{i=1}^{n} \frac{\rho + \lambda_i}{\rho} = \text{trace } (I + \alpha A) = n, \tag{12}$$

so that

$$\sum_{i=1}^{n} \frac{\rho}{\rho + \lambda_i} = \frac{1}{2} + \sum_{i=2}^{n} \frac{\rho}{\rho + \lambda_i} \ge \frac{1}{2} + \frac{(n-1)^2}{n-2}.$$

The lemma now follows and the proof of the theorem is complete. □

The bound given by Theorem 5 is in fact the best possible.

THEOREM 6   *In the case when the underlying graph is the complete graph on n vertices, K_n, we have $s(x, y) = (n-2)/(2n-3)$ (for $x \neq y$). Moreover, K_n is the only edge-transitive graph on n vertices satisfying Theorem 4 which attains the bound given by Theorem 5.*

PROOF   When the underlying graph is the complete graph, the distribution of $\hat{\eta}^{\{x,y\}}$, the number of steps taken by $\varepsilon_t^{\{x,y\}}$ before annihilation, is geometric. It is then straightforward to calculate

$$s(x, y) = P(\varepsilon_\infty^{\{x,y\}} \text{ even}) = (n-2)/(2n-3). \tag{13}$$

Equality in the bound (9) implies that the Cauchy–Schwartz inequality (11) must be satisfied with equality. This can only be true if $\lambda_2 = \lambda_3 = \ldots = \lambda_n$ and in this case we must have $\lambda_2 = \lambda_3 = \ldots = \lambda_n = -1$, from (12). We know already that $\lambda_1 = n - 1$. This is, incidentally, another way of finding the eigenvalues of the complete graph. (Of course, knowledge of the eigenvalues gives (13) after substitution in (10).)

Now suppose we have an edge-transitive graph G for which the bound (9) is attained. As before, equality in the Cauchy–Schwartz inequality (11), implies that $\lambda_2 = \lambda_3 = \ldots = \lambda_n$ and so the graph G has only two distinct eigenvalues. We know, however, [1, Corollary 2.7] that any connected graph of diameter d has at least $d + 1$ distinct eigenvalues. Hence G has diameter 1 and so is the complete graph on n vertices.   □

Thus we see that over the class of all graphs with certain symmetry (namely edge-transitivity) the "correlation" between neighbouring vertices is least for the complete graph (that is, $s(x, y)$ is closest to $1/2$), a result which seems intuitively reasonable. Similar heuristic considerations make it tempting to conjecture that this "nearest neighbour correlation" at equilibrium in the complete graph is minimum over the class of all connected graphs.

Monte Carlo simulations and intuition suggest that (9) holds for some graphs which are not edge-transitive, for example an $m \times n$ toroidal square lattice (with $m \neq n$). The example given before the theorem shows that the condition of vertex-transitivity is not sufficient for the result to hold. The finding of other sufficient conditions or an appropriate necessary condition represents an open problem.

5. Conclusion

Positive correlation in spatial processes is a relatively well understood concept. For example, Harris [4] has shown that the measures which arise as equilibria for the so-called attractive particle systems are positively correlated in the very precise sense that

$$E_\mu(fg) \ge E_\mu(f)E_\mu(g) \tag{14}$$

for any pair of functions f, g which are monotone in the same sense (where E_μ denotes expectation with respect to μ). A key result in this area is the FKG inequality which states that (14) holds for any μ satisfying the condition

$$\mu(X)\mu(Y) \le \mu(X \cup Y)\mu(X \cap Y) \qquad (X, Y \subseteq V). \tag{15}$$

The phenomenon of negative correlation seems much less well understood. In fact even a precise definition of the concept seems difficult to formulate. Theorem 5 says that the equilibrium distribution for the antivoter model on suitable graphs satisfies a "nearest neighbour negative correlation".

For the graphs for which we have been able to calculate the equilibrium distribution explicitly, namely the complete graphs, bipartite graphs and the odd circuits, it is a curiosity that relation (15) holds with the inequality reversed. However, it is impossible for (14) to hold with the reverse inequality for all monotone f and g and we can prove no result of a similar type for measures μ satisfying the reverse of (15).

Acknowledgement

We have pleasure in acknowledging helpful conversations with Anthony Mansfield, John Spouge, and in particular Peter Cameron, who suggested the proof of (6).

References

1. N. L. Biggs, *Algebraic Graph Theory*, Cambridge Tracts in Mathematics no. 67, Cambridge University Press, Cambridge, 1974.
2. P. Erdös and P. Ney, Some problems on random intervals and annihilating particles. *Ann. Prob.* **2** (1974), 828–839.
3. D. Griffeath, *Additive and Cancellative Interacting Particle Systems*, Lecture Notes in Mathematics no. 724, Springer Verlag, Berlin, 1979.
4. T. E. Harris, A correlation inequality for Markov processes in partially ordered state spaces. *Ann. Prob.* **5** (1977), 451–454.
5. L. Lovász, *Combinatorial Problems and Exercises*, North Holland, Amsterdam, 1979.

12

SCHEMES OF GRAPHS

Pierre Duchet

ABSTRACT The notion of scheme was introduced in an earlier paper, where a weakened form of Hadwiger's conjecture is proposed. This paper collects conjectures and results involving schemes, in relation to contraction and colouring problems. Namely, a p-scheme is proved to be contractible onto $K_{\lfloor (2p)^{1/2} \rfloor}$ and the colour-interchange approach of the scheme-conjecture is discussed.

1. Introduction

For the sake of clarity, graphs are supposed finite, simple and connected [3]. In general, a *scheme of a graph* $H = (X, F)$ or an *H-scheme* may be viewed as a family of sets $\Sigma = (C_e)_{e \in F}$, where C_e corresponds to an edge e of H, that satisfies the following *scheme-condition* for every pair e, f of H-edges:

$$C_e \cap C_f = \varnothing \Leftrightarrow e \cap f = \varnothing. \qquad (1.1)$$

The sets C_e are called the Σ-*links*. The elements of the links are called the Σ-*vertices*.

For various questions, it is useful to consider the graph H as drawn in another graph $G = (V, E)$; we will say G *contains an H-scheme*, and we will note $G \wr H$, when there exists an H-scheme formed with connected subsets of vertices of G (connected means inducing a connected subgraph).

It is easily seen that $G \wr H$ if and only if there exist $p = |X|$ vertices of G, say $v_1, \ldots, v_p$, in one-to-one correspondence with the p vertices $x_1, \ldots, x_p$ of H, with the following properties:

> For every edge $x_i x_j$ of H there is an undirected path in G connecting v_i and v_j; the set of vertices of such a path is denoted by P_{ij}. $\qquad (1.2)$

> The P_{ij}s satisfy the scheme-condition $P_{ij} \cap P_{kl} = \varnothing$ when the indices i, j, k, l are different. $\qquad (1.3)$

Such a scheme $\Sigma = (P_{ij})_{i,j \in F}$, with paths as links, is called a *simple*

GRAPH THEORY AND COMBINATORICS
ISBN 0-12-111760-X

scheme and the Σ-links are called the Σ-paths. The set $\{v_1, \ldots, v_p\}$ is called the *basis* of Σ. Σ is also denoted under the form $(v_1, \ldots, v_p; P_{ij})$.

When $P_{ij} \cap P_{ik}$ is reduced to $\{v_i\}$, for every pair of edges i, j and i, k of H, the subgraph of G constituted by the vertices and the edges of the Σ-paths is nothing but a *subdivision* of H.

By the way, subdivisions constitute a first example of schemes. Less trivial examples are exhibited in Figs 1, 2 and 3, where the basis-vertices are surrounded. In Figs 1 and 3, the Σ-paths are formed by the shortest paths (= geodesics) joining the basis vertices. Figure 1 shows $K_{1,3} \wp K_3$; Fig. 2 shows that the Petersen graph contains a simple scheme of K_6. In Fig. 3 we see that the complete bipartite graph $K_{2,3}$ contains a simple scheme of K_5^-, the complete graph K_5 minus one edge: no link connects a and b; the other Σ-paths are $ax, ay, az, bx, by, bz, xcy, ycz, zcx$.

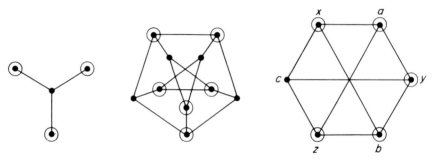

FIG. 1. $K_{1,3} \wp K_3$.   FIG. 2. $P_{10} \wp K_6$.   FIG. 3. $K_{3,3} \wp K_5^-$.

The proof of the following observation is omitted:

PROPOSITION 1.4   *If $G \wp H$ and $H \wp K$, then $G \wp K$.*   □

By a *p-scheme* we will mean a simple scheme of K_p. For a given graph G, the greatest integer p such that $G \wp K_p$ is called the *scheme number* of G and is denoted by $\sigma(G)$. Obviously $\sigma(K_p)$ is p.

THEOREM 1.5   *Let $p \leq q$ be positive integers. The scheme number of complete bipartite graphs $K_{p,q}$ is given by*

$$\sigma(K_{p,q}) = \begin{cases} p+1 & \text{for } K_{1,1}, K_{1,2}, K_{2,2}, K_{3,3}, \\ p+2 & \text{otherwise.} \end{cases}$$

PROOF   The small cases, and the inequality $\sigma(K_{p,q}) \geq p+2$ for $q \geq 4$ are easy and are left to the reader.

Conversely, let us suppose that $s = \sigma(K_{p,q}) \geq p + 3$. We prove the theorem by induction on p. If A, B is the bicolouring of $K_{p,q}$ with $|A| = p$ and $|B| = q$ and if S is the basis of an s-scheme Σ, put $a = |S \cap A|$ and $b = |S \cap B|$. Since every path of vertices in $S \cap B$ has to be joined by a Σ-path that avoids $S \cap A$, the induction hypothesis applied to the subgraph induced by the set $(A \backslash S) \cup B$ yields $b \leq p - a + 2$. $\square$

We complete this introduction by specifying some terminology. If $C_1, \ldots, C_p$ are connected subsets of a graph G, the *contraction* of G over the C_is is the graph whose vertices are the C_is and where C_i and C_j are joined by an edge if and only if they are adjacent in G. By a *subcontraction* of G, we mean a contraction of a partial subgraph of G. The vertices of a subcontraction are considered like the *contracted subsets* of G as well. We note that $G > H$ when H is isomorphic to a subcontraction of G. The *Hadwiger number* of G is the greatest integer p such that $G > K_p$ and is denoted by $\eta(G)$.

2. Topological schemes

The notions of "*subdivisions*" and "*subcontractions*" were inspired by considerations of a topological nature concerning connectedness, i.e. the problem of existence of paths between points. Thus, we have to look at how the notion of *scheme* works for topological graphs. We begin with

REMARK 2.1 *If $G > H$, then $G \,\rlap{\gt}{\wp}\, H$.*

PROOF The contracted subsets $V_1 \ldots V_p$ form a partition of the vertex set of G; by definition of a subcontraction, the V_is are connected and when V_i and V_j are adjacent vertices of H, $V_i \cup V_j$ is connected in G. For different indices, $(V_i \cup V_j) \cap (V_k \cup V_l)$ is empty. $\square$

Let M be a surface. Let $\mathscr{C}(M)$ denote the collection of all graphs that are subcontraction-minimal with the property not to be embeddable in M. The remark above implies that if a graph G is not embeddable in Σ, it must contain a scheme of some member of $\mathscr{C}(M)$. Conversely, I propose the following conjecture:

CONJECTURE 2.2 *Let M be a surface and G, H be graphs such that $G \,\rlap{\gt}{\wp}\, H$. If H is not embeddable in M, then G is not embeddable in M.*

The conjecture is true for the sphere S_2:

THEOREM 2.3 (P. Duchet, H. Meyniel)   *A graph is planar if and only if it does not contain a scheme of K_5 or $K_{3,3}$.*

PROOF   The "if" part is an immediate consequence of Remark 2.1 and of the Halin, Harary and Tutte [11, 12] version of Kuratowski's theorem [13]: *A graph is planar if and only if it does not admit K_5 or $K_{3,3}$ as subcontraction.*
    Conversely, a theorem of Tutte [15, Proposition 5.1] can be stated as follows: *If a graph G can be drawn on the plane in such a way that two non-adjacent edges are mapped into paths with an even number of crossings, the G is planar.* Our theorem follows.   □

3. Schemes and colourings

The following "*scheme-conjecture*" was raised in [8].

CONJECTURE 3.1 (Duchet and Meyniel [8])   *Every p-chromatic graph contains a p-scheme.*   □

Its solution for $p = 5$ would give a proof of the four-colour "compu-theorem" [1, 2].
    After Kempe's incorrect proof of the four-colour theorem, the idea of decreasing the number of colours needed in a graph colouring by the use of interchange of colours in a connected bicoloured component (see [14] for details concerning Kempe's exchanges), became the main angle of attack for various colouring problems (Brook's and Vizing's theorems, perfectness of Meyniel's graphs, etc.).
    More specifically, a *p-colouring* of a graph $G = (V, E)$ is a function Q: $V \to \{1, \ldots, p\}$ such that adjacent vertices have different images (= colours). If α and β are colours, a (α, β)-*bicoloured path* is a path using only vertices of colour α or β. We define a *chromatic p-scheme* in a graph G, by a pair (Q, Σ) with the following properties:

(a) Q is a p-colouring of G.
(b) Σ is a p-scheme in G with basis $\{v_1, \ldots, v_p\}$.
(c) $v_1, \ldots, v_p$ receive p different colours by Q.
(d) Every link $\Sigma(v_i, v_j)$ is constituted by the vertices of a bicoloured path joining v_i and v_j.

When we try to understand the fine topological structure of a p-

chromatic graph by the way of interchange of colours, the most natural conjecture we would like to solve may be stated as follows:

CONJECTURE 3.1′ *A p-chromatic graph G contains a chromatic p-scheme.* □

Perhaps Hajös had exactly the above statement in mind when he proposed his conjecture that *every p-chromatic graph contains a subdivision of K_p*: Hajös' conjecture was recently disproved by Catlin [5] and was proved false for almost every graph by Erdös and Fajtlowicz [9; see also 4]. Conjecture 3.1′ for $p = 4$ is also to be compared with a conjecture of Toft [15]: *a 4-chromatic graph contains a subdivision of K_4 in which all subdivided edges are paths of odd length*. (Note that a bicoloured path has an odd length.)

It is therefore of some importance to note the falsity of Conjecture 3.1′. This fact was pointed out in [8] and becomes more precise with

PROPOSITION 3.2 *Conjecture 3.1′ is false for every $p > 4$.*

PROOF Let us say that two vertices x and y of a graph are *twins* if they are adjacent and have exactly the same neighbours. We use the following lemma:

LEMMA 3.3 *If x and y are twins in a graph G and if x is a basis-vertex of a chromatic p-scheme (Q, Σ) in G, then y is also a basis-vertex of (Q, Σ).*

PROOF Suppose the contrary: x is the origin of $p - 1$ different bicoloured paths. None of these paths ends at y and, since every neighbour of y is adjacent to x, none of these paths can pass through y. Hence x has $p - 1$ neighbours with $p - 1$ different colours. Therefore y has the same colour as x, which is a contradiction; the lemma is thus proven. □

Now, consider the graph $G = L(kC_{2n+1})$, which appeared in Catlin's paper [5] as a counterexample to Hajös' conjecture, for $k > n \geqslant 2$. G is the line-graph of the multigraph kC_{2n+1} obtained from the odd cycle C_{2n+1} by replacing every edge by k parallel edges. We suppose that k, $n \geqslant 2$.

Two parallel edges of kC_{2n+1} constitute twins of G. Thus, by Lemma 3.3, every chromatic scheme of G must have λk basis-vertices for some integer λ. Since G has chromatic number $2k + \lceil k/n \rceil > 2k$ (see [5]), a chromatic scheme must have at least $3k$ colours, and in fact has exactly $3k$ colours, for a vertex of G has degree $3k - 1$. G is a counterexample to Conjecture 3.1′. □

The case $p = 4$ of the conjecture remains unsolved.

In spite of our negative answer to Conjecture 3.1′, the hope of proving the existence of a p-scheme in a p-chromatic graph via Kempe's style interchange method is not completely destroyed. For instance, I believe in the following stronger form of Conjecture 3.1:

CONJECTURE 3.4 (P. Duchet, H. Meyniel)   *Every p-chromatic graph G contains a partial subgraph G' with a chromatic p-scheme in G'* □

4. Schemes and contractions

By Remark 2.1, Conjecture 3.1 is weaker than Hadwiger's conjecture [**10**]: "*a p-chromatic graph is contractible onto K_p*".

By how much is Conjecture 3.1 weaker than Hadwiger's? In other words, what is the best function $g(p)$ such that the following implication holds for every graph G:

$$G \Vdash K_p \Rightarrow G > K_{g(p)}? \tag{4.1}$$

The problem first appeared in [**6**]; then, my guess was that $g(p) = (2p/3) + o(p)$ for large p. This is not true (see Corollary 4.5 below); I now conjecture that

CONJECTURE 4.2 $g(p) \geqslant \lceil p/2 \rceil$ and $g(p) = (p/2) + o(p)$ for large p. □

The best lower bound I know is given by the following result:

THEOREM 4.3   *For every positive integer p, $G \Vdash K_p$ implies $G > K_{\lfloor (2p)^{1/2} \rfloor}$.*

PROOF   Consider a graph G that contains a p-scheme Σ. Without real restriction, we suppose that every vertex of G is in Σ and that every Σ-link is reduced to a Σ-path. Put $q = \lfloor (2p)^{1/2} \rfloor$; we have $1 + (q(q-1)/2) \leqslant p$. In the basis of Σ we pick $1 + (q(q-1)/2)$ vertices and we partition them into q sets $A_1, A_2, \ldots, A_q$ with $|A_1| = 1$ and $|A_i| = i - 1$ for $i > 1$. Each A_i is considered with a total order, so we may speak of the kth element of A_i, for $1 \leqslant k < i$. First, we extend every A_i to a connected subset B_i: B_i will be the union of all links of Σ joining the first vertex of A_i to the other vertices of A_i. By the scheme condition, the B_is are pairwise disjoint.

For each $i > 1$, the unique vertex of $B_1 = A_1$, say a, is joined by a Σ-path of G to the first vertex of A_i, say b. We denote such a path by $a v_2 \ldots v_k b$. We add the vertices $v_2, v_3, \ldots$ to the set A_1 as long as the next vertex of the path is not in B_i. The set obtained in this way is connected and denoted by C_{1i}.

In a similar manner, for $j>i>1$, we form a connected set C_{ij} by adjoining to B_i the vertices of a Σ-path that connects the $(i-1)$th vertex of A_i to the ith vertex of A_j and stopping the augmentation just before we reach the set B_j.

We put

$$C_i = \bigcup_{j=i+1}^{q} C_{ij} \quad \text{for} \quad i<q,$$

and $C_q = B_q$.

The sets C_{ij} satisfy the scheme condition. Thus, the C_is are pairwise disjoint. Moreover, by definition of C_{ij}, the set $C_{ij} \cup B_j$ is connected. Hence $C_i \cup C_j$ is connected for every $i \neq j$: the contraction of G by the sets $C_1, \ldots, C_q$ produces K_q. $\square$

An upper bound for $g(p)$ can be obtained by considering the parameter α_d of affine spaces: α_d denotes the maximum cardinality of a set A of points of the affine space $AG(3, d)$ such that no line is included in A. Many authors have conjectured that the ratio $\alpha_d/3^d$ tends to 0 when the dimension d goes to infinity.

THEOREM 4.4   *For every integer $p>1$, $g(p) \leq p/2 + \alpha_d/2$ where $d =$* $\lceil \log_3 p \rceil$.

PROOF   We construct a bipartite graph G as follows:

P denotes a set of p points in the affine space $AG(3, d)$. L denotes the set of lines of $AG(3, d)$ that contains at least two points of P; a point $x \in P$ and a line $l \in L$ are joined by an edge of G if and only if $x \in l$. Let $l(a, b)$ denote the line of $AG(3, d)$ determined by two points a and b.

For a, b in P, put $\Sigma(a, b) = \{a, l(a, b), b\}$. The sets $\Sigma(a, b)$, for a, b running over P, form in G the links of a p-scheme Σ whose basis is P.

In order to demonstrate the theorem, we shall prove that $\eta(G) \leq p/2 + \alpha_d/2$. The inequality is clear for $p = 2, 3$. For $p \geq 4$, we have $d \geq 2$ and $\alpha_d \geq 4$, thus $p/2 + \alpha_d/2 \geq 4$. The inequality is therefore true if $\eta(G) \leq 4$.

So, assume that $q = \eta(G) \geq 5$. Denote by $C_1, \ldots, C_q$ the contraction classes for a contraction of G onto K_q. Every class contains at least one point; otherwise, the class would be reduced to a single line, which has at most three neighbours in G and q would be at most 4.

We may assume that the first α classes $C_1, \ldots, C_\alpha$ contain exactly one point, while the remaining classes possess at least two points. Let A be the set of points in $C_1 \cup \ldots \cup C_\alpha$. We observe that the only way in which two classes C_i and C_j, with $i, j \leq \alpha$, can be adjacent in G is that one of them possesses the line determined by the two points respectively in C_i and C_j. Therefore, three such classes can be pairwise adjacent only if the points belonging to each of them are not on the same line; hence, no line of $AG(3, d)$ is included in A and $\alpha \leq \alpha_d$.

The total number of points in the $q - \alpha$ remaining classes is $p - \alpha$. Thus, we have $2(q - \alpha) \leq p - \alpha$ or $q \leq p/2 + \alpha/2$. This yields the required inequality. $\square$

REMARK   It can be shown that $\eta(G)$ is at least $(3^d + 3)/2$.

COROLLARY 4.5   For $p \geq 81$, $g(p) \leq \frac{17}{27}p < \frac{2}{3}p$.

PROOF   A long case by case examination of the four-dimensional case yields $\alpha_4 \leq 21$. By an easy induction, we have $\alpha_d \leq 21 \times 3^{d-4}$, for $d \geq 4$. Applying Theorem 4.4, the corollary follows. $\square$

5. Convex schemes

A *convexity on a graph* $G = (V, E)$ is a set $\mathscr{C}$ of *connected subsets* of V which is preserved under intersection. The members of $\mathscr{C}$ are called the convex sets. By convention, $\varnothing$ and V are convex. The *convex hull* of a set A of vertices is the intersection of all convex sets containing A.

A *convex p-scheme* in G is defined to be a p-scheme $\Sigma = (v_1, \ldots, v_p; C_{ij})$, where C_{ij} is the convex hull of $\{v_i, v_j\}$.

The reader is referred to [7, 8] for the relationships between Radon's parameter for convexities and Hadwiger number. [8] contains a convex version of the scheme-conjecture 3.1. In [7] it is proved that any graph that contains a convex p-scheme is contractible onto $K_{\lceil (p+1)/2 \rceil}$.

I want to mention here that $p/2$ in this lower bound is probably the best possible order of magnitude. The reason is that the 3^d-scheme in the graph G, considered in the proof of Theorem 4.4, is a convex scheme for the usual convexity on G (a set is convex if and only if it contains all points belonging to the shortest paths between any two of its vertices).

References

1. K. Appel and W. Haken, Every planar map is four-colorable, I: Discharging. *Illinois J. Math.* **21** (1977), 429–490.
2. K. Appel, W. Haken and J. Koch, Every planar map is four-colorable, II: Reducibility. *Illinois J. Math.* **21** (1977), 191–567.
3. C. Berge, *Graphs and Hypergraphs*, North Holland, Amsterdam, 1973.

4. B. Bollobás and P. A. Catlin, Topological cliques of random graphs. *J. Combinat. Theory B* **30** (1981), 224–227.
5. P. A. Catlin, Hajós' graph-colouring conjecture: variations and counterexamples. *J. Combinat. Theory B* **26** (1979), 268–274.
6. P. Duchet, Problem 17. In *Proceedings of the Janos Bolyai International Conference, Eger,* 1981, to appear.
7. P. Duchet and H. Meyniel, Ensembles convexes dans les graphes, I: Théorèmes de Helly et de Radon pour graphes et surfaces. *Europ. J. Combinat.* **4** (1983), 127–132.
8. P. Duchet and H. Meyniel, Must a p-chromatic graph contain a p-scheme? Talk given at the Silver Jubilee Conference on Combinatorics and Optimization, Waterloo, 1982. Submitted for publication.
9. P. Erdös and S. Fajtlowicz, On the conjecture of Häjos. *Combinatorica* **1** (1981), 141–143.
10. H. Hadwiger, Ungelöste Problem. *Element. Math.* **13** (1958) 127–128.
11. R. Halin, Bemerkungen über ebene Graphen. *Math. Ann.* **53** (1964), 38–46.
12. F. Harary and W. T. Tutte, A dual form of Kuratowski's theorem. Mimeographed report.
13. K. Kuratowski, Sur le problème des courbes gauches en topologie. *Fund. Math.* **15** (1930), 271–283.
14. M. Las Vergnas and H. Meyniel, Kempe classes and the Hadwiger conjecture. *J. Combinat. Theory B* **31** (1981), 95–104.
15. B. Toft, Two theorems on critical 4-chromatic graphs. *Studia Sci. Math. Hungar.* **7** (1972), 83–89.
16. W. T. Tutte, Toward a theory of crossing numbers. *J. Combinat. Theory* **8** (1970), 45–53.

13

TREE–MULTIPARTITE GRAPH RAMSEY NUMBERS

P. Erdös, R. J. Faudree, C. C. Rousseau and R. H. Schelp

ABSTRACT   The Ramsey number $r(T, K(n, n))$ is studied in the case where T is a fixed tree of order m and n is large. In particular, we find that $r(K(1, m-1), K(n, n))$ is bounded above and below by $cmn/\log(m)$ where in each bound c is an appropriate positive constant.

1. Introduction

Given graphs $G_1, \ldots, G_k$, the *Ramsey number* $r(G_1, \ldots, G_k)$ is the smallest integer r so that, if we color the edges of K_r by k colors, then for some i the ith color class contains a copy of G_i. The study of $r(G_1, \ldots, G_k)$ or *generalized Ramsey theory* was popularized by Harary, although there were earlier papers on this subject, in particular that of Gerencsér and Gyárfás [4].

In [3] we considered Ramsey numbers of the form $r(H, G)$ where H is a fixed multipartite graph and G is a large sparse graph. The present paper is a companion to [3]. In it we focus on Ramsey numbers of the form $r(T, G)$ where T is a fixed tree and G is a large multipartite graph.

Before presenting these rather special results, we first shall review some of the problems of generalized Ramsey theory which have been of great interest to us. It would be very desirable to have an asymptotic formula for $r(K_3, K_n)$. At present, we only know that

$$c_1\left(\frac{n^2}{(\log n)^2}\right) < r(K_3, K_n) < c_2\left(\frac{n^2}{\log n}\right) \tag{1}$$

for all sufficiently large n. One would expect that, for $m \geqslant 4$ fixed and n sufficiently large,

$$r(K_m, K_n) < n^{m-1-\varepsilon}, \tag{2}$$

but this is open even for $m = 4$. Perhaps

$$r(C_4, K_n) < n^{2-\varepsilon}. \tag{3}$$

Erdös strongly believes this but others disagree. All agree that the

GRAPH THEORY AND COMBINATORICS
ISBN 0-12-111760-X

problem is likely to be difficult. No one doubts that

$$\lim_{n\to\infty} \frac{r(C_4, K_n)}{r(K_3, K_n)} = 0, \tag{4}$$

but even this is open at present. Szemerédi has observed that

$$r(C_4, K_n) < c\left(\frac{n^2}{(\log n)^2}\right), \tag{5}$$

which just fails to give (4). The argument is based on the following result, which is found in [1]. Let α, d and h denote the independence number, average degree and number of triangles respectively of a graph G of order N. Then

$$\alpha > c(N/d) \min\{\log(Nd^2/h), \log d\}. \tag{6}$$

(In (5) and (6) c stands for different absolute constants.) Now the desired result follows immediately by observing that in a graph G of order $N \geq c$ $(n/\log n)^2$ with no C_4 the average degree of G is $O(N^{1/2})$ and the number of triangles is at most as large as the number of edges, i.e. $Nd/2$.

Let G be a graph with q edges. Is it true that

$$r(K_3, G) \leq 2q + 1? \tag{7}$$

Equality holds in the case where G is a tree.

2. Results

Our first theorem gives a general upper bound for $r(T, K(n, n))$, where T is a tree of order m.

THEOREM 1   *Let T be a tree of order m. For all $n \geq 3m$,*

$$r(T, K(n, n)) \leq \lceil 4mn/\log(m) \rceil.$$

PROOF   As the result is trivial in the case $m \leq 3$, we may assume that $m > 3$. Let (red, blue) be a two-coloring of K_N where $N = \lceil 4mn/\log(m) \rceil$. If there is no red copy of T, then the number of red edges is at most $N(m-2)$. (This is a well-known result which is easily proved by induction.) Thus, we may assume that there are at least $\binom{N}{2} - N(m-2)$ blue edges, so that the average degree of the blue graph is at least $N - 2m + 3$. Let $d_1, d_2, \ldots, d_N$ be the degree sequence of the blue graph and let d denote the average degree of this graph. By a well-known argument, the inequality

$$\sum_{k=1}^{N} \binom{d_k}{n} > (n-1)\binom{N}{n}$$ (8)

implies that there is a blue copy of $K(n, n)$. By convexity, (8) will be satisfied if

$$N\binom{d}{n} > (n-1)\binom{N}{n},$$ (9)

and the latter certainly holds if

$$N\binom{N-2m}{n} > n\binom{N}{n}.$$ (10)

Note that (10) is equivalent to

$$N\binom{N-n}{2m} > n\binom{N}{2m}$$ (11)

and it certainly follows that there is a blue $K(n, n)$ if

$$\frac{N}{n}\left(1 - \frac{(n+2m)}{N}\right)^{2m} > 1.$$ (12)

With our choice of N and in view of the fact that $n \geqslant 3m$ we need only verify that

$$(4m/\log(m))\left(1 - \frac{5\log(m)}{12m}\right)^{2m} > 1$$ (13)

for all $m > 3$, and this is completely straightforward. □

REMARKS   Neither the constant 4 nor the inequality $n \geqslant 3m$ is a sharp condition. In fact, were we to set $N = \lceil cmn/\log(m) \rceil$ and assume n to be sufficiently large, then (11) would become

$$(cm/\log(m))\left(1 - \frac{\log(m)}{cm}\right)^{2m} > 1,$$ (14)

which is satisfied for all sufficiently large m by taking $c > 2$. Further, the critical value c_0 so that $c > c_0$ will ensure that (14) holds for *all* m is approximately $2 + 1/e$.

The complete r-partite graph having n vertices in each part will be denoted by $K_r(n, \ldots, n)$. In the following theorem, $\log^{(r)}(n)$ denotes the r-times iterated logarithm, i.e. $\log^{(1)}(n) = \log(n)$ and $\log^{(r)}(n) =$

$\log(\log^{(r-1)}(n))$, $r = 2, 3, \ldots$. The theorem is proved by induction, with Theorem 1 constituting the first step.

THEOREM 2   *Let T be a tree of order m. For each $r \geq 2$ there exists a constant c_r such that*

$$r(T, K_r(n, \ldots, n)) \leq \lceil c_r mn/\log^{(r-1)}(m) \rceil$$

whenever m is sufficiently large and $n \geq 3m$.

The proof of this result is very similar to the proof of Theorem 1 and so it will be omitted. Suffice it to say that using the strategy of the proof of Theorem 1 one can verify that the blue graph contains a $K(n, p)$, where $p = \lceil c_{r-1} mn/\log^{(r-2)}(m) \rceil$. This fact, together with the induction hypothesis, completes the proof.

The next result shows that the result of Theorem 1 is, within a constant factor, the correct magnitude in the case where T is a star.

THEOREM 3   *Let m be fixed. There exists a positive constant c such that*

$$r(K(1, m-1), K(n, n)) \geq \lfloor cmn/\log(m) \rfloor$$

for all sufficiently large n. If m is sufficiently large, $c = \frac{1}{6}$ will suffice.

PROOF   The proof uses the Lovász–Spencer method as developed in [7] and previously applied by the authors in [2]. We shall simply review the basic ideas of this method. Should additional details be needed, the reader is referred to the account given in [7]. Let $N = \lfloor cmn/\log(m) \rfloor$. We wish to show the existence of a two-colouring of the edges of K_N in which there is no red $K(1, m-1)$ and no blue $K(n, n)$. This will be accomplished by the probabilistic method, in particular by considering a random two-coloring in which each edge of the K_N is colored red with independent probability p. For each set S of m vertices of the K_N, let A_S denote the event that the red subgraph spanned by S contains $K(1, m-1)$. Similarly, for each set T of $2n$ vertices let B_T denote the event that the blue subgraph spanned by T contains $K(n, n)$. For a fixed A_S let N_{AA} denote the number of $S' \neq S$ such that A_S and $A_{S'}$ are dependent. Similarly, let N_{AB} denote the number of T such that A_S and B_T are dependent. In exactly the same way, define N_{BA} and N_{BB}. Letting A and B denote typical A_S and B_T respectively, the desired conclusion will follow from the fundamental lemma of Lovász if there exist constants a and b such that

$$aP(A) < 1, \qquad bP(B) < 1, \tag{15}$$

$$\log(a) > N_{AA} aP(A) + N_{AB} bP(B), \tag{16}$$

$$\log(b) > N_{BA} aP(A) + N_{BB} bP(B). \tag{17}$$

The following bounds are obvious:

$$N_{AA} \leqslant \binom{m}{2}\binom{N-2}{m-2}, \tag{18}$$

$$N_{AB}, N_{BB} \leqslant \binom{N}{2n}, \tag{19}$$

$$N_{BA} \leqslant \binom{2n}{2}\binom{N-2}{m-2}, \tag{20}$$

$$P(A) \leqslant mp^{m-1}, \tag{21}$$

$$P(B) \leqslant \binom{2n}{n}(1-p)^{n^2}. \tag{22}$$

With ε an appropriately small positive constant, set

$$p = (2+\varepsilon)\log(m)/n, \tag{23}$$

$$a = 1+\varepsilon, \tag{24}$$

$$b = m^{\varepsilon n}, \tag{25}$$

$$c = \tfrac{1}{6}. \tag{26}$$

Straightforward calculations verify that with these choices $N_{AA}aP(A)$ and $N_{AB}bP(B)$ tend to zero as $n \to \infty$ and that $\log(b)$ exceeds $N_{BA}aP(A)$, at least for all sufficiently large m. Thus with $n \to \infty$ and m taken to be sufficiently large, conditions (15)–(17) are satisfied and the proof is complete. $\square$

Although the bound of Theorem 1 is, in a certain sense, sharp in the case where T is a star, this is certainly not the case in general. In particular, the behavior of $r(T, K(n, n))$ is quite different in the case where T is a path. Häggkvist reports that he has proved the following result [5]:

THEOREM (Häggkvist) $r(P_m, K(n, k)) < m + n + k - 2$.

In any case, the crude upper bound $r(P_m, K(n, n)) \leqslant m + 4n$ follows from a simple argument using a result of Pósa [6]. Let (red, blue) be a two-coloring of the edges of K_N, where $N = m + 4n$. If there is no red P_m then Pósa's lemma yields a set of vertices X with its neighborhood in the red graph, $\Gamma(X)$, such that $|X| \leqslant m/3$ and $|\Gamma(X) \cup X| \leqslant 3|X|$. Repeated use of this result gives a set Y such that $n \leqslant |Y| \leqslant n + m/3$ and $|\Gamma(Y) \cup Y| \leqslant 3|Y| \leqslant 3n + m$. It follows that the blue graph contains a copy of $K(n, n)$.

3. Open questions and final remarks

What is the behavior of $r(T, K(n, n))$ when T has bounded degree? Perhaps the methods of Häggkvist will shed some light on this question. We have seen that for a tree, T, the Ramsey number $r(T, K(n, n))$ is linear in n. However, if T is replaced by a graph containing a cycle this is no longer true. In [7] Spencer showed that $r(C_m, K_n) \geqslant c(n/\log(n))^\alpha$, where $\alpha = (m-1)/(m-2)$. By the same method, one obtains the same bound for $r(C_m, K(n, n))$, except for the value of the positive constant c.

References

1. M. Ajtai, P. Erdös, J. Komlós and E. Szemerédi, On Turan's theorem for sparse graphs. *Combinatorica* **1** (1981), 313–317.
2. S. A. Burr, P. Erdös, R. J. Faudree, C. C. Rousseau and R. H. Schelp, An extremal problem in generalized Ramsey theory. *Ars Combinat.* **10** (1980), 193–203.
3. P. Erdös, R. J. Faudree, C. C. Rousseau and R. H. Schelp, Multipartite graph–sparse graph Ramsey numbers. To appear.
4. L. Geréncser and A. Gyárfás, On Ramsey-type problems. *Ann. Univ. Sci. Budapest. Eötvös Sect. Math.* **10** (1967), 167–170.
5. R. Haggkvist, Personal communication.
6. L. Pósa, Hamiltonian circuits in random graphs. *Discrete Math.* **14** (1976), 359–364.
7. J. Spencer, Asymptotic lower bounds for Ramsey functions. *Discrete Math.* **20** (1977), 69–76.

14

A COMBINATORIAL PROOF OF A CONTINUED FRACTION EXPANSION THEOREM FROM THE RAMANUJAN NOTEBOOKS

I. P. Goulden and D. M. Jackson

ABSTRACT   The paper concerns enumeration of paths on a restricted set of steps. As special cases of our results, we obtain natural proofs of three identities found in the Ramanujan notebooks.

1. Introduction

The following three results are taken from the Ramanujan notebooks [5] (see Vol. 2, Chap. XII, p. 145 ($=$ Vol. 1, Chap. XIII, p. 200)). We have retained his convention for the infinite sums.

ENTRY 17

$$\frac{1}{1+}\ \frac{a_1 x}{1\ +}\ \frac{a_2 x}{1\ +}\ \frac{a_3 x}{1\ +}\ \text{etc.} = 1 - A_1 x + A_2 x^2 - A_3 x^3 + \text{etc.}$$

Let $P_n = a_1 a_2 a_3 \ldots a_{n-1}(a_1 + a_2 + \ldots + a_n)$, then

$$P_n = \phi_0(n)A_n - \phi_1(n)A_{n-1} + \phi_2(n)A_{n-2} - \text{etc.},$$

where $\phi_r(n+1) - \phi_r(n) = a_{n-1}\phi_{r-1}(n-1)$. $\square$

In the above result we have suppressed Ramanujan's explicit listing of $P_1, \ldots, P_6$ in terms of the a_is and the A_is.

$$\frac{1}{1+}\ \frac{a_1 x}{1\ +}\ \frac{a_2 x}{1\ +}\ \frac{a_3 x}{1\ +}\ \text{etc.}$$

is the notation used for the continued fraction

$$\cfrac{1}{1 + \cfrac{a_1 x}{1 + \cfrac{a_2 x}{1 + \cfrac{a_3 x}{1 + \ldots}}}},$$

a convention which will be followed throughout this paper.

GRAPH THEORY AND COMBINATORICS
ISBN 0-12-111760-X

COROLLARY (i)   *If*

$$\frac{1}{1+b_1x+}\frac{a_1x}{1+b_2x+}\frac{a_2x}{1+b_3x+} \text{etc.} = 1 - A_1x + A_2x - \text{etc.},$$

then

$$P_n = a_1a_2a_3 \ldots a_{n-1}\overline{(a_1 + b_1 + \overline{a_2 + b_2} + \ldots + \overline{a_n + b_n})}$$

$$= \phi_0(n)A_n - \phi_1(n)A_{n-1} + \phi_2(n)A_{n-2} + \text{etc.},$$

where

$$\phi_r(n+1) - \phi_r(n) = b_n\phi_{r-1}(n) + a_{n-1}\phi_{r-1}(n-1). \quad \square$$

COROLLARY (ii)   *In the above results,*

$$D_{r-1} = \phi_0(r) + x\phi_1(r) + x^2\phi_2(r) + \text{etc.} \quad \square$$

D_k is Ramanujan's notation for the denominator polynomial of the kth convergent of a continued fraction.

Our interest in Entry 17 and its generalization, Corollary (i), started after reading Berndt's presentation, with proofs, of the material of Chapter XII of the second notebook [1]. Corollary (i) was the only result for which there was no proof and, as with many of Ramanujan's results, there were no clues to the way in which Ramanujan originally discovered it.

In this paper we show that the three results emerge quite naturally from a combinatorial context, namely the enumeration of paths on a restricted set of steps. Moreover, these results occur as elements of a family of related results.

2. Historical remarks

Entry 17 seems all the more interesting in the light of observations by De Morgan [3] and Rogers [7] that it is extremely difficult to obtain the A_ks from the continued fraction. Commenting on Entry 17 directly, and presumably on Corollary (i) as well because of its similarity to it, Wilson [6] writes (the square parentheses are ours) in his notes, "The reciprocal of Perron's corresponding C[ontinued] F[raction] (Ch. 8). But R[amanujan]'s results do not follow at all immediately." Although the continued fraction of Entry 17 is the subject of Chapter 8 of Perron [4], Ramanujan's result is not contained in it. The first proof of Entry 17 was evidently given by Andrews, who used a combinatorial argument, and is contained in [1], where Berndt also gives a proof of Corollary (ii).

There was in fact some difficulty in deciding precisely what Ramanujan meant in his statement of Corollary (i) since he did not include precise ranges of summations either here or in Entry 17. Berndt's original, and reasonable, assumption was that the range of summation for Corollary (i) was identical to that of Entry 17, which Andrews had proved. This, however, suggested that Corollary (i) was false. As we shall see, the ranges of summation are different and, with this understanding, Corollary (i) is in fact true.

It is always gratifying to a combinatorial theorist when he is able to shed some light on results which appear to have been unyielding to the more classical mathematical techniques. Corollary (i) is one such result. However, with insight drawn from the combinatorial proof, it is possible to construct a very short algebraic proof of the result, and this is given in the final section of this paper.

3. Combinatorial preliminaries

We begin by giving some facts about paths and their relationship to continued fractions. For further information see Goulden and Jackson [2].

If $\omega = \omega_1 \ldots \omega_n$ is a sequence over $\{-1, 0, 1\}$ with $n \geq 0$, then we say that $(\omega)_k$ is a *path* with initial altitude k and terminal altitude $k + \omega_1 + \ldots + \omega_n$. For $i \geq 1$, ω_i is a *step* in the path, and for $k + \omega_1 + \ldots + \omega_{i-1} = m$, then ω_i is a *rise*, *level* or *fall* at *altitude* m, if $\omega_i = 1, 0$, or -1 respectively. A path is *level-free* if it has no levels.

The product $(\sigma)_i(\rho)_j$ of the path $(\sigma)_i$ with the path $(\rho)_j$ is defined to be the path $(\sigma\rho)_i$, if $(\sigma)_i$ has terminal altitude j. If $\mathscr{P}$, $\mathscr{P}_1$ and $\mathscr{P}_2$ are sets of paths such that $\mathscr{P} = \{\pi_1 \pi_2 \mid (\pi_1, \pi_2) \in \mathscr{P}_1 \times \mathscr{P}_2\}$ then we write $\mathscr{P} = \mathscr{P}_1 \mathscr{P}_2$.

The path $(\omega_1 \ldots \omega_n)_k$ may be represented geometrically in the plane by a sequence $v_0 e_1 v_1 e_2 v_2 e_3 \ldots e_n v_n$ of vertices $v_j = (j, k + \omega_1 + \ldots + \omega_j)$ and edges (steps) $e_j = v_{j-1} v_j$. Thus a step from altitude m is represented by an edge whose initial vertex has ordinate m. The ordinate of the terminal vertex is equal to the terminal altitude of the path. The empty path at altitude k, denoted by ε_k, is represented by a single vertex, with ordinate k. This representation is useful for visualizing the action of the path decompositions given in this paper, and we shall not distinguish between a path and its geometrical representation in describing these decompositions.

Our first path decomposition is for the set $\mathscr{H}_i$ of all paths from altitude i to altitude i with no vertices at altitude less than i. We adopt the convention that $\varepsilon_i \in \mathscr{H}_i$. For example, $\pi_1 = (1\,0\,1\,0\,-1\,-1\,1\,1\,-1\,0\,0)_3$ and $\pi_2 = (1\,0\,1\,0\,-1\,-1)_3$, given in Fig. 1, are both in $\mathscr{H}_3$.

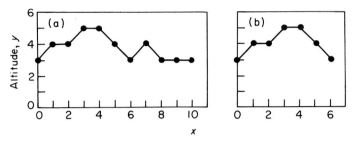

FIG. 1. Two paths in $\mathcal{H}_3$: (a) $\pi_1 = (1\ 0\ 1\ 0 - 1 - 1\ 1 - 1\ 0\ 0)_3$; (b) $\pi_2 = (1\ 0\ 1\ 0 - 1 - 1)_3$.

DECOMPOSITION 3.1

$$\mathcal{H}_i = \varepsilon_i \cup (\{(0)_i\} \cup \{(1)_i\} \mathcal{H}_{i+1}\{(-1)_{i+1}\}) \mathcal{H}_i.$$

PROOF Let $\sigma = \alpha\beta \in \mathcal{H}_i$ be an arbitrary non-empty path in $\mathcal{H}_i$, where the terminus of α is the vertex in σ (other than its origin) with altitude i and with the smallest positive abscissa of all vertices with altitude i in σ. Then $\beta \in \mathcal{H}_i$, and α is a non-empty path in $\mathcal{H}_i$ having no internal vertices at altitude less than or equal to i, so

$$\alpha \in \{(0)_i\} \cup \{(1)_i\} \mathcal{H}_{i+1}\{(-1)_{i+1}\},$$

and the result follows, since this process is reversible. $\square$

An example of the action of the above decomposition for $i = 3$ is provided by Fig. 1, where choosing $\sigma = \pi_1$ yields $\alpha = \pi_2$, in the notation of the above proof.

Let H_i be the ordinary generating function for $\mathcal{H}_i$ in which α_j, γ_j and β_j are indeterminates marking, respectively, rises, levels and falls at altitude j, for $j \geq 1$. Thus the number of paths in $\mathcal{H}_i$ with k_j rises, m_j levels and n_j falls at altitude j, for $j \geq 1$, is the coefficient of $\alpha_1^{k_1}\alpha_2^{k_2} \ldots \gamma_1^{m_1}\gamma_2^{m_2} \ldots \beta_1^{n_1}\beta_2^{n_2} \ldots$ in H_i.

LEMMA 3.2

$$H_i = \cfrac{1}{1 - \gamma_i -} \cfrac{\alpha_i\beta_{i+1}}{1 - \gamma_{i+1} -} \cfrac{\alpha_{i+1}\beta_{i+2}}{1 - \gamma_{i+2} -} \ldots \quad for \quad i \geq 1.$$

PROOF From Decomposition 3.1, we have

$$H_i = 1 + (\gamma_i + \alpha_i H_{i+1}\beta_{i+1})H_i,$$

so

$$H_i = (1 - \gamma_i - \alpha_i \beta_{i+1} H_{i+1})^{-1}.$$

The result follows by applying this result iteratively. $\square$

The following is a classical result on continued fractions (see, for example, Perron [**4**]).

LEMMA 3.3

$$\cfrac{1}{1 - \gamma_1 -} \cfrac{\alpha_1 \beta_2}{1 - \gamma_2 -} \cdots \cfrac{\alpha_{n-1} \beta_n}{1 - \gamma_n} = \frac{N_n}{D_n}$$

where N_n and D_n satisfy the recurrence equation

$$u_k = (1 - \gamma_k) u_{k-1} - \alpha_{k-1} \beta_k u_{k-2}, \qquad k \geq 2,$$

with initial conditions $N_0 = 0$, $N_1 = 1$ and $D_0 = 1$, $D_1 = 1 - \gamma_1$. $\square$

The continued fraction of Lemma 3.3 is called the nth *convergent* of

$$H_1 = \frac{1}{1 - \gamma_1 -} \frac{\alpha_1 \beta_2}{1 - \gamma_2 -} \cdots,$$

and N_n and D_n are, respectively, the *numerator* and *denominator* polynomials.

4. Entry 17 and path enumeration

Let $\mathcal{A}_n$ be the set of paths in $\mathcal{H}_1$ with a total of n rises and levels, and let $\mathcal{Q}_k(n)$ be the set of paths in $\mathcal{A}_n$ in which the first $k - 1$ steps are rises, for $1 \leq k \leq n + 1$. Note that $\mathcal{Q}_k(n)$ is contained in $\mathcal{Q}_i(n)$ for $i < k$, and that $\mathcal{Q}_{k-1}(n) - \mathcal{Q}_k(n)$ is the set of all paths in $\mathcal{Q}_{k-1}(n)$ whose $(k-1)$th step is either a level or fall (from altitude $k - 1$), if $3 \leq k \leq n + 1$. For example the paths $\pi_1 = (1\ 1\ -1\ 0\ 1\ 0\ -1\ -1\ 1\ 1\ -1)_1$ and $\pi_2 = (1\ 0\ 1\ 0\ -1\ -1\ 1\ 1\ -1)_1$, illustrated in Fig. 2, are in $\mathcal{Q}_3(6) - \mathcal{Q}_4(6)$ and $\mathcal{Q}_2(5) - \mathcal{Q}_3(5)$, respectively.

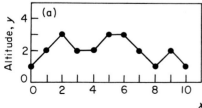 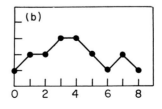

FIG. 2. Paths in $\mathcal{Q}_{k-1}(n) - \mathcal{Q}_k(n)$: (a) $\pi_1 = (1\ 1\ -1\ 0\ 1\ 0\ -1\ -1\ 1\ 1\ -1)_1$; (b) $\pi_2 = (1\ 0\ 1\ 0\ -1\ -1\ 1\ 1\ -1)_1$.

The following path decomposition for $\mathcal{D}_{k-1}(n) - \mathcal{D}_k(n)$ is obtained by considering separately the cases in which the $(k-1)$th step is a level or fall. The symbol $\leftrightsquigarrow$ denotes a one–one correspondence which preserves steps at each altitude.

DECOMPOSITION 4.1

(1) $\mathcal{D}_{k-1}(n) - \mathcal{D}_k(n) \leftrightsquigarrow \{(0)_{k-1}\} \times \mathcal{D}_{k-1}(n-1) \,\dot\cup\, \{(1)_{k-2}(-1)_{k-1}\}$
    $\times \mathcal{D}_{k-2}(n-1)$ for $3 \le k \le n+1$.

(2) $\mathcal{D}_1(n) - \mathcal{D}_2(n) \leftrightsquigarrow \{(0)_1\} \mathcal{D}_1(n-1)$, $n \ge 1$.

(3) $\mathcal{D}_1(n) = \mathcal{A}_n$, $n \ge 0$.

PROOF   (1) Let $\sigma \in \mathcal{D}_{k-1}(n) - \mathcal{D}_k(n)$, where $3 \le k \le n+1$. From the above discussion there are two cases.

Case 1   The $(k-1)$th step of σ is a level from altitude $k-1$. Let α and β be the subpaths of σ which precede and follow, respectively, the $(k-1)$th step, so $\sigma = \alpha(0)_{k-1}\beta$. Then $\alpha\beta \in \mathcal{D}_{k-1}(n-1)$ since $\alpha\beta$ has one fewer levels than σ, and the procedure is reversible.

Case 2   The $(k-1)$th step of σ is a fall from altitude $k-1$. The $(k-2)$th step is a rise from altitude $k-2$, by definition of $\mathcal{D}_{k-1}(n)$. Let α be the subpath of σ which precedes the $(k-2)$th step and β be the subpath of σ which follows the $(k-1)$th step, so $\sigma = \alpha(1)_{k-2}(-1)_{k-1}\beta$. Then $\alpha\beta \in \mathcal{D}_{k-2}(n-1)$, and the procedure is reversible.

The result follows by combining these two disjoint cases.

(2) Let $\sigma \in \mathcal{D}_1(n) - \mathcal{D}_2(n)$. The first step of σ must be a level from altitude 1, so $\sigma = (0)_1\beta$ where $\beta \in \mathcal{D}_1(n-1)$, and the result follows since this is reversible.

(3) Immediate.   □

This decomposition is illustrated for $k = 4$, Case 2 of part (1), in Fig. 2, where $\sigma = \pi_1$ and $\alpha\beta = \pi_2$, in the notation of the above proof.

Now let A_n and $Q_k(n)$, $1 \le k \le n+1$, be the ordinary generating functions for $\mathcal{A}_n$ and $\mathcal{D}_k(n)$, respectively, in which a_i and b_i are indeterminates marking rises and levels, respectively, at altitude i, for $i \ge 1$. We obtain recurrence equations for these generating functions directly from the above decomposition.

LEMMA 4.2

(1) $Q_k(n) = Q_{k-1}(n) - b_{k-1}Q_{k-1}(n-1) - a_{k-2}Q_{k-2}(n-1)$, $3 \le k \le n+1$.

(2) $Q_2(n) = Q_1(n) - b_1Q_1(n-1)$, $n \ge 1$.

(3) $Q_1(n) = A_n$, $n \ge 0$.

PROOF Immediate from Decomposition 4.1, by the sum and product lemmas for ordinary generating functions. □

Combining previous results, we obtain the main result of this paper.

THEOREM 4.3

(1) $\dfrac{1}{1+b_1x+}\dfrac{a_1x}{1+b_2x+}\dfrac{a_2x}{1+b_3x+}\cdots = \sum\limits_{n\geqslant 0} A_n(-x)^n.$

(2) $Q_k(n) = \sum\limits_{i=0}^{k-1} (-1)^i \phi_i(k)A_{n-i}, \ 1\leqslant k\leqslant n+1.$

(3) $\phi_i(k) - \phi_i(k-1) = b_{k-1}\phi_{i-1}(k-1) + a_{k-2}\phi_{i-1}(k-2), \ 1\leqslant i\leqslant k,$

where $\phi_0(k) = 1$ for $k\geqslant 1$, and $\phi_i(k) = 0$ for $i\geqslant k$.

PROOF (1) Follows from Lemma 3.2 with $i=1$ and $\alpha_j = a_j(-x)$, $\beta_j = b_j(-x)$, $\gamma_j = 1$ for $j\geqslant 1$.
(2) Follows immediately by repeated application of Lemma 4.2.
(3) Substituting (2) into Lemma 4.2 (1) we have

$$\sum_i (-1)^i A_{n-i}\{\phi_i(k) - \phi_i(k-1) - b_{k-1}\phi_{i-1}(k-1) - a_{k-2}\phi_{i-1}(k-2)\} = 0,$$

so (3) is clearly sufficient. The necessity of (3) follows from the initial conditions for $\phi_i(k)$ given by Lemma 4.2 (3). □

Among the results which can be deduced from Theorem 4.3 are the following:

(i) RAMANUJAN'S COROLLARY (i) *Let $k = n$. Then $Q_n(n) = a_1 \ldots a_{n-1}(a_1 + b_1 + \ldots + a_n + b_n)$ by direct calculation, and the result follows.*

(ii) RAMANUJAN'S ENTRY 17 *Let $k = n$ and $b_j = 0$ for $j\geqslant 1$. Setting all b_js equal to zero results in some compression of the range of non-zero values of $\phi_i(k)$, and a precise statement of Entry 17 as deduced from Theorem 4.3 is*

(1) $\dfrac{1}{1+}\dfrac{a_1x}{1}\dfrac{a_2x}{1}\cdots = \sum\limits_{n\geqslant 0} A_n(-x)^n;$

(2) $a_1 \ldots a_{n-1}(a_1 + \ldots + a_n) = \sum\limits_{0\leqslant i<n/2} (-1)^i \phi_i(n)A_{n-i}, \ n\geqslant 1;$

(3) $\phi_i(k) - \phi_i(k-1) = a_{k-2}\phi_{i-1}(k-2), \ 1\leqslant i<k/2;$

where $\phi_0(k) = 1$ for $k\geqslant 1$, and $\phi_i(k) = 0$ for $i\geqslant k/2$.

(iii) RAMANUJAN'S COROLLARY (ii)   *Let*

$$\Phi_{k-1} = \sum_{i \geq 0} \phi_i(k) x^i, \qquad k \geq 1.$$

Then $\Phi_0 = 1$, $\Phi_1 = 1 + b_1 x$ *and, multiplying both sides of* 4.3 (3) *by* x^i, *summing for* $i \geq 1$, *and replacing* $k - 1$ *by* k, *we get*

$$\Phi_k = (1 + b_k x)\Phi_{k-1} + a_{k-1} x \Phi_{k-2}, \qquad k \geq 2.$$

From Lemma 3.3, $\Phi_k = D_k$, *and the result follows.*

The choice $k = n + 1$ in Theorem 4.3 yields a result which seems simpler than Entry 17 or Corollary (ii) since $Q_{n+1}(n) = a_1 \ldots a_n$ (independent of $b_1, b_2, \ldots$!).

5. An algebraic proof

We now give a purely algebraic proof of Theorem 4.3, and hence of Ramanujan's results. The proof was arrived at by considering the combinatorial relationship between $\mathscr{A}_i$, $\mathscr{H}_i$ and $\mathscr{D}_k(n)$. Although we redefine A_n and $Q_k(n)$ algebraically in the following development, the reader may like to check that they are in fact identical with A_n and $Q_k(n)$ as combinatorially defined earlier.

Define F_i, A_n, q_k and $Q_k(n)$ as power series in $x, a_1, b_1, a_2, b_2, \ldots$ by

(i) $F_i = \dfrac{1}{1 + b_i x +} \dfrac{a_i x}{1 + b_{i+1} x +} \dfrac{a_{i+1} x}{1 + b_{i+2} x +} \cdots, \quad i \geq 1,$

(ii) $\displaystyle\sum_{n \geq 0} A_n(-x)^n = F_1,$

(iii) $\displaystyle\sum_{n \geq k-1} Q_k(n)(-x)^n = q_k = a_1 \ldots a_{k-1}(-x)^{k-1} F_1 \ldots F_k, \ k \geq 1.$

From (iii) we have

$$q_k - q_{k-1} = -a_1 \ldots a_{k-2}(-x)^{k-2} F_1 \ldots F_{k-1}\{a_{k-1} x F_k + 1\}, \qquad k \geq 2,$$

but (i) gives

$$a_{k-1} x F_k + 1 = F_{k-1}^{-1} - b_{k-1} x.$$

Combining these results yields

$$\left.\begin{aligned}
q_k - q_{k-1} &= a_{k-2} x q_{k-2} + b_{k-1} x q_{k-1}, \qquad k \geq 2, \\
q_2 - q_1 &= -1 + b_1 x q_1, \\
q_1 &= F_1.
\end{aligned}\right\} \qquad (*)$$

By equating the coefficients of $(-x)^n$ in the equations of (*), we obtain Lemma 4.2 for $Q_k(n)$ and A_n as redefined. But A_n is clearly identical with its original definition, so Theorem 4.3 and thus Ramanujan's results follow, by the proof given in Section 4.

Acknowledgements

We are indebted to B. C. Berndt for bringing the question of Entry 17 to our attention, and for furnishing us with the historical details. This research was supported by the Natural Sciences and Engineering Research Council of Canada (Grants U0073 and A8235).

References

1. B. C. Berndt, R. L. Lamphere and D. M. Wilson, Chapter XII of Ramanujan's second notebook: continued fractions.
2. I. P. Goulden and D. M. Jackson, *Combinatorial Enumeration*, John Wiley, New York, 1983.
3. A. De Morgan, On the reduction of a continued fraction to a series. *Phil. Mag.* (3) **24** (1844), 15–17.
4. O. Perron, *Die Lehre von den Kettenbruchen*, Leipzig and Berlin, 1929.
5. S. Ramanujan, *Notebooks of Srinivasa Ramanujan*, Vols 1 and 2, Facsimile edition, Tata Institute of Fundamental Research, Bombay, India, 1957.
6. Ramanujan Manuscripts, Add. ms. b1070[7]; *vide* B. M. Wilson's notes and proofs of results in Chapter 12 (of B), Trinity College Library, Trinity College, Cambridge. [*Authors' note*: B = handwritten copy of notebooks, by T. A. Satagopan.]
7. L. J. Rogers, On the representation of certain asymptotic series as convergent continued fractions. *Proc. London Math. Soc.* **4** (1907), 72–89.

15

ON INTERSECTIONS OF LONGEST CYCLES

Martin Grötschel

ABSTRACT Let G be a 2-connected graph, let $k \in \{2, 3, 4, 5\}$, and suppose that G has at least $k + 1$ nodes. We prove in this paper that whenever two different longest cycles of G meet in k nodes, then these nodes form an articulation set of G. We give some applications of this result, e.g. a lower bound on the circumference of vertex-transitive graphs, and conjecture possible generalizations.

1. Introduction and main results

A considerable amount of very successful research effort has been spent on the following type of problem: Given a graph (or digraph) with some properties (e.g. assumptions on the connectivity or degree sequence) and given a certain number of nodes or edges or paths, determine a lower (or upper) bound on the length of a longest cycle (or path) of the graph containing (not containing) the given nodes, edges, or paths; see for instance [3–6, 8, 9] and many others.

To our knowledge almost nothing has been done to determine the intersection patterns of longest cycles. Such results are of interest, for instance, in designing recursive algorithms (e.g. in combinatorial optimization) in which in every step a longest cycle is shrunk or deleted and one wants to know a bound on the length of the longest cycle in the resulting graph. The case where two longest cycles meet in two nodes has been analysed in [7] (to obtain a polynomial algorithm for the max-cut problem in graphs without long odd cycles).

In this section we state the main results of the paper. The (non-standard) notation can be found in Section 2, and the proofs in the subsequent sections. In this paper we only consider 2-*connected graphs* (the essential case) with at least four nodes. The circumference of such graphs is at least four and every two longest cycles meet in at least two nodes. The result of [7] can be stated as follows:

THEOREM 1.1 *Let C and D be two different longest cycles of a graph G meeting in exactly two nodes, say in u and v. Let $C = P_1 \cup P_2$ and $D = Q_1 \cup Q_2$, where P_i, Q_i, $i = 1, 2$ are $[u, v]$-segments of C (respectively*

GRAPH THEORY AND COMBINATORICS
ISBN 0-12-111760-X

D). Then the following hold:

(a) $\{u, v\}$ is an articulation set of G.

(b) $|P_1| = |P_2| = |Q_1| = |Q_2|$ and therefore the circumference is even.

(c) The truncated paths $\bar{P}_1$, $\bar{P}_2$, $\bar{Q}_1$, $\bar{Q}_2$ obtained by removing the end-nodes u, v of P_1, P_2, Q_1, Q_2 are in different components of $G - \{u, v\}$.

(d) Every longest cycle contains u and v.

(e) For every longest cycle B, the two paths of $B - \{u, v\}$ in $G - \{u, v\}$ belong to different components of $G - \{u, v\}$. $\square$

This result is generalized here to the following.

THEOREM 1.2   Let $k \in \{3, 4, 5\}$, and let $G = [V, E]$ be a graph with at least $k + 1$ nodes. Suppose that C and D are two different longest cycles meeting in a set W of exactly k nodes. Then the following hold:

(a) W is an articulation set of G.

(b) In the case $k = 3$, the paths obtained by removing W from C and D are in different components of $G - W$.

The proof of Theorem 1.2 will be given in Sections 3 and 4; conjectures about further generalizations in Section 5. We first show that not all cases of Theorem 1.1 are generalizable; this is, for instance, obvious for (b) of Theorem 1.1. Moreover, we discuss some consequences of Theorem 1.2.

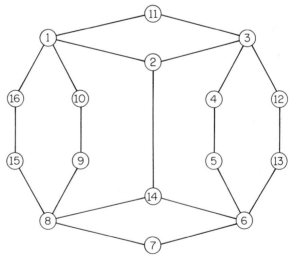

FIG. 1.

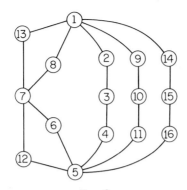

FIG. 2.

Theorem 1.2(b) does not hold for $k = 4$. Consider the graph G of Fig. 1. The circumference of G is 10. The cycle C through the nodes $1, 2, \ldots, 10$ and the cycle D through 1, 11, 3, 12, 13, 6, 14, 8, 15, 16 meet in $W = \{1, 3, 6, 8\}$. $G - W$, however, has only seven components, since node 2 of $C - W$ and node 14 of $D - W$ form one component of $G - W$. If Theorem 1.2(b) were to hold for $k = 4$, $G - W$ should have eight components.

Theorem 1.1(d) cannot be generalized to the case $k = 3$. Consider the graph G of Fig. 2. The circumference of G is 8, and the cycles C_1 through the nodes $1, 2, \ldots, 8$, C_2 through the nodes 1, 9, 10, 11, 5, 12, 7, 13, and C_3 through the nodes $1, 2, \ldots, 5$, 16, 15, 14 are longest cycles of G. C_1 and C_2 meet in $\{1, 5, 7\}$, while C_1 and C_3 meet in $\{1, 5\}$ only. Thus, if two longest cycles meet in three nodes, then not all longest cycles necessarily contain these three nodes.

To prove Theorem 1.2 we want to assume that the circumference is at least $k + 1$. This can be done because of the following observations. A 2-connected graph with at least four nodes has a cycle of length at least four. So this assumption is automatically fulfilled for $k = 3$. The 2-connected

those shown in Fig. 3. There may be additional parallel edges, but no "diagonals". Theorem 1.2 is obviously true for these graphs. The 2-connected graphs with at least six nodes and circumference five look like those shown in Fig. 4. The graph G' in Fig. 4 is a graph on four nodes containing a path of length three from u to v, and further parallel edges may exist. For the graphs shown in Fig. 4, Theorem 1.2 clearly holds as well.

We may therefore assume in the sequel that G has circumference at least $k + 1$, for $2 \le k \le 5$.

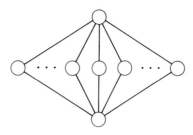

FIG. 3. The 2-connected graphs with circumference 4.

We shall now give an application of Theorem 1.2.

Babai [2] proved that every connected vertex-transitive graph with at least four nodes contains a cycle of length greater than $(3n)^{1/2}$. If a vertex-transitive graph is k-connected, $k \in \{4, 5, 6\}$, then Theorem 1.2 implies that every two different longest cycles meet in at least k nodes. Following the (nice and simple) proof of Babai [2], we can conclude

THEOREM 1.3   *If G is a k-connected vertex-transitive graph, $k \in \{4, 5, 6\}$, then G contains a cycle of length greater than $(kn)^{1/2}$.*   □

Note that vertex-transitive graphs are regular. By a result of Mader [10] and Watkins [12], the connectivity of a connected d-regular graph is at least $\frac{2}{3}(d + 1)$. Thus, if a connected vertex-transitive graph G is d-regular we get

G is Hamiltonian if $d = 2$ (obvious);

$$G \text{ has circumference at least } \begin{cases} (3n)^{1/2} & \text{if } d = 3, \\ (4n)^{1/2} & \text{if } d = 4, 5, \\ (5n)^{1/2} & \text{if } d = 6, \\ (6n)^{1/2} & \text{if } d \geqslant 7. \end{cases}$$

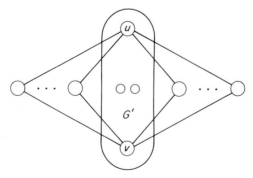

FIG. 4. The 2-connected graphs with circumference 5.

Although a number of people believe that all but four connected vertex-transitive graphs are Hamiltonian (see, for example, Alspach [1]), it seems that the lower bounds on the circumference given above are the best ones known at present.

The proofs of Theorem 1.2 for the various cases do not use the high symmetry of vertex-transitive graphs, and it seems to me that the constant $k^{1/2}$ in the bound can be considerably improved, but no improvement on the factor $n^{1/2}$ (the important one) can be obtained this way. For instance, if G has at least six nodes and is 3-regular, then one can obtain a lower bound of $2n^{1/2}$ for the circumference of G (see Alspach [1]).

2. Notation and some lemmas

Graphs are denoted by $G = [V, E]$ where V is the node set and E the edge set. Our graphs may have loops or multiple edges.

A path P with end-nodes u and v is called a $[u, v]$-*path*. The nodes of P different from u and v are called *internal nodes* of P. The path resulting from P by removing the two end-nodes of P is called the *truncation of* P and is denoted by $\bar{P}$. A truncated path $\bar{P}$ may be empty (if P is an edge), consist of one node (if P has two edges), or may be a path in the usual sense. A path P is called *internally disjoint from* a path or cycle C if none of the internal nodes of P is a node of C.

If P and Q are paths which are internally disjoint from each other and have at least one end-node in common, then $P \cup Q$ denotes the *concatenation* of P and Q which is either a path or a cycle. If C and D are paths or cycles and R is a path with one end-node contained in C, one end-node contained in D, and which is internally disjoint from C and D, then R is called a $[C, D]$-*path*. If we fix an orientation of a cycle C and have two nodes u, v of C, then we can speak of the $[u, v]$-*segment* of C which is the path from u to v consistent with the orientation.

If C and D are two cycles which both contain the nodes u and v and if P is a $[u, v]$-segment of C and Q a $[u, v]$-segment of D, then P and D are called *parallel* if P is internally disjoint from D and Q internally disjoint from C. We say that P and Q are *parallel paths on the cycles* C and D.

The length of a path or cycle C is the number of its edges and is denoted by $|C|$. The length of a longest cycle in a graph G is called the *circumference* of G.

An *articulation set* is a set of nodes whose removal results in a disconnected graph or the graph with one node.

The following simple lemmas help to reduce considerably the number of cases to be considered in the proof of our theorem.

LEMMA 2.1   *Let P and Q be two parallel paths on two longest cycles of a graph G. Then P and Q have the same length.*

PROOF   Let C and D be two longest cycles of G and let P and Q be parallel paths on C and D. Then C and D can be written as concatenations of paths as follows: $C = P \cup P'$ and $D = Q \cup Q'$. By assumption $P \cup Q'$ and $Q \cup P'$ are cycles, and since C and D have maximum length we get $|P \cup Q'| \leq |Q \cup Q'|$ implies $|P| \leq |Q|$, and $|Q \cup P'| \leq |P \cup P'|$ implies $|Q| \leq |P|$, which proves the claim.   □

LEMMA 2.2   *Let C and D be two longest cycles of G having a common node u. Let P be a segment on C and Q be a segment on D such that P and Q have u as one end-node and such that P is internally disjoint from D and Q is internally disjoint from C. Then G contains no $[\bar{P}, \bar{Q}]$ path internally disjoint from C and D.*

PROOF   Suppose that R is a path in G connecting an internal node of P, say x, to an internal node of Q, say y, such that R is internally disjoint from C and D. We may assume that the situation is as depicted in Fig. 5, where $C = P \cup P_1 = P' \cup P'' \cup P_1$, $D = Q \cup Q_1 = Q' \cup Q'' \cup Q_1$ and P, P_1 are $[u, v]$-segments of C, while Q, Q_1 are $[u, w]$-segments of D.

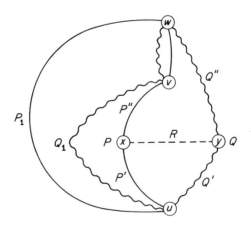

FIG. 5.

Clearly, $D_1 := Q_1 \cup P' \cup R \cup Q''$ and $C_1 := P_1 \cup Q' \cup R \cup P''$ are cycles of G, and hence $|D_1| \leq |D|$ implies $|P' \cup R| \leq |Q'|$ and $|C_1| \leq |C|$ implies $|Q' \cup R| \leq |P'|$. Therefore we have $|Q'| < |Q' \cup R| \leq |P'| < |P' \cup R| \leq |Q'|$, which is a contradiction.   □

LEMMA 2.3   *Let C and D be two longest cycles of a graph G which are concatenations of three paths, i.e.* $C = P_1 \cup P_2 \cup P_3$, $D = Q_1 \cup Q_2 \cup Q_3$. *Suppose that P_1 and Q_1 are parallel, and P_2 and Q_2 are internally disjoint respectively from D and C. Then G contains no* $[\bar{P}_1, \bar{P}_2]$-, $[\bar{P}_1, \bar{Q}_2]$-, $[\bar{Q}_1, \bar{Q}_2]$-, *and no* $[\bar{Q}_1, \bar{P}_2]$-paths internally disjoint from C and D.

PROOF   We show that G contains no $[\bar{P}_1, \bar{Q}_2]$-path internally disjoint from C and D. The other cases follow by symmetry.

Suppose that R is a path linking a node x of $\bar{P}_1$ to a node y of $\bar{Q}_2$ which is internally disjoint from C and D. We may assume that the situation is as depicted in Fig. 6, where $C = P_1' \cup P_1'' \cup P_2 \cup P_3$ (P_3 is a $[z, u]$-segment of C) and $D = Q_1 \cup Q_2' \cup Q_2'' \cup Q_3$ (Q_3 is a $[w, u]$-segment of D). We see that $C_1 := P_1' \cup R \cup Q_2' \cup P_2 \cup P_3$ and $D_1 := Q_1 \cup P_1'' \cup R \cup Q_2'' \cup Q_3$ are cycles and that $|C_1| + |D_1| = |C| + |D| + 2|R|$, which contradicts the fact that C and D have maximum length.   □

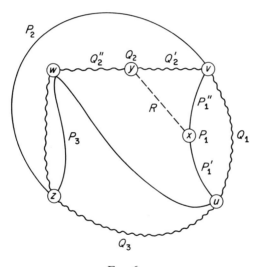

FIG. 6.

LEMMA 2.4   *Let P and Q be two parallel paths on two longest cycles C and D of a graph G. Let S be a segment on one of the cycles disjoint from the other. If there is a $[\bar{P}, S]$-path (respectively a $[\bar{Q}, S]$-path) internally disjoint from D, then there is no $[\bar{Q}, S]$-path (respectively no $[\bar{P}, S]$-path) internally disjoint from C and D.*

PROOF   Suppose that S is a segment on D and suppose that there are a

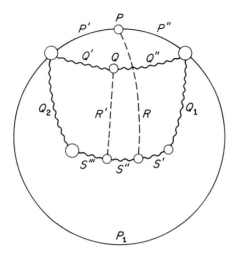

FIG. 7.

$[\bar{P}, S]$-path R and a $[\bar{Q}, S]$-path R'. Then, up to isomorphic rearrangements, the situation can be depicted as shown in Fig. 7. Here $S = S' \cup S'' \cup S'''$, $Q = Q' \cup Q''$, $P = P' \cup P''$, $C = P \cup P_1$, $D = Q \cup Q_1 \cup S \cup Q_2$. Clearly, $C_1 := P_1 \cup P'' \cup R \cup S'' \cup R' \cup Q'$ and $C_2 = P_1 \cup Q'' \cup R' \cup S'' \cup R \cup P'$ are cycles which are not longer than C, but $|C_1| + |C_2| = 2|C| + 2(|R| + |R'| + |S''|)$, a contradiction. □

3. The cases $k = 3, 4$

Note that the cases (a), (b) and (c) of Theorem 1.1 follow immediately from Lemmas 2.1, 2.2 and 2.3. The cases (d) and (e) need different arguments and are not proved here.

PROOF OF THEOREM 1.2(a), (b) FOR $k = 3$ Let the node set W in which the two longest cycles C and D intersect be $\{u, v, w\}$. Then C and D can be written as concatenations of three paths, say $C := P_1 \cup P_2 \cup P_3$ and $D := Q_2 \cup Q_3$, such that P_1, Q_1 are parallel $[u, v]$-segments, P_2, Q_2 are parallel $[v, w]$-segments and P_3, Q_3 are parallel $[u, v]$-segments. Since $|V| \geqslant 4$ and G is 2-connected, C and D have length at least 4, so at least one of the paths P_i, $i = 1, 2, 3$, has length at least 2, say P_1. By Lemma 2.1 Q_1 has the same length as P_1, and by Lemma 2.2, there is no $[\bar{P}_1, \bar{Q}_1]$-path in G disjoint from C and D.

Each of the paths P_2, P_3, Q_2, Q_3 satisfies the assumptions of Lemma 2.3 with respect to the parallel paths P_1 and Q_1. Therefore there are no $[\bar{P}_1, \bar{P}_2]$-, $[\bar{P}_1, \bar{P}_3]$-, $[\bar{P}_1, \bar{Q}_2]$-, $[\bar{P}_1, \bar{Q}_3]$-paths internally disjoint from C and D. This implies that the component of $G - W$ containing $\bar{P}_1$ does not contain any of the (non-empty) truncated paths $\bar{P}_2$, $\bar{P}_3$, $\bar{Q}_1$, $\bar{Q}_2$, $\bar{Q}_3$. Thus we have shown that $G - W$ has at least two components (the ones containing $\bar{P}_1$ and $\bar{Q}_1$) and that each of the paths $C - W$ and $D - W$ belongs to a different component of $G - W$, which finishes the proof. $\square$

PROOF OF THEOREM 1.2(a) FOR $k = 4$ Suppose the two longest cycles C and D meet in $\{u_1, u_2, u_3, u_4\}$. Let $C := P_1 \cup P_2 \cup P_3 \cup P_4$, where P_i is a $[u_i, u_{i+1}]$-segment of C, $i = 1, 2, 3$, and P_4 a $[u_4, u_1]$-segment. There are two possible intersection patterns for $D = Q_1 \cup Q_2 \cup Q_3 \cup Q_4$, namely, each of the four segments of D is parallel to a segment of C, or exactly two segments of C are parallel to segments of D.

Case 1: Each segment of D is parallel to a segment of C, say P_i is parallel to Q_i, $i = 1, 2, 3, 4$ By assumption one of the paths P_i has length at least two, say P_1. Using Lemmas 2.1, 2.2 and 2.3 we obtain, just as in the proof above for the case $k = 3$, that there can be no $[\bar{P}_1, \bar{Q}_1]$-, $[\bar{P}_1, \bar{Q}_2]$-, $[\bar{P}_1, \bar{Q}_4]$-, $[\bar{P}_1, \bar{P}_2]$-, $[\bar{P}_1, \bar{P}_4]$-paths in G internally disjoint from C and D. If the truncated path $\bar{P}_3$ (and hence $\bar{Q}_3$) is empty, then $\bar{Q}_1$ and $\bar{P}_1$ belong to different components of G–W. If $\bar{P}_3$ (and hence $\bar{Q}_3$) is non-empty, then by Lemma 2.2 there is no $[\bar{P}_3, \bar{Q}_3]$-path internally disjoint from C and D. By Lemma 2.2 there is a $[\bar{P}_1, \bar{P}_3]$-path (or a $[\bar{P}_1, \bar{Q}_3]$-path) internally disjoint from C and D, there can be no $[\bar{Q}_1, \bar{P}_3]$-path (or $[\bar{Q}_1, \bar{Q}_3]$-path) internally disjoint from C and D. Hence $\bar{P}_1$ and $\bar{Q}_1$ belong to different components of G–W in this case too, i.e. W is an articulation set.

Case 2: Two segments of C are parallel to segments of D, say P_1, Q_1 and P_3, Q_3 are parallel   One of the paths P_i has length at least two. There are two subcases to consider:

Case 2.1: One of the paths P_i parallel to a path Q_i has length at least two, say P_1   The proof of subcase 2.1 is word for word the same as the proof of Case 1.

Case 2.2: One of the paths P_2 or P_4 has length at least two, say P_2   By Lemma 2.2 there is no $[\bar{P}_2, \bar{Q}_i]$-path internally disjoint from C and D for $i = 1, \ldots, 4$. By Lemma 2.3 there is no such $[\bar{P}_2, \bar{P}_1]$-path and no such $[\bar{P}_2, \bar{P}_3]$-path. The only remaining possibility is that there is a $[\bar{P}_2, \bar{P}_4]$-path R internally disjoint from C and D. We may assume that the situation is as depicted in Fig. 8.

Then $D_1 := Q_1 \cup P_2' \cup R \cup P_4' \cup Q_3 \cup Q_4$ and $D_2 := Q_1 \cup Q_2 \cup Q_3 \cup P_2'' \cup R \cup P_4''$ are cycles which by assumption are not longer than D. How-

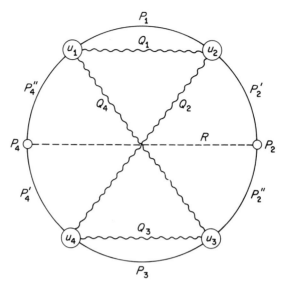

FIG. 8.

ever, $|D_1| + |D_2| = |D| + |Q_1| + |P_2| + |Q_3| + |P_4| + 2|R| = 2|D| + 2|R|$, since $|P_2| + |P_4| = |Q_2| + |Q_4|$, which is a contradiction. So there is no $[\bar{P}_2, \bar{P}_4]$-path internally disjoint from C and D either. This shows that the component containing $\bar{P}_2$ contains none of the other truncated paths and thus that W is an articulation set. ☐

REMARK 3.1 If $C = P_1 \cup P_2 \cup P_3 \cup P_4$ and $D = Q_1 \cup Q_2 \cup Q_3 \cup Q_4$ are longest cycles as above which meet in exactly four nodes u_1, u_2, u_3, u_4 and if the truncated paths $\bar{P}_i, \bar{Q}_i, i = 1, \ldots, 4$ are non-empty, then one can conclude from the proof above that $G - \{u_1, u_2, u_3, u_4\}$ has at least seven components. ☐

The graph G of Fig. 1 has two longest cycles meeting in four nodes W and where all truncated paths are non-empty. By Remark 3.1, $G - W$ has the minimum possible number of components. Moreover, one can show that there is no graph with these properties with fewer nodes than the graph of Fig. 1.

4. The case $k=5$

We shall now prove Theorem 1.2(a) for $k = 5$. Let $W = \{u_1, \ldots, u_5\}$ be the set of nodes in which the longest cycles C and D intersect, and assume that $C = P_1 \cup P_2 \cup P_3 \cup P_4 \cup P_5$, where P_i is the $[u_i, u_{i+1}]$-segment of C, $i = 1, \ldots, 4$ and P_5 the $[u_5, u_1]$-segment. Let $D = Q_1 \cup Q_2 \cup Q_3 \cup Q_4 \cup Q_5$. By enumerating the possible ways in which C and D can intersect, one can see that (up to renumbering) there are four cases to consider, which we depict graphically.

Case 1 All segments are parallel (Fig. 9).

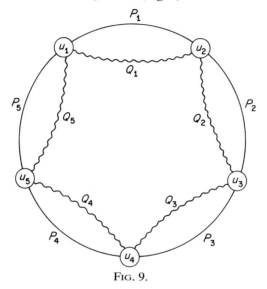

FIG. 9.

Case 2 Three segments are parallel (Fig. 10).

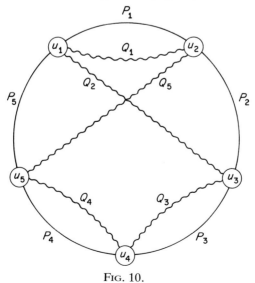

FIG. 10.

Case 3   Two segments are parallel (Fig. 11).

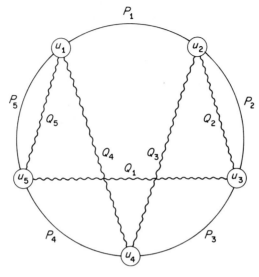

FIG. 11.

Case 4   No segments are parallel (Fig. 12).

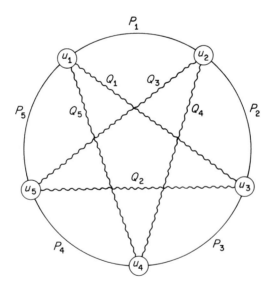

FIG. 12.

We have to discuss each of these cases separately. By going into detail one can also give lower bounds on the number of components of $G - W$ as in Remark 3.1, but we restrict ourselves to the proof that W is an articulation set. We assume in all cases that W is not an articulation set and derive a contradiction. As shown in Section 1, we can assume that the circumference of G is at least 6.

PROOF FOR CASE 1   Since the circumference of G is at least 6, one of the paths P_i has length at least 2. By symmetry we may assume that $|P_1| \geq 2$.

By Lemmas 2.2 and 2.3, G contains no $[\bar{P}_1, X]$-path internally disjoint from C and D for all $X \in \{\bar{Q}_1, \bar{Q}_2, \bar{Q}_5, \bar{P}_2, \bar{P}_5\}$. Suppose there is a path to one of the other truncated paths. By symmetry we may assume that there is a $[\bar{P}_1, \bar{Q}_3]$-path R in G internally disjoint from C and D. The end-nodes of R split P_1 into P_1' and P_1'' and Q_3 into Q_3' and Q_3''. (Here and in the sequel we assume that the splitting is "clockwise", i.e. the segment on C following P_5 is P_1', then P_1'', then P_2, and on D we go from Q_2 to Q_3' to Q_3'' to Q_4.) Then $C_1 := R \cup Q_3' \cup P_3 \cup P_4 \cup P_5 \cup Q_1 \cup P_1''$ is a cycle in G. $|C_1| \leq |C|$ implies $|R \cup Q_3' \cup P_1''| \leq |P_2|$.

Therefore $|P_2| \geq 3$, and if $G - W$ is connected then there must be a path linking $\bar{P}_2$ to one of the other truncated paths. As before, $\bar{P}_2$ can only be linked to one of the paths $\bar{P}_4$, $\bar{P}_5$, $\bar{Q}_4$, $\bar{Q}_5$. By symmetry we may assume that G contains a $[\bar{P}_2, \bar{Q}_5]$-path R' internally disjoint from C and D. R' splits P_2 into P_2' and P_2'' and Q_5 into Q_5' and Q_5''. Set

$$C_2 := P_1'' \cup P_2' \cup R' \cup Q_5'' \cup P_5 \cup P_4 \cup P_3 \cup Q_3' \cup R$$

and

$$C_3 := P_1' \cup Q_1 \cup Q_2 \cup P_2'' \cup R' \cup Q_5' \cup P_4 \cup Q_3'' \cup R,$$

then C_2 and C_3 are cycles satisfying $|C_1| + |C_2| = 2(|C| + |R| + |R'|)$, which contradicts the maximality of $|C|$.

PROOF FOR CASE 2   One of the paths P_i has length at least two. There are three subcases to consider.

Subcase 2.1 $|P_1| \geq 2$.
Subcase 2.2 $|P_2| \geq 2$ or $|P_5| \geq 2$, say $|P_2| \geq 2$.
Subcase 2.3 $|P_3| \geq 2$ or $|P_4| \geq 2$, say $|P_3| \geq 2$.

PROOF OF SUBCASE 2.1, $|P_1| \geq 2$ By Lemmas 2.2 and 2.3 G contains no $[\bar{P}_1, X]$-path internally disjoint from C and D for all $X \in \{\bar{Q}_1, \bar{Q}_2, \bar{Q}_5, \bar{P}_2, \bar{P}_5\}$. We now show that there are no such $[\bar{P}_1, X]$-paths for $X \in \{\bar{P}_3, \bar{P}_4, \bar{Q}_3, \bar{Q}_4\}$ as well. By symmetry it is sufficient to prove that

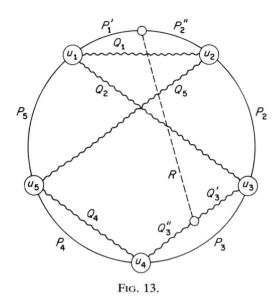

FIG. 13.

there is no $[\bar{P}_1, \bar{Q}_3]$-path internally disjoint from C and D. Suppose R is such a path.

R splits P_1 into $P_1' \cup P_1''$ and Q_3 into $Q_3' \cup Q_3''$ (see Fig. 13). Then we can construct the following cycles:

$$C_1 := R \cup P_1'' \cup Q_1 \cup P_5 \cup P_4 \cup P_3 \cup Q_3',$$

$$D_1 := R \cup P_1' \cup Q_2 \cup P_2 \cup Q_5 \cup Q_4 \cup Q_3'',$$

which satisfy $|C_1| + |D_1| = |C| + |D| + 2\,|R|$, a contradiction.

PROOF OF SUBCASE 2.2, $|P_2| \geqslant 2$ By Lemmas 2.2 and 2.3 there is no $[\bar{P}_2, X]$-path internally disjoint from C and D for all $X \in \{\bar{P}_1, \bar{P}_3, \bar{Q}_1, \bar{Q}_2, \bar{Q}_3, \bar{Q}_5\}$. So there may be such paths for $X \in \{\bar{Q}_4, \bar{P}_4, \bar{P}_5\}$. The cases $X = \bar{Q}_4$ and $X = \bar{P}_4$ are symmetric, so we have to consider two further subcases.

(a) Suppose there is a $[\bar{P}_2, \bar{P}_5]$-path R internally disjoint from C and D. R splits P_2 into $P_2' \cup P_2''$ and P_5 into $P_5' \cup P_5''$. Consider the cycles

$$C_1 := P_1 \cup P_2' \cup R \cup P_5' \cup P_4 \cup P_3 \cup Q_2,$$

$$D_1 := Q_1 \cup Q_5 \cup Q_4 \cup Q_3 \cup P_2'' \cup R \cup P_5'',$$

then $|C_1| + |D_1| = |C| + |D| + 2\,|R|$, a contradiction.

(b) Suppose that there is a $[\bar{P}_2, \bar{Q}_4]$-path R internally disjoint from C and D and that R splits P_2 into $P_2' \cup P_2''$ and Q_4 into $Q_4' \cup Q_4''$. (Such a path in fact may exist!) Since $|P_4| = |Q_4| \geqslant 2$ and we assume that $G - W$ is connected, there must be a $[\bar{P}_4, X]$-path internally disjoint from C and D for some of the truncated paths X. By Lemmas 2.2 and 2.3 there can be no such path for $X \in \{\bar{P}_3, \bar{P}_5, \bar{Q}_3, \bar{Q}_4, \bar{Q}_5\}$, by Lemma 2.4 no such path for $X = \bar{P}_2$, and by Subcase 2.1 no such path for $X \in \{\bar{P}_1, \bar{Q}_1\}$. The only possibility remaining is a $[\bar{P}_4, \bar{Q}_2]$-path R' internally disjoint from C and D splitting P_4 into $P_4' \cup P_4''$ and Q_2 into $Q_2' \cup Q_2''$ (see Fig. 14). The cycles

$$C_1 := P_1 \cup P_2' \cup R \cup Q_4' \cup P_3 \cup Q_2' \cup R' \cup P_4'' \cup P_5,$$

$$D_1 := Q_1 \cup Q_5 \cup Q_4'' \cup R \cup P_2'' \cup Q_3 \cup P_4' \cup R' \cup Q_2''$$

satisfy $|C_1| + |D_1| = |C| + |D| + 2(|R| + |R'|)$, a contradiction. This finishes the proof of Subcase 2.2.

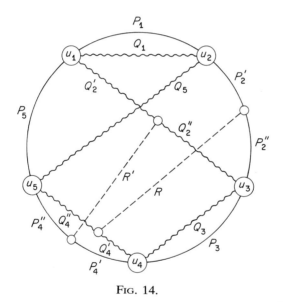

FIG. 14.

PROOF OF SUBCASE 2.3, $\|P_3\| \geqslant 2$ By Lemmas 2.2 and 2.3 there can be no $[\bar{P}_3, X]$-path internally disjoint from C and D for $X \in \{\bar{Q}_2, \bar{Q}_3, \bar{Q}_4, \bar{P}_2, \bar{P}_4\}$. Moreover, by Subcase 2.1 there can be no such path for $X \in \{\bar{Q}_1, \bar{P}_1\}$. What remains is the possibility of a $[\bar{P}_3, \bar{P}_5]$-or a $[\bar{P}_3, \bar{Q}_5]$-path internally disjoint from C and D. Note that these two cases are symmetric, and in fact the existence of a $[\bar{P}_3, \bar{P}_5]$-path is equivalent to the existence of a $[\bar{P}_2, \bar{Q}_4]$-path which has been discussed in Subcase 2.2(b). This proves Subcase 2.3.

PROOF OF CASE 3   As in Case 2, we also have to discuss the three subcases $|P_1| \geq 2$, $|P_2| \geq 2$ and $|P_3| \geq 2$. The proof runs along the same lines as that of Case 2.

PROOF OF CASE 4   One can show that there is no $[\bar{P}_i, \bar{Q}_j]$-path internally disjoint from C and D for all $i, j \in \{1, \ldots, 5\}$ at all. Namely, suppose that $|P_1| \geq 2$, then by Lemmas 2.2 and 2.3 there are no $[\bar{P}_1, X]$-paths internally disjoint from C and D for all $X \in \{\bar{Q}_1, \bar{Q}_3, \bar{Q}_4, \bar{Q}_5\}$. Suppose that there is such a $[\bar{P}_1, \bar{Q}_2]$-path R splitting P_1 into $P_1' \cup P_1''$ and Q_2 into $Q_2' \cup Q_2''$, then one of the cycles $C_1 := R \cup P_1'' \cup Q_4 \cup P_3 \cup Q_1 \cup P_5 \cup Q_2''$ and $D_1 := R \cup P_1' \cup Q_5 \cup P_4 \cup Q_3 \cup P_2 \cup Q_2'$ must be longer than C, a contradiction. This shows that $G - W$ is disconnected. This finishes the proof of the case $k = 5$.   □

5. Some Conjectures

One can of course ask what is the largest number k for which Theorem 1.2(a) holds. In fact, $k = 5$ is best possible. Consider the graph G with $n \geq 8$ nodes shown in Fig. 15. G has circumference 6 and contains two different cycles of length 6 meeting in the six nodes $1, 2, \ldots, 6$ such that the removal of these six nodes leaves a connected graph.

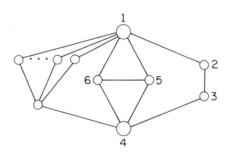

FIG. 15.

It seems, however, sensible to add to Theorem 1.2 the condition that the circumference of G should be at least $k + 1$ (for $k \leq 5$ this slightly weakens the theorem but is the only interesting case). Then the graph of Fig. 15 is no longer a counterexample. Indeed I believe that this version of Theorem 1.2(a) is also true for $k = 6$ and 7, but I have resisted any inclination to try the enumerative proof technique used above to solve these cases.

For $k = 8$ the Petersen graph P gives a counterexample. If u and v are two adjacent nodes of P, then $P-u$ (respectively $P-v$) contains a Hamiltonian cycle C (respectively D). The intersection of C and D is the eight nodes of P different from u and v. These eight nodes do not form an articulation set by construction.

I do not know of an infinite number of graphs in which two longest cycles meet in a set of eight nodes which is not an articulation set, but I know of such a class for $k = 9$. Consider Fig. 16. The graph shown therein has $n \geqslant 12$ nodes and circumference 10. The cycles C through the nodes $1, 2, \ldots, 10$ and D through 2, 3, 4, 11, 8, 9, 10, 5, 6, 7 are longest cycles of length 10 and meet in the set $W = \{2, 3, \ldots, 10\}$. $G - W$ consists of one component.

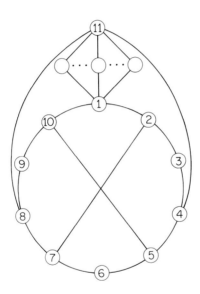

Fɪɢ. 16.

It should be clear that connectivity plays a role in the problem we consider. So one can state the questions above in more generality.

For every integer $k \geqslant 2$ let $f(k)$ denote the largest integer with the following property: If G is a k-connected (non-Hamiltonian) graph and if two longest cycles of G meet in at most $f(k)$ nodes then these nodes form an articulation set.

Let $f'(k)$ be defined as $f(k)$ above, making the additional assumption that the k-connected graph G has circumference larger than $f'(k)$. Clearly, we have $f'(k) \geq f(k)$.

Theorem 1.2(a) and the graph of Fig. 15 show that $f(2) = 5$. The Petersen graph shows that $f'(3) \leq 7$, and we conjecture that $f'(2) = f'(3) = 7$. It seems to be a very hard problem to determine the functions f and f'. I believe however that the following is true:

CONJECTURE 5.1 $f(k) \geq k$. □

In other words, if G is k-connected and two longest cycles meet in k nodes, then these nodes form an articulation set. Conjecture 5.1 – if true – would imply

CONJECTURE 5.2 *In a k-connected graph two longest cycles meet in at least k nodes.* □

I have been told by R. Häggkvist and A. Bondy that Conjecture 5.2 was first conjectured in 1979 by Scott Smith (then a high-school student), and that the truth of the conjecture probably has been verified up to $k = 10$ (up to $k = 6$ it follows from Theorem 1.2(a)). The general case is still unsettled.

Another interesting problem coming up in certain applications is to determine the intersection pattern of two longest odd cycles or a longest odd and longest even cycle. This will be investigated in a forthcoming paper.

References

1. B. Alspach, The search for long paths and cycles in vertex-transitive graphs and digraphs. In *Combinatorial Mathematics VII* (K. L. Alavi, ed.), Lecture Note in Mathematics no. 884, Springer Verlag, Berlin, 1981, pp. 14–22.
2. L. Babai, Long cycles in vertex-transitive graphs. *J. Graph Theory* **3** (1979), 301–304.
3. J. C. Bermond and C. Thomassen, Cycles in digraphs – a survey. *J. Graph Theory* **5** (1981), 1–43.
4. B. Bollobás, *Extremal Graph Theory*, Academic Press, London, 1978.
5. J. A. Bondy and L. Lovász, Cycles through specified vertices of a graph. *Combinatorica* **1** (1981), 117–140.
6. M. Grötschel, Graphs with cycles containing given paths. *Ann. Discrete Math.* **1** (1977), 233–245.
7. M. Grötschel and G. L. Nemhauser, A polynomial algorithm for the max-cut problem on graphs without long odd cycles. *Math. Programming* **29** (1984) (in press).

8. R. Häggkvist and C. Thomassen, Circuits through specified edges. *Discrete Math.* **41** (1982) 29–34.
9. D. A. Holton and M. D. Plummer, Cycles through prescribed and forbidden point sets. *Ann. Discrete Math.* **16** (1982), 129–148.
10. W. Mader, Eine Eigenschaft der Atome endlicher Graphen. *Arch. Math.* (*Basel*) **22** (1971), 333–336.
11. H. Walter and H.-J. Voß, *Über Kreise in Graphen*, VEB Deutscher Verlag der Wissenschaften, Berlin, 1974.
12. M. E. Watkins, Connectivity of transitive graphs. *J. Combinat. Theory* **8** (1970), 23–29.

16

SIMPLICIAL DECOMPOSITIONS AND TRIANGULATED GRAPHS

R. Halin

ABSTRACT   This paper surveys applications of simplicial decompositions to problems on triangulated graphs, especially infinite ones, in conjunction with perfect graphs, interval graphs and a generalization of the latter. It also contains some recent results relating to these problems.

1. Introduction

The triangulated graphs (also called rigid-circuit graphs [3] or chordal graphs [5]) form a class of graphs as important to real world applications (see [6]) as to questions regarding the fundamentals of mathematics (see [15], also, for further references). A graph is called *triangulated* if every circuit of length not less than 4 in G has a chord (or, in other words, if there is no induced cycle of length not less than 4 in G).

The purpose of this note is to show how simplicial decompositions of graphs can be used to study triangulated graphs; infinite graphs will be considered in particular. We shall be concerned with the following topics: (a) the question of perfectness of infinite triangulated graphs; (b) interval graphs; (c) the possibility of representing a triangulated graph as the intersection graph of certain subtrees of a tree.

The graphs considered in this paper are simple, i.e. they are undirected and do not contain loops or multiple edges, but they may be infinite. Instead of "complete graph" we shall say "simplex". The cliques of a graph G are the maximal (with respect to inclusion) simplices of G.

2. Simplicial decompositions of graphs

The main tool of this note is introduced in the following.

DEFINITION   Let G be a graph and let $\mathfrak{D} = (G_\lambda)_{\lambda < \sigma}$ be a family of subgraphs of G where σ is an ordinal greater than 0 and λ runs through all ordinals less than σ. Then this family $\mathfrak{D}$ is said to form a *simplicial decomposition* of G, if G is the union of the G_λ and for each τ, $0 < \tau < \sigma$,

GRAPH THEORY AND COMBINATORICS
ISBN 0-12-111760-X

$(\bigcup_{\lambda < \tau} G_\lambda) \cap G_\tau = S_\tau$ is a simplex properly contained in both $\bigcup_{\lambda < \tau} G_\lambda$ and G_τ.

The S_λ will be refered to as the *simplices of attachment* of $\mathfrak{D}$. G is called *prime* if G does not have a simplicial decomposition with at least two members (or, equivalently, if G does not have a separating simplex). A simplicial decomposition G_λ $(\lambda < \sigma)$ of G is called a *prime-graph decomposition* (PGD) of G if all its members G_λ are prime.

We then have the *decomposition theorem*:

THEOREM 1   *Every graph G without an infinite simplex has a PGD G_λ $(\lambda < \sigma)$. This decomposition can be chosen reduced (i.e. $G_\lambda \subseteq G_\tau$ implies $\lambda = \tau$); then the G_λ are exactly all the maximal prime induced subgraphs of G.*

See [9] for a proof.

If in a simplicial decomposition as above an S_τ is finite, then there is a $\lambda < \tau$ such that $S_\tau \subseteq G_\lambda$. This is in general not true if S_τ is infinite. Therefore the following definition makes sense:

DEFINITION   A simplicial decomposition $\mathfrak{D} = (G_\lambda)_{\lambda < \sigma}$ is called *strict* if for every τ, $0 < \tau < \sigma$, there is an $f(\tau) < \tau$ such that S_τ is contained in $G_{f(\tau)}$.

In the case of strictness we get a tree $T(\mathfrak{D}, f)$, associated with the simplicial decomposition $\mathfrak{D}$, whose vertices are the ordinals less than σ and whose edges are the unordered pairs $[f(\tau), \tau]$ $(0 < \tau < \sigma)$.

3. Prime-graph decompositions of triangulated graphs

It is easy to characterize the triangulated graphs as follows (see [9, p. 195]):

THEOREM 2   *A graph G is triangulated if and only if the cliques of G are exactly the maximally prime induced subgraphs of G.*

By Theorem 1 we have

THEOREM 2′   *A graph G not containing an infinite simplex is triangulated if and only if it has a PGD consisting of its cliques.*

REMARK There exists a countable triangulated graph with only one infinite clique which does not have a PGD. Further, the comparability graph of the regular tree of degree 3 with one vertex selected as a root is a triangulated graph which allows uncountably many distinct PGDs, having pairwisely no member in common. So, if the absence of an infinite simplex is not stipulated, the decomposition theorem (Theorem 1) becomes wrong, as regards both its statement of existence and its uniqueness.

4. The perfectness of triangulated graphs

The following is a restatement of a result in Berge's book [1, p. 349]:

THEOREM 3 *A graph with a simplicial decomposition having a finite number of members is perfect if and only if each of these members is perfect.*

By Theorems 3 and 2' every finite triangulated graph is perfect (this was first proved by Hajnal and Surányi [13]). From this, by the colouring theorem of de Bruijn and Erdös in connection with Lovász' well-known theorem on perfect graphs, it follows easily that every countable triangulated graph is perfect.

However, Theorem 3 becomes wrong if simplicial decompositions with infinitely many members are considered. Laver constructed an example of an uncountable triangulated graph which has a PGD containing countable simplices as its members and which is $\bar{\chi}$-perfect but not χ-perfect (see [9, p. 210]). On the other hand, Hajnal and Surányi [13] showed that every triangulated graph whose complement does not contain an infinite simplex must be perfect. The counterpart of Hajnal and Surányi's result was conjectured by Wagon [15]:

THEOREM 4 *Every triangulated graph without an infinite simplex is perfect.*

By applying Theorem 1 I have been able to prove this conjecture without difficulty [10].

5. Interval graphs

The interval graphs form an especially important class of triangulated graphs. (By an *interval graph* we understand any graph isomorphic to the intersection graph of a finite family of intervals on the real line.)

Among the known characterizations of the interval graphs that by Lekkerkerker and Boland [14] is the deepest and the most beautiful. In fact they proved two criteria for interval graphs: in their main theorem they characterize the interval graphs as those triangulated graphs which do not possess an asteroidal triple of vertices; from this result they then derive a characterization by a class of forbidden induced subgraphs. (An *asteroidal triple* in a graph G is a triple of vertices x, y, z of G such that any two of them can be connected by a path which does not meet the third vertex nor one of neighbours.)

By using the PGDs of triangulated graphs I have been able to give a new proof of Lekkerkerker and Boland's main theorem which is much simpler than the original one [11]. My proof is based on the following characterization of interval graphs, which is closely related to Fulkerson and Gross's criterion [4] (the latter is formulated in terms of matrices):

THEOREM 5   *A finite graph G is an interval graph if and only if it has a PGD $G_0, \ldots, G_k$ consisting of its cliques which is* consecutive *in the following sense: For each j, $0 < j \leq k$, the simplex of attachment S_j is contained in the preceding clique G_j.*

The latter condition can also be expressed as follows: *There is a PGD $\mathfrak{D}$ of G and an f such that $T(\mathfrak{D}, f)$ is a (finite) path.*

It may be of interest to mention the following characterization of interval graphs which seems, in this form, to be new:

THEOREM 6   *A graph G is an interval graph if and only if among any three cliques of G there is at least one which separates the two others.*

This also holds for infinite graphs G if the notion of an interval graph is extended to the infinite case in the natural way (see [11]).

6. On the representation of triangulated graphs in trees

We call a graph G *tree-representable* if there is a tree T and a family of subtrees of T such that G is isomorphic to the intersection graph of this family. If T is restricted to finite paths, then just the interval graphs are obtained; therefore the tree-representable graphs appear as a natural generalization of the interval graphs. It is easy to see that every tree-representable graph must be triangulated. In the finite case the converse is also true. So we get the following theorem, which was obtained independently by Surányi (see [**7**]), Gavril [**5**], Buneman [**2**] and Walter [**16**]:

THEOREM 7   *The finite tree-representable graphs coincide with the finite triangulated graphs.*

In the proof that every finite triangulated graph G is tree-representable we can make use of a tree structure $T(\mathfrak{D}, f)$ associated with some PGD $\mathfrak{D}$ of G. Let $\mathfrak{D}$ consist of the cliques $G_0, \ldots, G_k$ of G and f be a function as considered in Section 2. It can be shown without difficulties that for each vertex v of G the indices i with $v \in V(G_i)$ span a subtree T_v of $T(\mathfrak{D}, f)$ and that, by letting $v \to T_v$, a representation of G in $T(\mathfrak{D}, f)$ is given.

In [**12**] I considered the question of tree-representability for infinite triangulated graphs. The preceding idea of construction can be carried over to an infinite triangulated graph G, provided, however, that such a tree $T(\mathfrak{D}, f)$ exists, which means that G has a strict PGD. Vice versa it can also be shown that every tree-representable graph must have a strict PGD consisting of simplices. In the finite case the proof is based on the fact that every family $(B_i)_{i \in I}$ of subtrees of a finite tree T must have the so-called *Helly property*, i.e.

If $B_i \cap B_j \neq \varnothing$ for all pairs $i, j \in I$, then $\bigcap_{i \in I} B_i \neq \varnothing$.

This statement does not hold in the infinite case, as (for example) the family of infinite subpaths of a one-sided infinite path shows. However, we get a *modified Helly property* if we complete the underlying tree T by adding the so-called ends of T as figurative vertices and allow them to be elements of the intersections in question. (Two one-sided infinite paths in T are equivalent if they have an infinite subpath in common; to each equivalence class arising in this way there corresponds an end of T.)

Thus we may state

THEOREM 8   *A graph G is tree-representable if and only if it has a strict PGD containing simplices as its members.*

In particular, by Theorem 1, we have the following sharpening of Theorem 7:

THEOREM 8′   *A graph without an infinite simplex is tree-representable if and only if it is triangulated.*

References

1. C. Berge, *Graphes et hypergraphes*, 2nd edn, Dunod, Paris, 1973.
2. P. Buneman, A characterization of rigid circuit graphs. *Discrete Math.* **9** (1974), 205–212.
3. G. A. Dirac, On rigid circuit graphs. *Abh. Math. Sem. Univ. Hamburg* **25** (1961), 71–76.
4. D. R. Fulkerson and O. A. Gross, Incidence matrices and interval graphs. *Pacific J. Math.* **15** (1965), 835–855.
5. F. Gavril, The intersection graphs of subtrees in trees are exactly the chordal graphs. *J. Combinat. Theory B* **16** (1974), 47–56.
6. M. C. Golumbic, *Algorithmic Graph Theory and Perfect Graphs*, Academic Press, New York, 1980.
7. A. Gyárfas, and J. Lehel, A Helly type problem in trees. In *Combinatorial Theory and Applications* (P. Erdös, A. Rényi and V. T. Sós eds), Coll. Math. Soc. J. Bolyai no. 4, North Holland, Amsterdam 1970, pp. 571–584.
8. R. Halin, *Graphentheorie I*, Wiss. Buchges., Darmstadt, 1980.
9. R. Halin, *Graphentheorie II*, Wiss. Buchges., Darmstadt, 1981.
10. R. Halin, A note on infinite triangulated graphs. *Discrete Math.* **31** (1980), 325–326.
11. R. Halin, Some remarks on interval graphs. *Combinatorica* **2** (1982), 297–304.
12. R. Halin, On the representation of triangulated graphs in trees. *Europ. J. Combinat.* (to appear).
13. A. Hajnal and J. Surányi, Über die Auflösung von Graphen in vollständige Teilgraphen. *Ann. Univ. Budapestinensis* **1** (1958), 113–121.
14. C. Lekkerkerker and J. Boland, Representation of a finite graph by a set of intervals on the real line. *Fundamenta Math.* **51** (1962), 45–64.
15. S. Wagon, Infinite triangulated graphs. *Discrete Math.* **22** (1978), 183–189.
16. J. R. Walter, Representations of chordal graphs as subtrees of a tree. *J. Graph Theory* **2** (1978), 265–267.

17

INDEPENDENT CYCLES IN GRAPHS OF GIVEN MINIMUM DEGREE

A. J. Harris

ABSTRACT In this paper, we find an essentially best possible lower bound on the number of long vertex-disjoint cycles in large graphs of given minimum degree.

There are a number of results concerning the number of independent cycles in a graph whose size or degree sequence satisfy certain constraints. For an account of many of these results, see Bollobás [1, Chap. 3]. For example, Erdös and Pósa [4] (see also Bollobás [1, p. 114]) gave a sharp lower bound on the number of independent cycles in large graphs. Corrádi and Hajnal [2] showed that if G is a graph with $\delta(G) \geq 2k$ and $|G| \geq 3k$, then G contains k independent cycles. Bollobás posed the following more general question: Let $g(n, \delta, l)$ be the largest integer for which every graph of order n and minimum degree δ contains at least $g(n, \delta, l)$ independent cycles of length at least l. Can we accurately estimate $g(n, \delta, l)$? For $l = 3$, there is no restriction on the length of the cycles, and it follows from the Corrádi–Hajnal theorem that $g(n, 2k, 3) = k$ provided that $n \geq 3k$. The main aim of this paper is to give a reasonably sharp lower bound on $g(n, \delta, l)$.

We shall define a function $f(\delta, l)$ for $l \geq 3$, $\delta \geq l + 1$ and show that all sufficiently large graphs of minimum degree δ contain $f(\delta, l)$ independent cycles of length at least l, where $f = g$ for infinitely many values of δ and l, and $f = g - 1$ otherwise. The proof proceeds by successively deleting cycles from G; in order to show that this process gives the desired number of cycles, we shall prove a series of technical lemmas.

THEOREM *Let* $3 \leq l \leq \delta - 1$. *Define*

$$N(\delta, l) = 24\delta^4 + 35\delta^3 (l-2)^{\lceil l/2 \rceil - 1},$$

$$f(\delta, l) = \begin{cases} \left\lfloor \dfrac{2(\delta+1)}{l} \right\rfloor - 1 & \text{if } l \text{ is even,} \\[4mm] \left\lfloor \dfrac{2(\delta+2)}{l+1} \right\rfloor - 1 & \text{if } l \text{ is odd.} \end{cases}$$

Then $g(n, \delta, l) \geq f(\delta, l)$ *provided that* $n \geq N(\delta, l)$.

GRAPH THEORY AND COMBINATORICS
ISBN 0-12-111760-X

PROOF   Let l be fixed. We use induction on δ. Let $\mathcal{G}(\delta, n, b)$ denote the set of all graphs of order at least n such that all but at most b vertices have degree at least δ. A cycle is said to be *short* if its length is at most $l-1$, otherwise it is *long*. Furthermore, given a graph G, let $L(G)$ denote the length of the shortest long cycle in G. Notice that the theorem is trivially true for $l-1 \leqslant \delta \leqslant \lceil 3l/2 \rceil - 2$, since in this case, we merely require one cycle of length at least l. The theorem of Corrádi and Hajnal [2] mentioned above implies the assertion for $l = 3$. We therefore suppose that $l \geqslant 4$.

In our proof we may assume that the vertices of degree at least $\delta + 1$ form an independent set. We shall refer to vertices of degree less than δ as *bad* vertices, and the rest as *good* vertices.

Our first lemma is of some interest by itself.

LEMMA 1   *Suppose that* $l \geqslant 4$, $b \geqslant 0$ *and*

$$n \geqslant 5b(l-2)^{\lceil l/2 \rceil - 1},$$

Then every graph of order n with at most b vertices of degree less than $l-1$ has circumference at least l.

PROOF   Suppose that G is a smallest graph forming a minimal counter-example. Then G is clearly connected. Also, $\delta(G) \geqslant 2$, since otherwise the deletion of a vertex of degree 1 would give a smaller counter-example.

We begin by proving that $b \geqslant 3$. Suppose that $b \leqslant 2$. Let x be a vertex of minimum degree and let P be a longest path containing x as an interior vertex. Since there is at most one other vertex of degree less than $l-1$, at least one end-vertex of P, say y, has degree at least $l-1$. Then an edge incident with y and a segment of P give a cycle of length at least l. Hence $b \geqslant 3$.

Next we show that G contains a cutvertex. Suppose G is 2-connected. By a theorem of Erdös and Gallai [3] (see Bollobás [1, p. 137]), the absence of a cycle of length at least l implies that G has at most $(n-1)(l-1)/2$ edges, i.e. has total degree at most $(n-1)(l-1)$. The proof proceeds by finding a lower bound on the diameter of G. The total degree of the good vertices is at least $(n-b)(l-1)$, and of the bad vertices at least $2b$. Hence, not many of the good vertices can have degrees much greater than $(l-1)$. More precisely, if, for every vertex x, we define the *surplus degree* $\sigma(x)$ to be $\max\{\delta(x)-(l-1), 0\}$, then

$$\sum_{x \in G} \sigma(x) \leqslant (l-1)(b-1) - 2b$$

$$\leqslant (l-3)(b-1)$$

It is easily seen that, for a given diameter d, the number of vertices such that G satisfies this constraint is maximized if one vertex has surplus degree $(l-3)(b-1)$, i.e. has degree $\Delta = (l-3)(b-1)+(l-1)$, and the rest have surplus degree zero. Hence:

$$n \leq 1 + \Delta \sum_{i=0}^{d-1} (l-2)^i$$
$$= 1 + [(b-1)(l-3)+(l-1)][(l-2)^d - 1]/(l-3)$$
$$= 1 + \left(b + \frac{2}{l-3}\right)[(l-2)^d - 1]$$
$$< (b+2)(l-2)^d - 1.$$

So, if $n \geq (b+2)(l-2)^{\lceil l/2 \rceil - 1}$, the diameter of G is at least $\lceil l/2 \rceil$. Pick two vertices x and y at this distance. Since G is supposed to be 2-connected, x and y must lie on a cycle of length at least l, contradicting our assumption. Hence, G contains a cutvertex.

Consider an end-block B of G, and let v be the cutvertex of B. Suppose that B contains b' bad vertices of G. Since v may have fewer than $l-1$ neighbours in B, B considered as a graph in its own right contains at most $b'+1$ bad vertices and, since it itself contains no cycle of length at least l, at least 3. Hence, $b' \geq 2$ and B satisfies

$$|B| < (b'+3)(l-2)^{\lceil l/2 \rceil - 1}.$$

So

$$|(G \backslash B) \cup \{v\}| \geq (5b - b' - 3)(l-2)^{\lceil l/2 \rceil - 1}$$
$$\geq 5(b - b' + 1)(l-2)^{\lceil l/2 \rceil - 1}.$$

But $(G \backslash B) \cup \{v\}$ contains $\leq b - b' + 1$ bad vertices. So since, by our assumption, G is a minimal counterexample, $(G \backslash B) \cup \{v\}$ must itself contain a cycle of length at least l, contrary to our assertion. This proves the lemma. $\square$

Note that if the constraint on l in Lemma 1 is omitted then, as shown by a path, the assertion is false.

LEMMA 2   *Suppose that $G \in \mathcal{G}(\delta, n, b)$ and contains an l-cycle C. Then the graph $G' = G \backslash C$ belongs to the set*

$$\mathcal{G}\left(\delta - \left\lceil \frac{l}{2} \right\rceil, n - l, b + \left\lfloor \frac{\delta^2 l}{2} \right\rfloor\right).$$

PROOF   Suppose l is even. Let S be the set of good vertices not lying on C having at most $\delta - l/2 - 1$ neighbours in $G \backslash C$. There are at most $l/2$

vertices of degree greater than δ on C, since such vertices form an independent set. Each of these vertices has at most $|S|$ neighbours in S, whereas each of the vertices of degree δ has at most $(\delta-2)$ neighbours in S. The number of edges e joining S to C thus satisfies

$$e \leqslant \frac{l}{2}(\delta-2)+\frac{l}{2}|S|.$$

However, since each vertex of S has degree at least δ, and has at most $\delta-l/2-1$ neighbours in $G\backslash C$, it has at least $l/2+1$ neighbours on C. Hence e satisfies

$$e \geqslant \left(\frac{l}{2}+1\right)|S|.$$

Thus,

$$\frac{l}{2}(\delta-2)+\frac{l}{2}|S| \geqslant \left(\frac{l}{2}+1\right)|S|;$$

i.e.

$$|S| \leqslant \frac{l}{2}(\delta-2).$$

However, since each vertex of S has at most $\delta-l/2-1$ neighbours in $G\backslash C$, the number of vertices of degree less than $\delta-l/2$ in $G\backslash C$ is at most

$$|S|\left(\delta-\frac{l}{2}-1\right)+|S|+b \leqslant \frac{l}{2}(\delta-2)\left(\delta-\frac{l}{2}-1\right)+\frac{l}{2}(\delta-2)+b$$

$$=\frac{l}{4}(2\delta^2-4\delta-\delta l+2l)+b$$

$$\leqslant \left\lfloor\frac{\delta^2 l}{2}\right\rfloor+b$$

So

$$G\backslash C \in \mathscr{G}\left(\delta-\frac{l}{2}, n-l, b+\left\lfloor\frac{\delta^2 l}{2}\right\rfloor\right)$$

as required.

Suppose that l is odd. This case is entirely analogous to the case for l even. Let S be the set of good vertices not lying on C having at most $\delta-(l+1)/2$ neighbours in $G\backslash C$. On C there are at most $(l-1)/2$ vertices of degree greater than δ, and letting e be the number of edges joining S to C, we see that

$$e \leqslant \left(\frac{l+1}{2}\right)(\delta-2)+\left(\frac{l-1}{2}\right)|S|.$$

Also

$$e \geq \left(\frac{l+1}{2}\right)|S|.$$

Whence

$$|S| \leq \tfrac{1}{2}(l\delta + \delta - 2l - 2).$$

Hence, the number of vertices of degree less than $\delta - (l-1)/2$ in $G \backslash C$ is at most

$$|S|\left(\delta - \frac{l-1}{2}\right) + |S| + b \leq \tfrac{1}{4}(l\delta + \delta - 2l - 2)(2\delta - l + 1) + b$$

$$\leq \left\lfloor \frac{\delta^2 l}{2}\right\rfloor + b$$

So

$$G\backslash C \in \mathcal{G}\left(\delta - \frac{l-1}{2}, n - l, b + \left\lfloor \frac{\delta^2 l}{2}\right\rfloor\right)$$

as required. □

LEMMA 3   Suppose that $G \in \mathcal{G}(\delta, n, b)$ contains a cycle of length greater than l, but no cycle of length precisely l. Let $G' = G \backslash C$. Let C be a cycle of length $L(G)$. Then

$$G\backslash C \in \mathcal{G}\left(\delta - \left\lfloor \frac{l+1}{2}\right\rfloor, n - L(G), b\right).$$

PROOF   Suppose that there is a vertex $u \in G\backslash C$ with at least $\lceil l/2 \rceil + 1$ neighbours on C. We show that there exists a shorter long cycle of length at least l through u.

By deleting some edges connecting u to C, we may suppose that u has precisely $\lceil l/2 \rceil + 1$ neighbours on C. If C is of length greater than $l+1$, successively contract edges of C which do not have both ends adjacent with u, until a cycle C' of length precisely $l + 1$ is obtained. Denote the vertices of C' by $x_1, x_2, \ldots, x_{l+1}$. It is now sufficient to show that there exist two neighbours of u at distance precisely three on C', since if x_i and x_{i+3} are neighbours of u, then $x_1 x_2 \ldots x_i u x_{i+3} \ldots x_{l+1} x_1$ is an l-cycle, and by expanding the edges previously contracted, we find a long cycle which is shorter than C.

Consider each of the $\lceil l/2 \rceil + 1$ neighbours of u on C'. Call each vertex at distance three on C from any of these vertices *forbidden*. Allowing multiplicity, $2\lceil l/2 \rceil + 2$ vertices of C' are thus called *forbidden*. However, each vertex of C' can have been called forbidden at most twice (as a result of the two vertices at distance three on C' being neighbours of u).

Hence, at least $\lceil l/2 \rceil + 1$ are forbidden. But there are only $\lfloor l/2 \rfloor$ vertices on C' which are not adjacent to u, so some forbidden vertex is adjacent to u. Hence, there exist two vertices at distance three apart on C', both of which are adjacent to u, and we may shorten the cycle C. This contradicts our definition of C as a shortest long cycle in G, proving the lemma. $\square$

This lemma shows that if $G \in \mathcal{G}(\delta, n, b)$, and $L(G) > l$, then deleting a cycle of length $L(G)$ gives a graph in $\mathcal{G}(\delta - \lfloor (l + 1)/2 \rfloor, n - L(G), b)$. This is sufficient to prove the result by induction on δ provided that $L(G)$ is small compared to n. Before we can complete the proof of the theorem, it remains to find a reasonable upper bound for $L(G)$. In fact, we show that if $G \in \mathcal{G}(\delta, n, b)$, and G contains a long cycle of length at least l, then

$$L(G) \leqslant \max\{l^2, [4l \log(n)/\log(2\delta - 2l + 1)] + 2l(b + 1)\}. \tag{1}$$

Since inequality (1) holds if $L(G) \leqslant l^2$, in the proof of the next six lemmas we may and shall assume that $L(G) \geqslant l^2$.

Define a *local block* B of G to be one of the following: (i) a maximal subgraph such that any two vertices lie on a short cycle; (ii) two vertices joined by an edge not lying on any short cycle; (iii) an isolated vertex. Thus, the set of all local blocks covers G.

Our proof of (1) is based on some properties of the local blocks.

LEMMA 4   *No local block contains a long cycle.*

PROOF   Suppose that B is a local block containing a long cycle. Let C be a shortest long cycle in B, and choose two diametrically opposite vertices u and v of C. (If C is of odd order, we let u and v be almost opposite.) For each vertex x on C, define the function $d_C(x)$ to be the distance of x from u on C.

In B, there exists a path $P = uu_1u_2 \ldots u_{k-1}v$ of length $k < l/2$ from u to v (since u and v lie on a short cycle). Suppose that P respectively meets C in the vertices $u = u_{i_0}, u_{i_1}, \ldots, u_{i_s} = v$.

We show that $d_C(u_{i_j}) < jl$.

Suppose not. Then let u_{i_t} be the first of these vertices (other than $u = u_{i_0}$) such that $d_C(u_{i_t}) \geqslant tl$. Clearly, u_{i_t} and $u_{i_{t-1}}$ are not adjacent on C else $u_{i_{t-1}}$ would do. However, then $u_{i_{t-1}} \text{ - - - } u_{i_t} \ldots u_{i_{t-1}}$ is a cycle of length at least l, but shorter than C. (Here, we let "- - -" denote part of the path P, and "$\ldots$" denote the shorter arc of C from $u_{i_{t-1}}$ to u_{i_t}.) This contradicts the minimality of C as a long cycle in B. So $d_C(u_{i_t}) \geqslant jl$ and we have

$$\left\lfloor \frac{l^2}{2} \right\rfloor \leqslant d_C(v) = d_C(u_{i_s}) < sl \leqslant kl \leqslant \left\lfloor \frac{l^2}{2} \right\rfloor,$$

which is a contradiction, proving the lemma. □

LEMMA 5   *No local block contains a path of length* $\lfloor l^2/2 \rfloor$.

PROOF   Suppose that $P = x_0 x_1 \ldots x_r$ is a maximal path in some local block B. For a vertex x on P, define $d_P(x)$ to be the distance from x_0 to x on P. Let $x_0 u_1 u_2 \ldots u_{k-1} x_r$ be a path of length $k < l/2$ from x_0 to x_r in B, and let the common vertices be $x_0, u_{i_1}, u_{i_2}, \ldots, u_{i_{t-1}}, x_r$. By Lemma 4, every cycle in B is short. Hence, it is clear that $d_P(u_{i_t}) < tl$, and that

$$d(x_r) < kl \leqslant \left\lfloor \frac{l^2}{2} \right\rfloor. \quad \square$$

LEMMA 6   *Each pair of local blocks meets in at most one vertex.*

PROOF   Suppose that there exist two local blocks B and B' having two vertices u and v in common. Neither of B and B' is properly contained in the other, or else one of them would not be maximal. Pick a vertex $z \in B - B'$. Then there exists a vertex $z' \in B' - B$ such that z and z' do not both lie on a short cycle (since otherwise B would not be maximal). Since z, u and v lie in B and B is 2-connected, there exists a fan from z to u and v. By Lemma 5, the total length of the fan is less than $\lfloor l^2/2 \rfloor$. We similarly find a fan from z' to u and v with total length less than $\lfloor l^2/2 \rfloor$. The union of these fans contains a cycle through z and z'. This cycle has length less than l^2, so, by hypothesis, it is a short cycle. This contradicts the fact that z and z' belong to distinct local blocks. □

If two local blocks have non-empty intersection (which, by Lemma 6, is a single vertex), then the blocks are said to be *incident* with one another. The common vertex is said to be a *local cutvertex*.

The *local block-cutvertex* graph of G, denoted lbc(G), is defined to be the bipartite graph whose vertices are the local blocks and local cut-vertices of G, two vertices being joined if the local block contains the local cutvertex.

The above lemmas concerning local blocks enable us to use the local block-cutvertex graph to estimate $L(G)$.

LEMMA 7   *Let* $g(\text{lbc}(G))$ *denote the girth of* lbc(G). *Then*

$$L(G) < \frac{\lg(\text{lbc}(G))}{4}.$$

PROOF   Pick a shortest cycle $C = x_1 x_2 \ldots x_g x_1$ in lbc(G). Since lbc(G) is bipartite, g is even. Suppose that x_1 and x_3 are two vertices of C which are local cutvertices of G. Then, in G, these vertices lie on a cycle of length less than l in their common local block, and hence are at distance less than $l/2$ apart in this block. By continuing this argument for all pairs of vertices x_{2i-1}, x_{2i+1} (mod($g + 1$)) ($i = 1, 2, \ldots, g/2$), we find a cycle in G corresponding to C of length $k < lg/4$.

It remains to show that $k \geqslant l$. Observe that if $k < l$, then every vertex on this cycle in G lies in a short cycle, and hence is contained in some local block, contradicting the fact that the cycle passes through several local blocks. This proves the lemma.   □

We next examine the structure of lbc(G) in more detail to find an upper bound on its girth. For each local block of G containing a bad vertex, we call this vertex in lbc(G) *bad*. Similarly, for each bad local cutvertex, we call the vertex in lbc(G) *bad*. All other vertices of lbc(G) are called *good*.

LEMMA 8   *Let $G \in \mathcal{G}(\delta, n, b)$, then given a good vertex $v \in$ lbc(G) corresponding to a local block B_v in G, either there are at least $2(\delta - l + 1)$ vertices at distance at most 4 from v in lbc(G), or in lbc(G) there is a bad vertex at distance at most 4 from v.*

PROOF   Suppose that in lbc(G), neither v nor any other vertex at distance at most 4 is bad. Let $|B_v| = i$. We split the proof into two cases according to whether i is larger or smaller than δ.

Case 1: $i > \delta$ Since B_v itself contains no cycle of length at least l, then by a theorem of Erdös and Gallai [3] (see also Bollobás [1, p. 137]), the number of edges in B_v satisfies $e(B_v) \leqslant (l - 1)(i - 1)/2$. But $\delta(G) = \delta$, so every vertex which is not a local cutvertex has degree at least δ in B_v. By counting degrees, we see that the number of vertices in B_v which are not local cutvertices is at most $(l - 1)(i - 1)/\delta \leqslant li/\delta - 1$. Therefore, the number of local cutvertices in B_v is at least $(1 - l/\delta)i + 1$. So the number of vertices in lbc(G) adjacent to v is at least $(1 - l/\delta)i + 1$. Since each local cutvertex is, by definition, adjacent to another local block, there are at least $(1 - l/\delta)i + 1$ vertices corresponding to local blocks at distance 2 from v. So, in lbc(G), even the number of vertices within distance 2 from v is at least $2(1 - l/\delta)i + 2 \geqslant 2(\delta - l + 1)$.

Case 2: $1 \leqslant i \leqslant \delta$ In this case, since $\delta(G) = \delta$, and B_v contains at most δ vertices, every vertex in B_v is a local cutvertex with at least $\delta - i + 1$ neighbours in other local blocks. Suppose that there exists a vertex in B_v incident with a local block B of size greater than δ. Then, by the reasoning of Case 1, the vertex B of $\mathrm{lbc}(G)$ has at least $2(\delta - l + 1)$ vertices within distance 2. So in $\mathrm{lbc}(G)$, there are certainly at least $2(\delta - l + 1)$ vertices within distance 4 from v. If there is no such block B, then all the blocks incident with B_v are themselves of size at most δ. All the vertices of the neighbouring local blocks must themselves be local cutvertices. There are at least $i(\delta - i + 1)$ such vertices, and since $L(G) \geqslant l^2$, it is clear that all these other local blocks are distinct. Hence, in $\mathrm{lbc}(G)$ there are at least $i(\delta - i + 1)$ vertices at distance 3 from v, and at least $i(\delta - i + 1)$ vertices at distance 4. So the total number of vertices at distance at most 4 is at least $2i(\delta - i + 1) \geqslant 2\delta > 2(\delta - l + 1)$ and the lemma is proved. $\square$

LEMMA 9 $L(G) \leqslant [4l \log(n)/\log(2\delta - 2l + 1)] + 2l(b + 1)$.

PROOF   By assumption, G contains a long cycle, so let G_0 be a component of G containing a long cycle. We suppose that $|G_0| = n_0$, and that $|\mathrm{lbc}(G_0)| = m_0$. Let v be a vertex of $\mathrm{lbc}(G_0)$ corresponding to a local block in G_0. If v and all vertices at distance at most 4 from v are good, then, by Lemma 8, there are at least $2(\delta - l + 1)$ vertices at distance at most 4. Recall that g is the girth of $\mathrm{lbc}(G)$. It is clear by convexity that the minimum number of vertices at distance $\lfloor (g - 1)/8 \rfloor$ from v in $\mathrm{lbc}(G_0)$ is found when all the bad vertices are close to v, thereby minimizing the number of vertices at any specified distance from v. Since each edge of G lies in precisely one local block,

$$n_0^2 \geqslant e(G_0) + |G_0| \geqslant m_0.$$

So, by Lemma 8, if $d = \lfloor (g - 1)/8 \rfloor$, then certainly

$$n^2 \geqslant n_0^2 \geqslant m_0 \geqslant 1 + 2(\delta - l + 1) \sum_{j=0}^{d-b-1} [(2\delta - 2l) + 1]^j$$

$$\geqslant 1 + 2(\delta - l + 1)(2\delta - 2l + 1)^{d-b-1}$$

$$\geqslant (2\delta - 2l + 1)^{d-b}.$$

So

$$2 \log(n) \geqslant \left(\frac{g}{8} - b - 1 \right) \log(2\delta - 2l + 1)$$

$$\geqslant \left(\frac{L(G)}{2l} - b - 1 \right) \log(2\delta - 2l - 1),$$

where the second inequality follows from Lemma 7. So

$$L(G) \leqslant [4l \log(n)/\log(2\delta - 2l + 1)] + 2l(b + 1)$$

and the lemma is proved. $\square$

It follows from Lemmas 3–9 that, given $l \geqslant 4$ and a graph $G \in \mathcal{G}(\delta, n, b)$ containing a long cycle, we have

$$L(G) \leqslant \max\{l^2, [4l \log(n)/\log(2\delta - 2l + 1)] + 2l(b + 1)\}.$$

The graph obtained from G by deleting a cycle of length $L(G)$ belongs to the set

$$\mathcal{G}\left(\delta - \left\lfloor \frac{l+1}{2} \right\rfloor, n - L(G), b\right).$$

Subject to finding some lower bound on n, we are now able to prove the main theorem.

Define δ_i as follows:

$$\delta_i = \delta - i \left\lfloor \frac{l+1}{2} \right\rfloor.$$

Given a graph $G_0 \in \mathcal{G}(\delta, n, 0)$, we successively define a sequence of graphs $G_1, G_2, \ldots, G_k$, where G_i is obtained by deleting a shortest long cycle from G_{i-1}, and then deleting one by one edges joining vertices of degree greater than δ_i. We show that if $n \geqslant N(\delta, l)$ then $k \geqslant f(\delta, l)$.

Define $\Lambda_0 = 0$, and for $1 \leqslant i \leqslant f(\delta, l) - 1$, define Λ_i by

$$\Lambda_i = \Lambda_{i-1} + \max\left\{l^2, \left\lfloor 4l \log(n)/\log(2\delta_{i-1} - 2l + 1) + 2l\left(\frac{(i-1)\delta^2 l}{2} + 1\right)\right\rfloor\right\}.$$

It is then clear by induction on $L(G_i)$ that, for $1 \leqslant i \leqslant f(\delta, l) - 1$,

$$G_i \in \mathcal{G}\left(\delta_i, n - \Lambda_i, i\left\lfloor \frac{\delta^2 l}{2} \right\rfloor\right). \tag{2}$$

We shall show that if $n \geqslant N(\delta, l)$, then $i \leqslant f(\delta, l) - 1$ implies that G_i contains a long cycle.

For $i \leqslant f(\delta, l) - 1$ and $l \geqslant 4$ it is easily checked that $\delta_{i-1} \geqslant \lceil 5l/4 \rceil$, so

$$2\delta_{i-1} - 2l + 1 \geqslant 3,$$

$$L_i \leqslant \Lambda_{i-1} + \max\left\{l^2, 4l \log(n)/\log(3) + 2l\left(\left(\frac{i\delta^2 l}{2}\right) + 1\right)\right\}$$

$$\leqslant \Lambda_{i-1} + 4l \log(n)/\log(3) + 2\delta^3 l.$$

But $\Lambda_0 = 0$ and $i \leqslant f(\delta, l) - 1 < 2\delta/l$, so certainly

$$L_i \leqslant 8\delta \log(n)/\log(3) + 4\delta^4. \tag{3}$$

We are now in a position to prove that $k \geqslant f(\delta, l)$. It is required to show that provided that n is large, for all $1 \leqslant i \leqslant f(\delta, l) - 1$, the graph G_i

contains a long cycle. By (3),

$$|G_i| \geq n - 8\delta \log(n)/\log(3) - 4\delta^4.$$

Also, by (2), the number of bad vertices in G_i is at most δ^3 and $\delta_i \geq l - 1$, so if

$$n - 8\delta \log(n)/\log(3) - 4\delta^4 \geq 5\delta^3(l-2)^{\lceil l/2 \rceil - 1}$$

then, by Lemma 1, G_i contains the required cycle. For this to be satisfied, it is clearly sufficient that

$$n - 7.3\delta\sqrt{n} - 4\delta^4 \geq 5\delta^3(l-2)^{\lceil l/2 \rceil - 1}.$$

Very crude calculations show that

$$n \geq 24\delta^4 + 35\delta^3(l-2)^{\lceil l/2 \rceil - 1}$$

will do. This completes the proof of the theorem. □

Interestingly, the above result is very close to being best possible. Pick $l \geq 3$ and $\delta \geq l$. Let G^δ denote an arbitrary graph on δ vertices. Then if $n \geq 2\delta$, the graph

$$G = G^\delta + E^{n-\delta}$$

is a graph on n vertices with minimum degree δ which contains only $\lfloor 2\delta/l \rfloor$ long cycles if l is even and $\lfloor 2\delta(l+1) \rfloor$ long cycles if l is odd. This shows that

$$f(\delta, l) \geq g(n, \delta, l) - 1$$

provided that $n \geq 2\delta$. Thus, for $n \geq N(\delta, l)$, the result is at most one short of best possible. Furthermore, if $n \geq N(\delta, l)$ and either l is even and δ is of the form $kl/2 - 1$, or l is odd and δ is of the form $k(l+1)/2 - 1$ or $k(l+1)/2 - 2$, then $f = g$.

References

1. B. Bollobás, *Extremal Graph Theory*, Academic Press, London, 1978.
2. K. Corrádi and A. Hajnal, On the maximal number of independent circuits in a graph. *Acta Math. Acad. Sci. Hungar.* **14** (1963), 423–439.
3. P. Erdös and T. Gallai, On maximal paths and circuits of graphs. *Acta Math. Acad. Sci. Hungar.* **10** (1959), 337–356.
4. P. Erdös and L. Pósa, On the number of disjoint circuits of a graph. *Publ. Math. Debrecen* **9** (1962), 3–12.

18

CONNECTED DOMINATION IN GRAPHS

S. T. Hedetniemi and Renu Laskar

ABSTRACT   A set D of vertices in a graph $G = (V, E)$ is a connected dominating set if every vertex in $V - D$ is adjacent to at least one vertex in D and the subgraph $\langle D \rangle$ induced by the vertices in D is connected. The connected domination number $\gamma_c(G)$ is the minimum number of vertices in a connected dominating set in G. In this paper we extend the original work on connected domination by Sampathkumar and Walikar by establishing several new upper and lower bounds for $\gamma_c(G)$ and by establishing a Nieminen-type identity for connected domination of the form $\gamma_c + \varepsilon_T = p$, where p denotes the number of vertices in G and ε_T denotes the maximum number of end-vertices in a spanning tree of G.

We also show that the problem of determining $\gamma_c(G)$ for an arbitrary graph G is NP-complete.

1. Introduction

A *dominating set* in a graph $G = (V, E)$ is a set D of vertices such that every vertex in $V - D$ is adjacent to at least one vertex in D. The domination number of a graph G, denoted $\gamma(G)$, is the minimum number of vertices in a dominating set.

Although notions of dominating sets of queens on chessboards date back at least to the 1800s (cf. [3]), the modern study of domination can be attributed initially to such people as Berge [4], Ore [14], Vizing [16] and Liu [12]. For a survey of results on domination see Cockayne and Hedetniemi [7], Cockayne [8] or Laskar and Walikar [11]. More recently the notions of total domination [1, 6] and connected domination [15] have been introduced. A dominating set D is a *total dominating set* if the subgraph $\langle D \rangle$ induced by D has no isolates. A dominating set D is a *connected dominating set* if $\langle D \rangle$ is connected. The *total (connected) domination number*, denoted by $\gamma_t(G)$ ($\gamma_c(G)$), is the minimum number of vertices in a total (connected) dominating set.

For a connected graph G, the following inequalities are self-evident:

$$\gamma \leqslant \gamma_t \leqslant \gamma_c \quad \text{for connected graphs } G \text{ with } \Delta(G) < p - 1, |V(G)| = p. \tag{1}$$

The inequalities in (1) suggest several avenues for exploration. In

GRAPH THEORY AND COMBINATORICS
ISBN 0-12-111760-X

particular, for any known bounds of the form $\gamma \leq \alpha$ or $\gamma_t \leq \alpha$ it is possible that in fact $\gamma_c \leq \alpha$, and conversely, any new result of the form $\gamma_c \leq \beta$ automatically provides a new upper bound for γ and γ_t.

In this paper we review the preliminary results of Sampathkumar and Walikar [15] on connected domination and then provide several extensions of their results along with several new results of the type mentioned above.

2. Known results on connected domination

In the first paper published on connected domination, Sampathkumar and Walikar [15] presented the following results:

(a) Let e denote the number of end-vertices in a tree with $p > 2$ vertices. Then $\gamma_c(T) = p - e$.

(b) Let H be a connected spanning subgraph of a graph G. Then $\gamma_c(G) \leq \gamma_c(H)$.

(c) For any connected graph G with $|V(G)| \geq 3$, $\gamma_c(G) \leq p - 2$.

(d) Let G be a connected graph with p vertices, q edges and maximum degree Δ, then

$$p/(\Delta + 1) \leq \gamma_c(G) \leq 2q - p.$$

(e) $\gamma_c(G) = p/(\Delta + 1)$ if and only if $\Delta = p - 1$; i.e. $\gamma_c(G) = 1$.

(f) $\gamma_c(G) = 2q - p$ if and only if G is a path.

(g) $\gamma_c(G) + \gamma_c(\bar{G}) \leq p(p - 3)$, where $\bar{G}$ is the complement of G.

(h) For both G and $\bar{G}$ connected $\gamma_c(G) + \gamma_c(\bar{G}) = p(p - 3)$ if and only if $G = P_3$.

In the next section we obtain a variety of new results on connected domination including an extension of (a) and improvements of (g) and the upper bound in (d) above.

3. New results

The literature on domination in graphs contains many results which relate γ to other parameters of a graph. In this section we obtain a series of similar results for γ_c.

Let ε_F denote the maximum number of pendant edges (an edge containing a vertex of degree 1) in any spanning forest of G. In 1974 Nieminen [13] demonstrated that a close relationship exists between ε_F and the domination number γ.

THEOREM 1 (Nieminen [13]) *For any connected graph G with p vertices,*

$$\gamma + \varepsilon_F = p. \quad \square$$

A similar equality holds for connected domination. Let ε_T denote the maximum number of end-vertices (equivalently, pendant edges) in a spanning tree of a connected graph G.

THEOREM 2 *For any connected graph G with p vertices,*

$$\gamma_c + \varepsilon_T = p.$$

PROOF Let T be a spanning tree of G with ε_T end-vertices and let F denote the set of end-vertices. Then $T - F$ is a connected dominating set, having $p - \varepsilon_T$ vertices, i.e. $\gamma_c \leq p - \varepsilon_T$.

In order to show that $\gamma_c \geq p - \varepsilon_T$ let D be a connected dominating set with γ_c vertices. Since $\langle D \rangle$ is a connected subgraph of G, let T_D be any spanning tree of $\langle D \rangle$. We can now form a spanning tree T of G by adding the remaining $p - \gamma_c$ vertices of G to T_D, joining each of these vertices to one vertex of D to which it is adjacent. In this way T will have at least $p - \gamma_c$ end-vertices; i.e. $\varepsilon_T \geq p - \gamma_c$ or $\gamma_c \geq p - \varepsilon_T$. $\quad \square$

COROLLARY 2a (Sampathkumar and Walikar [15]) *For any tree T with p vertices and e end-vertices,* $\gamma_c(T) = p - e$. $\quad \square$

COROLLARY 2b *The problem of determining $\gamma_c(G)$ for an arbitrary connected graph G is NP-complete.*

PROOF In [9] it is shown that the problem of determining ε_T for an arbitrary connected graph is NP-complete. Thus, from the fact that $\gamma_c + \varepsilon_T = p$, it must follow that the problem of determining γ_c is also NP-complete. $\quad \square$

The domination and total domination numbers have previously been related to $\Delta(G)$, the maximum degree of a vertex in G. A similar result is easy to establish for connected domination which "improves" these earlier results.

THEOREM 3 *For any connected graph G with p vertices, q edges and maximum degree Δ,*

(i) $p - q \leq \gamma \leq p - \Delta$ (Berge [4]);
(ii) $\gamma_t \leq p - \Delta + 1$ (Cockayne, Dawes and Hedetniemi [6]);
(iii) $\gamma_c \leq p - \Delta$.

PROOF OF (iii)   Let v be a vertex in G of maximum degree Δ. Then a spanning tree T of G can be formed in which v is adjacent to each of its neighbors in G, i.e. T has a vertex of degree Δ and hence has at least Δ end-vertices. Thus, by Theorem 2, $\gamma_c \leq p - \Delta$.   □

As observed by Sampathkumar and Walikar [15], the connected domination number of a tree T is simply the number of vertices which are not end-vertices of T. Based on this observation we can determine the trees for which γ_c achieves the upper bound in Theorem 3.   □

PROPOSITION 4   *For any tree* T, $\gamma_c = p - \Delta$ *if and only if* T *is a spider* (*a tree having at most one vertex of degree not less than* 3).

PROPOSITION 5   *For any connected graph* G, *let* diam(G) *denote the diameter of* G. *Then*

$$\text{diam}(G) - 1 \leq \gamma_c.$$

PROOF   Let diam(G) $= k$ and let u and v be two vertices such that the distance from u to v is k. Let D be a connected dominating set and consider whether u and/or v are in D.
   Case 1 $u, v \in D$; then since $\langle D \rangle$ is a connected graph, $|D| \geq k + 1$.
   Case 2 $u \in D$, $v \notin D$; then since v must be adjacent to at least one vertex in D, it follows that $|D| \geq k$.
   Case 3 $u \notin D$, $v \notin D$. In this case it is easy to see that $|D| \geq k - 1$.   □

Let $\beta_1(G)$ denote the maximum number of edges in a set S such that no two edges in S have a vertex in common; β_1 is called the matching number of G.
   It is easy to see that for any connected graph G, $\gamma(G) \leq 2\beta_1$, i.e. the set of vertices contained in any maximum matching forms a dominating set of G. This result can be strengthened, however, by selecting a maximum matching the vertices of which form a connected subgraph.

THEOREM 6   *Every connected graph* G *contains at least one* β_1-*set* M *such that* $\langle V(M) \rangle$ *is a connected subgraph.*

PROOF   Let M be a maximum matching of G that maximizes the cardinality of a largest connected component C of $\langle V(M) \rangle$. Suppose $\langle V(M) \rangle$ is not connected; if T is a spanning tree of G there exists another connected component of $\langle V(M) \rangle$, say C', which is connected to C by a pair of T-edges, vu and uv' with $v \in C$, $v' \in C'$. We have $u \notin V(M)$.

Replacing in M the edge containing v' by the edge uv', we obtain a matching M'. $\langle C \cup \{u, v'\}\rangle$ is connected, hence $\langle V(M')\rangle$ has a connected component larger than C: Contradiction with our choice of M and C. $\square$

COROLLARY 6   *For any connected graph* G, $\gamma_c \leqslant 2\beta_1$. $\square$

There are a large number of results in the graph theory literature of the form

$$\alpha + \bar{\alpha} \leqslant p \pm \varepsilon.$$

Results of this form have previously been obtained for γ and γ_t, and can easily be extended to include γ_c.

THEOREM 7   *For any graph* G *such that* G *and* $\bar{G}$ *are connected,*

(i) $\gamma + \bar{\gamma} \leqslant p + 1$. (Jaeger and Payan [10]);
(ii) $\gamma_t + \bar{\gamma}_t \leqslant p + 2$ *for graphs with at least two vertices, with equality if and only if* G *or* $\bar{G} = mK_2$ (Cockayne, Dawes and Hedetniemi [6]);
(iii) $\gamma_c + \bar{\gamma}_c \leqslant p + 1$.

PROOF OF (iii)   From Proposition 4,

$$\gamma_c \leqslant p - \Delta \quad \text{and} \quad \bar{\gamma}_c \leqslant p - \bar{\Delta}.$$

Therefore,

$$\gamma_c + \bar{\gamma}_c \leqslant (p - \Delta) + (p - \bar{\Delta})$$
$$= 2p - (\Delta + \bar{\Delta})$$
$$= 2p - (\Delta + p - 1 - \delta), \quad \text{where } \delta \text{ is the minimum degree in } G,$$
$$= p + 1 + \delta - \Delta$$
$$\leqslant p + 1,$$

since $\delta - \Delta \leqslant 0$. $\square$

Notice that for the cycle C_5 of length 5, if $G = C_5$ then $\bar{G} = C_5$ and $\gamma_c(G) + \gamma_c(\bar{G}) = 3 + 3 = 6 = p + 1$. Thus the upper bound in Theorem 7 (iii) is best possible. However, for nearly all trees a slight improvement is possible.

COROLLARY 7   *For any tree* T *with* $p > 2$ *vertices,* $T \neq K_{1,n}$, $\gamma_c(T) + \gamma_c(\bar{T}) \leqslant p$.

PROOF   Every nontrivial tree has at least two end-vertices, say u and v.

But u and v form a connected dominating set for $\bar{T}$. Thus $\gamma(\bar{T}) = 2$. But since for every tree T, $\gamma_c(T) \leq p - 2$, the result follows. $\square$

In [7] Cockayne and Hedetniemi defined the *domatic number* of a graph $G = (V, E)$, denoted $d(G)$, to equal the maximum order of a partition of $V = \{V_1, V_2, \ldots, V_k\}$ such that for every i, $1 \leq i \leq k$, V_i is a dominating set. In much the same way we can define the *total domatic* and *connected domatic numbers* $d_t(G)$ and $d_c(G)$, respectively, to be the largest order of a partition of V into total and connected dominating sets.

The next result on d_c extends in the natural way previous results on d and d_t.

THEOREM 8   *For any graph G such that G and $\bar{G}$ are connected,*

(i) $d + \bar{d} \leq p + 1$ (Cockayne and Hedetniemi [7]);
(ii) $d_t + \bar{d}_t \leq p - 2$ (Cockayne, Dawes and Hedetniemi [6]);
(iii) $d_c + \bar{d}_c \leq p - 1$.

PROOF OF (iii)   Since $G \neq K_n$ and $\bar{G} \neq K_n$, we know that $d_c \leq \delta$, and

$$\bar{d}_c \leq \bar{\delta} = p - 1 - \Delta \leq p - 1 - \delta.$$

Hence, $d_c + \bar{d}_c \leq p - 1$. $\square$

The next two results combine γ_c and d_c.

THEOREM 9   *For any connected graph G with p vertices, $\gamma_c + d_c \leq p + 1$, with equality if and only if $G = K_p$.*

PROOF   Let $k = d_c$ and let $D_1, D_2, \ldots, D_k$ be a partition of $V(G)$, where each D_i is a connected dominating set. It follows immediately that

$$p = |D_1| + |D_2| + \ldots + |D_k| \geq \gamma_c d_c;$$

i.e. $d_c \leq p/\gamma_c$. Also $d_c \leq \delta + 1$, where δ denotes the minimum degree of G. Thus, $d_c \leq \min\{p/\gamma_c, \delta + 1\}$. Now $\gamma_c \leq p - \Delta \leq p - \delta$, and therefore

$$\gamma_c + d_c \leq p - \delta + \delta + 1 = p + 1.$$

Now suppose $\gamma_c + d_c = p + 1$. Then $\gamma_c = p - \delta$ and $d_c = \delta + 1$. Now $d_c = \delta + 1$ implies that there exist a set D_i which contains only a minimum degree vertex x. But then, since each D_j is a dominating set, x must be joined to all vertices of D_j, for each $j \neq i$. In other words, $\delta = p - 1$ and hence $G = K_p$. Also, $\gamma_c(K_p) = 1$, $d_c(K_p) = p$, and therefore $\gamma_c(K_p) + d_c(K_p) = p + 1$. $\square$

Notice that $d_c = 1 + \delta$ if and only if $G = K_p$.

4. Some open problems

As we have seen, domination, total domination and connected domination fit well together in forming the beginnings of a unified general theory of domination. Many results on domination and total domination generalize immediately to connected domination and in turn new results on connected domination shed new light on properties of domination and total domination.

A number of interesting conjectures remain concerning possible generalizations to connected domination of previous results on domination. We mention just a few in closing.

Walikar, Sampathkumar and Acharya [17] have shown that

$$\gamma \leq p - \kappa, \tag{2}$$

where κ is the connectivity of G.

Jaegar and Payan [10] showed that

$$\gamma \bar{\gamma} \leq p; \tag{3}$$

$$\gamma \leq \bar{d}. \tag{4}$$

Allan and Laskar [2] and Bollobás and Cockayne [5] demonstrated that

$$\gamma \leq 2 \mathrm{ir} - 1 \tag{5}$$

where ir is the irredundance number of G. Irredundance number of a graph G was defined by Cockayne and Hedetniemi [7] as follows. A set $S \subseteq V(G)$ is an irredundant set if, for every $x \in S$,

$$N[x] \nsubseteq \bigcup_{y \in S - x} N[y].$$

The irredundance number of G, denoted by ir, is the minimum cardinality of all maximal irredundant sets.

Allan, Hedetniemi and Laskar [1] proved that

$$i + \gamma_t \leq p, \tag{6}$$

where i is the minimum cardinality of all maximal independent sets.

In each case (i.e. (2)–(6)) one can ask whether similar results hold if γ_c or d_c is substituted for γ, γ_t or d.

Note added in proof

The contraction number (or Hadwiger number) $h(G)$ of a connected graph G is the maximum number of vertices of a complete graph obtained from G by iterated contractions of edges. Duchet and Meyniel

[18] note that the deletion of a connected dominating set decreases the value of $h(G)$.

Duchet and Meyniel also proved in the above-mentioned paper that

$$\gamma_c \leqslant 2\beta_0 - 1 \quad \text{and} \quad \gamma_c \leqslant 3\gamma - 2.$$

Acknowledgement

This research was supported in part by grant no. ISP-8011451 (EPSCoR) from the National Science Foundation.

References

1. R. B. Allan, S. T. Hedetniemi and R. Laskar, A note on total domination. *Discrete Math.* (to appear).
2. R. B. Allan and R. Laskar, On domination and some related concepts in graph theory. Congressus Numerantium XXI, Utilitas Mathematicae, Winnipeg, 1978, pp. 43–56.
3. W. W. Ball Rouse, *Mathematical Recreations and Problems of Past and Present Times*, Macmillan, London, 1892.
4. C. Berge, *The Theory of Graphs and Its Applications*, Dunod, Paris, 1958; English translation published by Methuen, London, 1962.
5. B. Bollobás, and E. J. Cockayne, Graph theoretic parameters concerning domination, independence and irredundance. *J. Graph Theory* **3** (1979), 241–249.
6. E. J. Cockayne, R. Dawes and S. T. Hedetniemi, Total domination in graphs. *Networks* **10** (1980), 211–219.
7. E. J. Cockayne and S. T. Hedetniemi, Towards a theory of domination in graphs. *Networks* **7** (1977), 247–261.
8. E. J. Cockayne, Domination of undirected graphs: a survey. In *Theory and Applications of Graphs*, Lecture Notes in Mathematics no. 642, Springer Verlag, Berlin, 1978, pp. 141–147.
9. M. R. Garey and D. S. Johnson, *Computers and Intractability: A Guide to the Theory of NP-Completeness*, Freeman, San Francisco, 1978.
10. F. Jaegar and C. Payan, Relations du type Nordhaus–Gaddum pour le nombre d'absorption d'un graphe simple. *Compt. Rend. Acad. Sci. Paris A* **274** (1974), 728–730.
11. R. Laskar and H. B. Walikar, On domination related concepts in graph theory. In *Proceedings of the International Symposium, Indian Statistical Institute, Calcutta, 1980*, Lecture Notes in Mathematics no. 885, Springer-Verlag, Berlin, 1981, pp. 308–320.
12. C. L. Liu, *Introduction to Combinatorial Mathematics*, McGraw-Hill, New York, 1968.
13. J. Nieminen, Two bounds for the domination number of a graph. *J. Inst. Maths Applics* **14** (1974), 183–187.
14. O. Ore, *Theory of Graphs*, American Mathematical Society, Providence, R.I., 1962.

15. E. Sampathkumar and H. B. Walikar, The connected domination number of a graph. *Math. Phys. Sci.* **13** (1979), 607–613.

16. V. G. Vizing, A bound on the external stability number of a graph. *Dokl. Akad. Nauk* **164** (1965), 729–731.

17. H. B. Walikar, E. Sampathkumar and D. Acharya, *Recent Developments in the Theory of Graphs and its Applications*, MRI Lecture Notes in Mathematics No. 1, 1978.

18. P. Duchet and H. Meyniel, On Hadwiger's number and stability number. *Ann. Discrete Math.* **13** (1982), 71–74.

19

SHELLING STRUCTURES, CONVEXITY AND A HAPPY END

Bernhard Korte and László Lovász

ABSTRACT   Greedoids were introduced by the authors as structures with an exchange property generalizing matroid exchange. Another special class of greedoids arises from various "shelling" procedures. These structures have occurred in the literature under various names: abstract convexity, anti-exchange structures, alternative precedence structures, discs, etc. In many respect they are analogous to matroids. We study some of their properties, in particular circuits in them, and generalize the "happy end theorem" of Erdös and Szekeres to such structures.

1. Introduction

Matroids can be characterized as sets E endowed with a "closure" operation $\sigma: 2^E \to 2^E$ which satisfies the usual requirements of closure, i.e.

(C1) $\qquad\qquad \sigma(X) \supseteq X,$

(C2) $\qquad\quad X \subseteq Y$ implies $\sigma(X) \subseteq \sigma(Y),$

(C3) $\qquad\qquad \sigma\sigma(X) = \sigma(X),$

and, in addition, the Steinitz–McLane exchange property:

(SM) $\quad$ If $\;y, z \notin \sigma(X)\;$ and $\;z \in \sigma(X \cup y)\;$ then $\;y \in \sigma(X \cup z).$

The linear closure operation on a set of vectors in (say) $\mathbb{R}^n$ is the most important example of such a closure operator. It was noted by Edelman [3] and Jamison-Waldner [5] that the convex hull operator, defined on any set of vectors in $\mathbb{R}^n$, satisfies a rather similar but in a sense "dual" exchange property (besides, of course, (C1)–(C3)):

(ASM) If $\;y, z \notin \sigma(X)\;$ and $\;z \in \sigma(X \cup y)\;$ then $\;y \notin \sigma(X \cup z).$

They called a set E endowed with a closure operator σ satisfying (ASM) an *anti-exchange structure.*

It turns out, however, that if we consider exchange properties of independent sets rather than of the closure, then a natural common generalization of matroids and anti-exchange structures can be found: *greedoids*, as they were introduced by Korte and Lovász [6]. We are

GRAPH THEORY AND COMBINATORICS
ISBN 0-12-111760-X

indebted to A. Björner, who first pointed out to us this connection between anti-exchange structures and greedoids. One way to define them is in terms of their feasible sets. Let E be a finite set and $\mathcal{F} \subseteq 2^E$. Then the pair $(E, \mathcal{F})$ is a *greedoid* if and only if the following hold:

(H1) $\varnothing \in \mathcal{F}$.

(H4) $\begin{cases} \text{If } X, \ Y \in \mathcal{F} \text{ and } |X| < |Y| \text{ there exists an element } y \in Y - X \\ \text{such that } X \cup y \in \mathcal{F}. \end{cases}$

Note that if, in addition to these, subclusiveness of $\mathcal{F}$ was assumed, then we would get just matroids. Thus, matroids are special greedoids. It is less apparent, but the reader may verify it with little effort, that the complements of closed sets in an anti-exchange structure form a greedoid. In this case $\mathcal{F}$ is not subclusive but closed under taking unions.

In this paper we shall study various structural properties of anti-exchange structures, which will also be called shelling structures for reasons to be discussed later. Some overlap with the papers of Edelman [3], Jamison-Waldner [5], Björner [1] and Crapo [2] is inevitable; but most of the results, including the notion and properties of circuits, are new.

In Section 2 we give some preliminaries; in particular we collect some definitions and results on greedoids which will be needed. A more detailed study of structural aspects of greedoids can be found in Korte and Lovász [7, 8]. In Section 3, various definitions and examples of shelling structures are given. Section 4 contains a discussion of circuits in shelling structures. Finally, in Section 5 we formulate one more exchange property, which holds for shelling structures associated with a convex hull, but not for all shelling structures. It is natural to extend various combinatorial properties of convexity to such shelling structures. As a first step in this direction, we prove a generalization of the "happy end theorem" due to Erdös and Szekeres [4].

2. Preliminaries: greedoids

A *language* over a finite ground set E (which is called the *alphabet*) is a collection of *finite sequences* of elements of E. We call these sequences *strings* or *words* and denote them by small Greek letters. E^* is the collection of all possible strings over E. The underlying set of a word is denoted by $\tilde{\alpha}$. $\tilde{\mathcal{L}} \subseteq 2^E$ is the collection of all underlying sets of $\mathcal{L}$. Similarly to the cardinality symbol, we use $|\alpha|$ to denote the length of a string α. The notation $x \in \alpha$ means $x \in \tilde{\alpha}$. The concatenation of two words α, β is denoted by $\alpha \cdot \beta$, i.e. the string α followed by the string β.

A language is called *simple* if no letter is repeated in any of its words, i.e. $|\alpha| = |\tilde{\alpha}|$ for every $\alpha \in \mathcal{L}$. It is called *normal* if it has no dummy letters, i.e. every element of E appears in at least one word. $(E, \mathcal{L})$ is called a *simple hereditary language* if

(G1) $\qquad\qquad\qquad \varnothing \in \mathcal{L}.$

(G2) $\quad \begin{cases} \text{If } \alpha \in \mathcal{L} \text{ and } \alpha = \beta \cdot \gamma, \text{ then } \beta \in \mathcal{L}, \text{ i.e. every beginning section} \\ \text{of a word belongs to the language.} \end{cases}$

A simple hereditary language is a *greedoid* if in addition the following holds:

(G3)   If $\alpha, \beta \in \mathcal{L}$ and $|\alpha| > |\beta|$, then there exists an $x \in \alpha$ such that $\beta \cdot x \in \mathcal{L}$.

We observe that (G1)–(G3) are ordered analogues of the independence axioms of matroids. Apart from this definition of hereditary languages and greedoids as a collection of *ordered* sets, we can also define them in an *unordered* version by considering the underlying sets of strings. Then an *accessible set-system* $(E, \mathcal{F})$ is a set system $(\mathcal{F} \subseteq 2^E)$ with

(H1) $\qquad\qquad\qquad \varnothing \in \mathcal{F};$

(H2) $\qquad$ for all $X \in \mathcal{F}$ there exists a $x \in X$ such that $X - x \in \mathcal{F}.$

For a greedoid we have in addition the following:

(H3) $\quad \begin{cases} \text{If } X, Y \in \mathcal{F} \text{ and } |X| = |Y| + 1 \text{ then there exists a } x \in X - Y \text{ such} \\ \text{that } Y \cup x \in \mathcal{F}. \end{cases}$

It follows immediately that for a simple hereditary language (greedoid) $(E, \mathcal{L})$ defined by (G1)–(G3), the system $(E, \tilde{\mathcal{L}})$ satisfies (H1)–(H3). Conversely, it was shown in Korte and Lovász [6] that given a system $(E, \mathcal{F})$ satisfying (H1)–(H3) there exists a unique ordered system $(E, \mathcal{L})$ with $\tilde{\mathcal{L}} = \mathcal{F}$ which fulfils (G1)–(G3).

Elements of $\mathcal{F}(\mathcal{L})$ are called *feasible* or *independent sets* (*words*). A maximal feasible subset of a set $X \subseteq E$ is called a *basis* of X. A *basic word* of X is a maximal feasible word which uses only letters from X. A greedoid is called *full* if E is feasible.

It is an easy observation that (H2) and (H3) are equivalent to (H4) as stated in the introduction. Therefore, greedoids can be considered either as simple languages satisfying ordered versions of the matroid axioms or as unordered set systems satisfying (H1) and (H4), i.e. direct relaxations of matroids. In the following we will use both definitions of greedoids.

3. Shelling structures: alternative definitions and examples

There are many equivalent ways to define shelling structures; one was given in the introduction. Here we start with another definition, which is very often the way such structures are given or can be recognized.

Let E be a finite set and let, for each $e \in E$, a set system $H_e \subseteq 2^{E-e}$ be given. The sets in H_e are considered as *alternative precedences* for e. Let

$$\mathscr{L} = \{x_1 \ldots x_k \mid \forall 1 \leq i \leq k,\ x_i \in E \text{ and there is a set } U \in H_{x_i}$$
$$\text{such that } U \subseteq \{x_1, \ldots, x_{i-1}\}\}.$$

(in words, x_i may occur only if at least one "alternative precedence set" occurs before x_i).

To exclude trivial cases, we assume that $\mathscr{L}$ is normal. Then the pair $(E, \mathscr{L})$ is called an *alternative precedence structure* or *shelling structure*. This second name is motivated by the following examples.

EXAMPLE 1   Let T be a tree and $E = V(T)$. Let $\mathscr{L} = \{x_1 \ldots x_k \mid x_i$ is an end-point of the tree $T - x_1 - \ldots - x_{i-1}\}$. Then $(E, \mathscr{L})$ is a shelling structure. As alternative precedences, we can consider, for $e \in E$, the components $T_1, \ldots, T_r$ of $T - e$, and let $H_e = \{E - e - V(T_i): 1 \leq i \leq r\}$.

EXAMPLE 2   Let E be a finite set of vectors in $\mathbb{R}^n$. Let $\mathscr{L} = \{x_1 \ldots x_k \mid x_i$ is a vertex of the convex hull of $E - x_1 - \ldots - x_{i-1}$, for $i = 1, \ldots, k\}$. Then $(E, \mathscr{L})$ is a shelling structure. As alternative precedences for a point $e \in E$, consider all closed half-spaces with e on the boundary, and let H_e consist of the intersections of $E - e$ with these half-spaces.

EXAMPLE 3   Let $(P, \leq)$ be a finite poset and let $\mathscr{L} = \{x_1 \ldots x_k \mid \{x_1, \ldots, x_k\}$ is an order ideal in $(P, \leq)$ and its ordering is compatible with $\leq\}$. Then $(P, \mathscr{L})$ is a shelling structure. For every point $e \in P$, we take just one alternative precedence set $\{x \in P : x < e\}$.

Several other examples can be found in Korte and Lovász [7], but these three will suffice to illustrate the notions and results that follow.

A simple intrinsic characterization of shelling structures was observed by Björner [1]:

LEMMA 3.1   A normal hereditary language $(E, \mathscr{L})$ is a shelling structure if and only if the following axiom holds:

(S)   If $\alpha, \beta \in \mathscr{L}$, $\tilde{\alpha} \nsubseteq \tilde{\beta}$ then there exists an $x \in \alpha$ such that $\beta x \in \mathscr{L}$.

PROOF   For sake of completeness, we give the simple proof. Suppose first that $(E, \mathscr{L})$ is a shelling structure, and let $\alpha, \beta \in \mathscr{L}$ with $\tilde{\alpha} \nsubseteq \tilde{\beta}$. Let x be the first element in α not in $\tilde{\beta}$, and write $\alpha = \alpha_1 x \alpha_2$. Then x has an alternative precedence set $U \subseteq \tilde{\alpha}_1$. But then $U \subseteq \tilde{\alpha}_1 \subseteq \tilde{\beta}$ and so $\beta x \in \mathscr{L}$.

Conversely, suppose that $(E, \mathscr{L})$ is a normal hereditary language satisfying (S). Define, for each $e \in E$, $H_e = \{\tilde{\alpha}: \alpha e \in \mathcal{L}\}$. Let the alternative precedences H_e define a shelling structure $(E, \mathcal{L}')$. We show that $\mathcal{L} = \mathcal{L}'$. It is obvious that $\mathcal{L} \subseteq \mathcal{L}'$, so consider any $\beta \in \mathcal{L}'$. We want to show that $\beta \in \mathcal{L}$. Write $\beta = \beta'x$. Using induction on $|\beta|$, we may assume that $\beta' \in \mathcal{L}$. Furthermore, $\beta \in \mathcal{L}'$ implies that there is a set $U \in H_x$ such that $U \subseteq \tilde{\beta}'$. This means that there exists a word α such that $\alpha x \in \mathcal{L}$ and $\tilde{\alpha} \subseteq \tilde{\beta}'$. But then (S) implies that $\beta = \beta'x \in \mathcal{L}$. $\square$

If $\alpha \in \mathcal{L}$ in a shelling structure $(E, \mathcal{L})$ then, by (S), α can be extended to a basic word containing all elements of E. Thus the set of basic words determines the shelling structure.

The previous lemma also shows that every shelling structure is a greedoid. Let $(E, \mathcal{L})$ be a shelling structure and let $\mathcal{F}$ be the set of underlying sets of words in $\mathcal{L}$. Then $(E, \mathcal{F})$ determines $(E, \mathcal{L})$ uniquely; this is true for every greedoid. We shall sometimes speak about the shelling structure $(E, \mathcal{F})$ instead of the shelling structure determined by $(E, \mathcal{F})$.

LEMMA 3.2 *A normal accessible set-system* $(E, \mathcal{F})$ *is the system of feasible sets of a shelling structure if and only if* $E \in \mathcal{F}$ *and* $\mathcal{F}$ *is closed under union.*

PROOF Suppose first that $\mathcal{F}$ is the set of feasible sets of a shelling structure $(E, \mathcal{L})$ and let $X, Y \in \mathcal{F}$. Then $X = \tilde{\alpha}$ and $Y = \tilde{\beta}$ for some $\alpha, \beta \in \mathcal{F}$. Repeated application of (S) yields a word $\beta\alpha' \in \mathcal{L}$ such that $\widetilde{\beta\alpha'} = X \cup Y$.

Conversely, let us assume that $\mathcal{F}$ is closed under union, and let

$$\mathcal{L} = \{x_1 \ldots x_k \mid \{x_1, \ldots, x_i\} \in \mathcal{F} \quad \text{for all} \quad i\}.$$

We claim that (S) holds, i.e. $(E, \mathcal{L})$ is a shelling structure. Obviously, $(E, \mathcal{L})$ will have the sets in $\mathcal{F}$ as feasible sets. Let $\alpha, \beta \in \mathcal{L}$ with $\tilde{\alpha} \not\subseteq \tilde{\beta}$ and let x be the first element in α not in β. Write $\alpha = \alpha_1 x \alpha_2$. Then $\widetilde{\alpha_1 x} \in \mathcal{F}$ and $\tilde{\beta} \in \mathcal{L}$ and so $\widetilde{\alpha_1 x} \cup \tilde{\beta} = \widetilde{\beta x} \in \mathcal{F}$. But then $\beta x \in \mathcal{L}$. $\square$

As a corollary, we see that every set $X \subseteq E$ contains a unique largest feasible subset, which we call the *basis of X*. (In a general greedoid, only the cardinality of maximal feasible subsets of X is uniquely determined by X.)

We call the complements of feasible sets *convex sets*. It follows by Lemma 3.2 that the intersection of convex sets is convex. Consequently, every set is contained in a unique minimal convex set, called the *convex hull.* We denote by $\tau(X)$ the convex hull of X. Clearly, $X - \tau(E - X)$ is the basis of X.

The convex hull operator $\tau: 2^E \to 2^E$ determines the shelling structure uniquely: feasible sets are just the complements of convex sets, and a set X is convex if and only if $\tau(X) = X$.

LEMMA 3.3 *A mapping* $\tau: 2^E \to 2^E$ *is the convex hull operator of a shelling structure on E and if and only if the following hold:*

(C0) $\tau(\varnothing) = \varnothing$.
(C1) $\tau(X) \supseteq X$ *for all* X, $Y \subseteq E$.
(C2) $X \subseteq Y \Rightarrow \tau(X) \subseteq \tau(Y)$ *for all* X, $Y \subseteq E$.
(C3) $\tau\tau(X) = \tau(X)$ *for all* $X \subseteq E$.
(ASM) *If* y, $z \notin \tau(X)$ *and* $z \in \tau(X \cup y)$ *then* $y \notin \tau(X \cup z)$.

PROOF   Suppose first that τ is the convex hull operator of a greedoid. (C0)–(C3) are obvious. To prove that (ASM) holds, let B be the basis of $E - X$. Then y, $z \in B$. Furthermore, let B_1 be the basis of $E - X - y$, then by hypothesis $z \notin B_1$. Trivially, $B_1 \subseteq B - y - z$. By (S), there is a $u \in B$ such that $B_1 \cup u \in \mathscr{F}$. Since B_1 is the basis of $E - X - y$, we must have that $u = y$. Thus $B_1 \cup y$ is a feasible subset of $E - X - z$ and so $y \notin \tau(X \cup z)$.

Conversely, suppose that (C0)–(C3) and (ASM) hold. Then let

$$\mathscr{F} = \{X \subseteq E \mid \tau(E - X) = E - X\}.$$

We claim that $(E, \mathscr{F})$ is a shelling structure. Clearly $\varnothing \in \mathscr{F}$ by (C0).

Let $X \in \mathscr{F}$. Then $\tau(E - X) = E - X$ by definition. Choose an element $x \in X$ such that $\tau(E - X \cup x)$ is (inclusionwise) minimal; we claim that $X \cup x \in \mathscr{F}$, i.e. $\tau(E - X \cup x) = E - X \cup x$. Assume indirectly that $\tau(E - X \cup x) \neq E - X \cup x$. Then, by (C1), $\tau(E - X \cup x) \supseteq E - X \cup x$. Let $x' \in \tau(E - X \cup x) - (E - X \cup x)$. Then $E - X \cup x' \subseteq \tau(E - X \cup x)$ and so, by (C2) and (C3),

$$\tau(E - X \cup x') \subseteq \tau(\tau(E - X \cup x)) = \tau(E - X \cup x).$$

Furthermore, $x \notin \tau(E - X \cup x')$, by (ASM). This contradicts the choice of x. Thus we have shown that $\mathscr{F}$ is accessible.

The fact that $\mathscr{F}$ is closed under union follows by a standard argument on closure operations. Let X, $Y \in \mathscr{F}$, then

$$\tau(E - (X \cup Y)) \supseteq E - (X \cup Y)$$

by (C1). Further,

$$\tau(E - (X \cup Y)) \subseteq \tau(E - X) = E - X$$

by (C2), similarly

$$\tau(E - (X \cup Y)) \subseteq E - Y.$$

Hence $\tau(E - (X \cup Y)) \subseteq (E - X) \cap (E - Y) = E - (X \cup Y)$ and so $\tau(E - (X \cup Y)) = E - (X \cup Y)$, which shows that $X \cup Y \in \mathscr{F}$. □

This last characterization of shelling structures is a first example of the phenomenon that these structures are in many respects "dual" to matroids. Another such example is the following.

A greedoid $(E, \mathscr{F})$ is an *interval greedoid* if for all A, B, $C \in \mathscr{F}$ with $A \subseteq B \subseteq C$ and $x \in E - C$ such that $A \cup x \in \mathscr{F}$ and $C \cup x \in \mathscr{F}$, it follows that $B \cup x \in \mathscr{F}$. $(E, \mathscr{F})$ is called an *interval greedoid without lower (upper) bounds* if for all A, $B \in \mathscr{F}$ with $A \subseteq B[B \subseteq A]$ and $x \in E - B[x \in E - A]$ such that $B \cup x \in \mathscr{F}$, it follows that $A \cup x \in \mathscr{F}$. We do not go into the discussion of the interval property, which is a notion of central importance for greedoids; see Korte and Lovász [8, 9] and Crapo [2] (here greedoids with the interval property are called *selectors*). But the two stronger interval properties characterize two major classes of greedoid:

LEMMA 3.4 (a) *Greedoids with the interval property without lower bounds are exactly the matroids.*

(b) *Greedoids with the interval property without upper bounds are exactly the shelling structures.*

PROOF (a) The fact that matroids have the interval property without lower bounds is obvious. Conversely, if a greedoid has this property, then $\mathscr{F}$ is subclusive. Let $A \in \mathscr{F}$ and $x \in A$; we then claim that $A - x \in \mathscr{F}$. Since $(E, \mathscr{F})$ is a greedoid, we have a $y \in A$ such that $A - y \in \mathscr{F}$. If $y = x$ we are done, so suppose $y \neq x$. Using induction, we may assume that $A - y - x \in \mathscr{F}$. But then, by the interval property without lower bounds, $(A - y - x) \cup y = A - x \in \mathscr{F}$. Thus $\mathscr{F}$ is subclusive and the greedoid is a matroid.

(b) The fact that a shelling structure has the interval property without upper bounds is trivial by its representation as an alternative precedence structure.

Consider a greedoid $(E, \mathscr{F})$ with the interval property without upper bounds. Let $\alpha_1\beta \in \mathscr{L}$ and $\tilde{\alpha} \not\subseteq \tilde{\beta}$. Let x be the first element of α not in $\tilde{\beta}$, and write $\alpha = \alpha_1 x \alpha_2$. Then $\tilde{\alpha}_1 \subseteq \tilde{\beta}$, $\tilde{\alpha}_1$, $\tilde{\beta} \in \mathscr{F}$, $x \in E - \tilde{\beta}$ and $\tilde{\alpha}_1 \cup x \in \mathscr{F}$. By the interval property without upper bounds, this implies that $\tilde{\beta} \cup x \in \mathscr{F}$ and hence $\beta x \in \mathscr{L}$. So $(E, \mathscr{L})$ satisfies (S) and hence it is a shelling structure. □

4. Circuits

For general greedoids, we introduced restrictions and contractions in Korte and Lovász [8]. These operations, however, are rather uninteresting for shelling structures: they do not preserve fullness and hence lead out of the class of shelling structures. A minor-producing operation more suited to shelling structures is the following. Let $(E, \mathscr{F})$ be a shelling structure, and $T \subseteq E$. Define

$$\mathscr{F}: T = \{A \cap T: A \in \mathscr{F}\}.$$

Then $(T, \mathscr{F}: T)$ is called the *trace of* $(E, \mathscr{F})$ *on* T.
It is obvious that if $T_1 \subseteq T_2$ then $(\mathscr{F}: T_2): T_1 = \mathscr{F}: T_1$.

LEMMA 4.1   *The trace* $(T, \mathscr{F}: T)$ *of* $(E, \mathscr{F})$ *on any subset* T *is a shelling structure.*

PROOF   It is easy to verify that $\mathscr{F}: T$ is accessible and closed under union. $\square$

A set $T \subseteq E$ is called *free* in the shelling structure $(E, \mathscr{F})$ if $(T, \mathscr{F}: T)$ is the free greedoid, i.e. if $\mathscr{F}: T = 2^T$. A set C is called a *circuit* if it is minimal non-free, i.e. it is not free but every proper subset of it is free.

EXAMPLE 1   Let $E = V(G)$ where G is a tree. Then the free sets are sets of end-points or subtrees of G. Moreover, circuits are triples of points lying on a path.

EXAMPLE 2   Let $E \subseteq \mathbb{R}^2$ and let $(E, \mathscr{F})$ be the shelling structure on E. Then for any $T \subseteq E$, $(T, \mathscr{F}: T)$ is the shelling structure on T. T is free if and only if every point in T is a vertex of the convex hull of T. The circuits are triples of colinear points as well as quadruples in which one point is in the interior of the triangle spanned by the other three.

EXAMPLE 3   Let $(P, \leqslant)$ be a poset and let $(P, \mathscr{L})$ be the schedule greedoid on this poset. Then the free sets are just the antichains. Moreover, the circuits are pairs $\{a, b\}$ with $a < b$.
   It is easy to see that a pair $\{a, b\}$ is either free or a circuit; and it is a circuit if and only if every feasible ordering of E contains a and b in the same order, or, equivalently, if every feasible set containing b contains a (or vice versa).

The following lemmas characterize free sets and circuits. For each $A \in \mathscr{F}$, let $\Gamma(A) = \{x \in E - A: A \cup x \in \mathscr{F}\}$ denote the set of continuations.

LEMMA 4.2   Let $(E, \mathscr{F})$ be a shelling structure and $A \in \mathscr{F}$. Then $\Gamma(A)$ is free. Conversely, let T be a free set and A the basis of $E - T$. Then $T = \Gamma(A)$.

PROOF.   Let $A \in \mathscr{F}$. It follows from the interval property without upper bounds that for each $X \subseteq \Gamma(A)$, $A \cup x \in \mathscr{F}$. Hence $\mathscr{F}: \Gamma(A) = 2^{\Gamma(A)}$ and so $\Gamma(A)$ is free.

Let T be free and let A be the basis of $E - T$. Trivially $\Gamma(A) \subseteq T$. Let $y \in T$. Then, by the freeness of T, there is a set $B \in \mathscr{F}$ with $B \cap T = \{y\}$. Then $B \not\subseteq A$ and so, by (S), there is a $z \in B$ such that $A \cup z \in \mathscr{F}$. But A is the basis of $E - T$ and so $z \notin E - T$. Hence $z = y$ and so $y \in \Gamma(A)$.   □

LEMMA 4.3   Let $(E, \mathscr{F})$ be a shelling structure and C a circuit in $(E, \mathscr{F})$. Let A be the basis of $E - C$. Then $\Gamma(A) \subseteq C$ and $|C - \Gamma(A)| = 1$.

PROOF   It is obvious that $\Gamma(A) \subseteq C$, and that $\Gamma(A) \neq C$ since then C should be free by Lemma 4.2. Let $x \in C - \Gamma(A)$. Then A is the basis of $E - C \cup x$ as well, and since $C - x$ is free, we have $C - x = \Gamma(A)$.   □

We shall call the element of $C - \Gamma(A)$ the *root of* C. Note that we do not claim that every set C with $|C - \Gamma(A)| = 1$ is a circuit. Counter-examples are found in the examples discussed below, e.g. in posets.

LEMMA 4.4   Let $(E, \mathscr{F})$ be a shelling structure, $C \subseteq E$ and $a \in C$. Then C is a circuit with root a if and only if $\mathscr{F}: C = 2^C - \{a\}$.

PROOF   Suppose that C is a circuit with root a. First we show that $\{a\} \notin \mathscr{F}: C$. Otherwise, there exists a set $B \in \mathscr{F}$ with $B \cap C = \{a\}$. But then $B \not\subseteq A$ and so, by (S), there is a $z \in B - A$ such that $A \cup z \in \mathscr{F}$, i.e. $z \in \Gamma(A)$. But this is impossible as $\Gamma(A) = C - a$ is disjoint from B.

Secondly, we note that every for $b \in C - a$, $\{b\} \in \mathscr{F}: C$. This is clear since $b \in C - a = \Gamma(A)$ and so $A \cup b \in \mathscr{F}$ and $(A \cup b) \cap C = \{b\}$.

Third, for all $b \in C - a$, $\{a, b\} \in \mathscr{F}: C$. For suppose that $\{a, b\} \notin \mathscr{F}: C$, then $a \notin \mathscr{F}: (C - b) = \mathscr{F}: C: (C - b)$, contradicting the hypothesis that C is minimal non-free.

Finally, $\mathscr{F}: C = 2^C - \{a\}$ follows on noticing that $\mathscr{F}: C$ is closed under union.

If $\mathscr{F}: C = 2^C - \{a\}$ then trivially C is minimal non-free, i.e. a circuit. It is clear that its root is a.   □

Our next lemma shows that the circuits of a shelling structure determine it uniquely.

LEMMA 4.5   *Let* $(E, \mathscr{F})$ *be a shelling structure and* $A \subseteq E$. *Then* $A \in \mathscr{F}$ *if and only if for each circuit* C, $A \cap C$ *is not the root of* C. *An ordering of* E *is a word in* $\mathscr{L}$ *if and only if it orders every circuit so that it starts with a non-root element.*

PROOF   It is clear by the definition of circuits that if A is feasible then it has to satisfy this condition. Conversely, suppose that $A \subseteq E$ and $A \notin \mathscr{F}$. Consider the basis B of A, and let $x \in A - B$. Then $\Gamma(B) \cup x$ is non-free by Lemma 4.2 and hence it contains a circuit C. Trivially, $x \in C$, and in fact x is the root of C. Also trivially, $\Gamma(B) \cap A = \varnothing$. Hence $a \cap C = \{x\}$, i.e. A does not satisfy the condition formulated in the lemma.   □

The structure of circuits in shelling structures is (contrary to first appearances) simpler than in matroids. This is due to the fact that any collection of rooted sets gives rise to a shelling structure. More precisely, a *rooted set* is a pair (C, a) where $a \in C$. Let $\mathscr{H} = \{(C_i, a_i): i = 1, \ldots, m\}$ be any collection of rooted subsets of a set E, and let

$$\mathscr{L}_{\mathscr{H}} = \{x_1 \ldots x_k \in E^*: \forall 1 \leq i \leq m \quad \text{and} \quad 1 \leq j \leq k,$$

$$\text{if} \quad a_i = x_j \quad \text{then} \quad C_i \cap \{x_1, \ldots, x_{j-1}\} \neq \varnothing\}$$

be the *language determined by* $\mathscr{H}$.

LEMMA 4.6   *If every element of* E *occurs in a word in* $\mathscr{L}_{\mathscr{H}}$ *then* $(E, \mathscr{L}_{\mathscr{H}})$ *is a shelling structure.*

PROOF   One may consider the non-empty subsets of $C_i - a_i$ as alternative precedences for a_i.   □

Every shelling structure can be determined by a system of rooted sets, namely by its circuits, by Lemma 4.5. Such a system of rooted sets is not unique, and it does not even need to contain all circuits. But there is always a unique minimal system of circuits determining a shelling structure, as we are going to show.

Let $(E, \mathscr{F})$ be a shelling structure, $C \subseteq E$, $a \in C$, and let A be the basis of $E - C$. We say that the rooted set (C, a) is a *critical circuit* if $A \cup a \notin \mathscr{F}$ but $A \cup a \cup c \in \mathscr{F}$ for all $c \in C - a$.

LEMMA 4.7   *Every critical circuit is a circuit.*

PROOF C is not free by Lemma 4.2; in fact, $\Gamma(A) \neq C$. The further hypotheses imply immediately that $\{c\} \in \mathscr{F}$: C and $\{a, c\} \in \mathscr{F}$: C for all

$c \in C - a$, and hence $X \in \mathscr{F}$: C for all $X \in 2^C - \{a\}$. Hence, by Lemma 4.4, C is a circuit. $\square$

EXAMPLE 1   If $E = V(G)$ for a tree G, then critical circuits are 3-paths.

EXAMPLE 2   If $E \subseteq \mathbb{R}^2$ then a critical circuit is either a triple of colinear points of E which are consecutive on the line (i.e. no further point of E belongs to their convex hull) or a quadruple consisting of three points forming a triangle and a fourth point in the interior of this triangle, such that this triangle contains no other point of E.

EXAMPLE 3   If $(P, \leqslant)$ is a poset, then, in its schedule greedoid, critical circuits are the pairs (a, b) where b covers a.

THEOREM 4.8   Let $(E, \mathscr{L})$ be a shelling structure. Then:

(a) $(E, \mathscr{L})$ is determined by its critical circuits.
(b) Every system of circuits determining $(E, \mathscr{L})$ contains all critical circuits.

PROOF   (a) Let $\mathscr{H}_0$ be the system of critical circuits. Trivially $\mathscr{L}(\mathscr{H}_0) \supseteq \mathscr{L}$. Suppose that $\mathscr{L}(\mathscr{H}_0) \neq \mathscr{L}$, and let $x_1 \ldots x_k$ be a shortest word in $\mathscr{L}(\mathscr{H}_0) - \mathscr{L}$. Then $x_1 \ldots x_{k-1} \in \mathscr{L}$. Let α be a longest continuation of $x_1 \ldots x_{k-1}$ with $\alpha x_k \notin \mathscr{L}$, and set $C = \{x_k\} \cup \Gamma(\alpha)$. Then $A = \tilde{\alpha}$ is the basis of $E - C$. Further, $\alpha x_k \notin \mathscr{L}$ and hence $A \cup x_k \notin \mathscr{F}$, but $\alpha y x_k \in \mathscr{L}$ for all $y \in \Gamma(A)$ and so $A \cup y \cup x_k \in \mathscr{F}$. So (C, x_k) is a critical circuit and so $(C, x_k) \in \mathscr{H}_0$. But $C \cap \{x_1, \ldots, x_k\} = \{x_k\}$, which contradicts the hypothesis that $x_1 \ldots x_k \in \mathscr{L}(\mathscr{H}_0)$.

(b) Let $\mathscr{H}$ be any family of circuits determining $(E, \mathscr{L})$ and (C, a) any critical circuit. Let A be the basis of $E - C$. Then $A \in \mathscr{F}$ but $A \cup a \notin \mathscr{F}$; hence there is a $(D, a) \in \mathscr{H}$ such that $D \cap A = \varnothing$. Let $y \in C - a$. Then $A \cup a \cup y \in \mathscr{F}$ by the definition of critical circuits, and so $D \cap (A \cup a \cup y) \neq \{a\}$. Thus $y \in D$ and so $C \subseteq D$. Since both C and D are circuits, this implies $C = D$, and so $(C, a) \in \mathscr{H}$. $\square$

THEOREM 4.9   A shelling structure is a poset greedoid if and only if all circuits have cardinality 2.

PROOF   We have seen before that poset greedoids have the pairs $(\{a, b\}, b)$ with $a < b$ as circuits.

Conversely, let $(E, \mathscr{L})$ be a shelling structure such that all circuits have cardinality 2. Let G be the directed graph on E with edges ab for which

$(\{a, b\}, b)$ is a circuit. Since $(E, \mathscr{L})$ contains at least one feasible ordering of all elements of E, and in this ordering a must anticipate b for every edge ab, G must be acyclic. Furthermore, if $ab \in E(G)$ and $bc \in E(G)$, then every feasible set containing c also contains b and so also contains a, and thus ac is a circuit. Hence G is a poset and, by Lemma 4.5, $(E, \mathscr{L})$ is just the shelling structure associated with G. $\square$

5. Closer to convexity

The anti-exchange property does not imply all simple combinatorial properties of convexity. Using the terminology introduced in the preceding sections, we formulate the following property:

(C4) $\begin{cases} \text{If } (A \cup x, x) \text{ and } (A \cup y, y) \text{ are circuits, then there is a unique} \\ \text{subset } A' \subseteq A \text{ such that } (A' \cup x \cup y, y) \text{ is a circuit.} \end{cases}$

This holds for the convex hull shellings of a set. In fact, A spans a simplex and x, y are interior points of A. Projecting y from x on the boundary of A, the resulting point y' will be contained in a unique minimal face A', and then $(A' \cup x, y)$ is the circuit.

Property (C4) does not hold for all shelling structures, e.g. not for posets.

It is a natural question to ask at this point: Is (C4), or maybe a longer list of similar combinatorial properties, enough to characterize convex hull shellings among all shelling structures? The answer to the question is probably "no", since one can generalize convex hull shellings to oriented matroids and so it seems to be related to the representability of matroids.

In this paper we achieve a much more moderate goal by showing that one can generalize for shelling structures with property (C4) at least one important property of convexity, namely a Ramsey-type result of Erdös and Szekeres [4], often called the "happy end theorem":

THEOREM 5.1 (Erdös and Szekeres [4])   *There exists a function $f: \mathbb{N} \to \mathbb{N}$ such that every set of $f(n)$ points in the plane such that no three are on a line contains all the vertices of some convex n-gon.*

In fact, the following bounds hold true for f:

$$2^{n-2} + 1 \leqslant f(n) \leqslant \binom{2n-4}{n-2} + 1.$$

We prove a generalization of this result:

THEOREM 5.2   *There exists a function* $f: \mathbb{N} \to \mathbb{N}$ *such that every shelling structure* $(E, \mathscr{F})$ *with property* (C4) *such that* $|E| \geq f(n)$ *and every circuit has at least four elements contains a free set of size* n.

Berore proving this, let us first state and prove a lemma.

LEMMA 5.3   *Let* $(E, \mathscr{F})$ *be a shelling structure with property* (C4) *such that every circuit has at least four elements and let* $X \subseteq E$. *Then there is a point* $y \in X$ *such that* $X - y$ *is not a circuit.*

PROOF   Suppose, by way of contradiction, that $X - y$ is a circuit for every $y \in X$. Let A be a basis of $E - X$, then $\Gamma(A) \neq X$ as X is non-free. Let $x_0 \in X - \Gamma(A)$. Then $X - x_0$ is a circuit, and A is obviously a basis of $E - (X - x_0)$. So $\Gamma(A) = X - x_0 - x_1$ for some $x_1 \in X - x_0$. But then, $X - x_1$ is also a circuit, A is a basis of $E - (X - x_1)$ and x_0 is the root of $X - x_1$.

By (C4), $X - x_0 - x_1$ has a unique subset Y such that $Y \cup x_1 \cup x_0$ is a circuit with root x_0. Hence $|Y \cup x_1 \cup x_0| = |X| - 1$ and so $Y = X - x_1 - x_0 - y_0$ for some $y_0 \in X - x_1 - x_0$. Thus $X - y_0$ is a circuit with root x_0. Similarly, there is a point $y_1 \in X - x_1 - x_0$ such that $X - y_1$ is a circuit with root x_1.

Now by the hypothesis that every circuit has at least four elements, we have $|X| \geq 5$ and so we can find an element $y_3 \in X - x_0 - x_1 - y_0 - y_1$. Consider $X - y_3$. This is a circuit. The root of $X - y_3$ cannot be any element of $X - x_0 - x_1 = \Gamma(A)$, since for every $y \in X - x_0 - x_1 - y_3$, $A \cup y \in \mathscr{F}$ and $(A \cup y) \cap (X - y_3) = \{y\}$, contradicting Lemma 4.5. Hence the root of $X - y_3$ is either x_0 or x_1, say the first. But then we get a contradiction with the uniqueness requirement in (C4). $\square$

PROOF OF THEOREM 5.2   Let us 2-colour all subsets of E as follows: let the free subsets be red and the non-free one blue. By a version of Ramsey's theorem, if $|E|$ is sufficiently large then there exists a subset $X \subseteq E$ such that $|X| = n + 1$ and, for all $1 \leq k \leq n$, all k-subsets of X have the same colour.

Let k be the minimum cardinality of a blue subset of X (if any). Then all subsets of X with cardinality k are non-free while all subsets of cardinality less than k are free. Thus all subsets of cardinality k are circuits. By Lemma 5.3, this is impossible unless $k = n + 1$. But then any n-element subset of X is free. $\square$

REMARK   The function $f(n)$ obtained from this proof grows horribly fast.

One may hope that using more of the special structure of shelling structures, better bounds could be derived, maybe even attaining the Erdös–Szekeres bound.

Acknowledgements

The work reported in this paper was supported by the joint research project "Algorithmic Aspects of Combinatorial Optimization" of the Hungarian Academy of Sciences (Magyar Tudományos Akadémia) and the German Research Association (Deutsche Forschungsgemeinschaft, SFB 21). The paper was originally published as report no. 83274–OR of the Institut für Ökonometrie und Operations Research, Bonn.

References

1. A. Björner, On matroids, groups and exchange languages. Preprint, Department of Mathematics, University of Stockholm, 1983.
2. H. Crapo, Selectors. A theory of formal languages, semimodular lattices and shelling processes. *Advanc. Math.* (in press).
3. P. H. Edelman, Meet-distributive lattices and the anti-exchange closure. *Algebra Universalis* **10** (1980), 290–299.
4. P. Erdös and G. Szekeres, A combinatorial problem in geometry. *Composito Math.* **2** (1935), 463–470.
5. R. E. Jamison-Waldner, A perspective on abstract convexity: classifying alignments by varieties. In *Convexity and Related Combinatorial Geometry* (D. C. Kay and M. Breem, eds), Marcel Dekker, New York, 1982, pp. 113–150.
6. B. Korte and L. Lovász, Mathematical structures underlying greedy algorithms. In *Fundamentals of Computation Theory* (F. Gécseg, ed.), Lecture Notes in Computer Science no. 117, Springer Verlag, New York, 1981, pp. 205–209.
7. B. Korte and L. Lovász, Greedoids – a structural framework for the greedy algorithm. In *Progress in Combinatorial Optimization: Proceedings of the Silver Jubilee Conference on Combinatorics, Waterloo, June 1982* (W. R. Pulleyblank, ed.), Academic Press, New York, 1984 (in press).
8. B. Korte and L. Lovász, Structural properties of greedoids. *Combinatorica* **3** (1984), 359–374.
9. B. Korte and L. Lovász, A note on selectors and greedoids. *Europ. J. Combinat.* (in press).

A CORRESPONDENCE BETWEEN SPANNING TREES AND ORIENTATIONS IN GRAPHS

Michel Las Vergnas

ABSTRACT   In a previous paper we have established a simple relation between the number of spanning trees and the number of
orientations of given activities of a graph with a total ordering of its
edge-set. We exhibit here a canonical correspondence between spanning trees and orientations consistent with this relation.

The theorems presented in this paper have several levels of generality:
graphs, vector spaces over the reals, oriented matroids, oriented matroid
perspectives [2]. We will consider here the case of graphs.

Let G be a (finite) undirected graph with edge-set E. Suppose E is
totally ordered. Let T be a spanning tree of G (without loss of generality
we may suppose G connected). An edge $e \in E \backslash T$ is *externally active* with
respect to T (and the ordering) if e is the smallest edge of the unique
cycle contained in $T \cup \{e\}$. Similarly an edge $e \in T$ is *internally active* if e
is the smallest edge of the cocycle of G determined by $T \backslash \{e\}$. The
spanning tree T has *internal activity* $\iota(T) = i$ and *external activity* $\varepsilon(t) = j$
if there are exactly i internally active and j externally active edges with
respect to T. These definitions are due to Tutte [4]. Let t_{ij} be the number
of spanning trees of G with internal activity i and external activity j.
Tutte has shown that t_{ij} does not depend on the ordering of E [4].
Furthermore,

$$\sum_{i,j} t_{ij} x^i y^j = t(G; x, y)$$

is the dichromatic polynomial or Tutte polynomial of G.

Consider now an orientation $\vec{G}$ of G. The following definitions are
introduced in [1]: An edge $e \in E$ is *externally active* with respect to the
orientation (and the given ordering) if there is some directed cycle of $\vec{G}$
with smallest edge e. Similarly e is *internally active* if there is some
directed cocycle of $\vec{G}$ with smallest edge e. Let $o(\vec{G})$ be the number of
externally active edges of $\vec{G}$ and $o^*(\vec{G})$ be the number of internally active
edges. Let o_{ij} denote the number of orientations $\vec{G}$ of G with $o^*(\vec{G}) = i$
and $o(\vec{G}) = j$. (In order that the expression of $t(G; x, y)$ below be valid

GRAPH THEORY AND COMBINATORICS
ISBN 0-12-111760-X

when G has loops, we have to make the convention that there are two ways to direct a loop.) In [1] we prove that o_{ij} does not depend on the ordering of G and that

$$t(G; x, y) = \sum_{i,j} 2^{-i-j} o_{ij} x^i y^j.$$

The proof is by deletion/contraction of the greatest edge in G. This theorem generalizes Stanley's well known theorem on the number of acyclic orientations of a graph [3]. In its full generality it contains several other counting results on orientations of graphs, regions determined by hyperplanes in spaces $\mathbb{R}^n$, oriented matroids, etc. (see [1, Section 4] for a discussion of these applications).

By comparing the two above expressions of $t(G; x, y)$ we get the remarkable relation

$$o_{ij} = 2^{i+j} t_{ij}.$$

This relation suggests a question: Is there a natural correspondence between orientations and spanning trees of a graph which preserves activities and has the right multiplicity? Explicitly: given an ordering of the edges of a graph G, we want to associate with any orientation $\vec{G}$ of G a spanning tree $T = f(\vec{G})$ of G by some canonical procedure in such a way that (a) $\iota(T) = o^*(\vec{G})$ and $\iota(T) = o(\vec{G})$, and (b) there are exactly $2^{\iota(T)+\varepsilon(T)}$ orientations $\vec{G}$ of G such that $T = f(\vec{G})$.

We describe below a correspondence with the desired properties. Proofs and generalizations will appear elsewhere [2]. On the relationship between the present work and an earlier paper of G. Berman, see [1, Section 3].

An edge of a directed graph $\vec{G}$ is either in a directed cycle or in a directed cocycle but not in both. It follows that internal and external activities can be dealt with separately. The general case follows easily by duality from the case of orientations and spanning trees of internal activities zero. In this case, if no confusion results, we say active instead of externally active.

Algorithm 1

Let $\vec{G}$ be a connected directed graph with $o^*(G) = 0$ (i.e. a strongly connected graph) with a total ordering of its edge-set.

Let a be the greatest active non-loop edge of $\vec{G}$ and $\gamma_1 < \gamma_2 < \ldots$ be the list in lexicographical order of the directed cycles of $\vec{G}$ activated by a (i.e. with least edge a). Let z_i be the greatest edge of γ_i, $i = 1, 2, \ldots$. Let k be the smallest index i such that, in $\vec{G}/z_i$ (the graph obtained from $\vec{G}$ by contracting z_i), the only new active edges (if any) are loops greater than z_i. Set $b(\vec{G}) = z_k$. Define $\vec{G}_1 = \vec{G}$, $b_1 = b(\vec{G})$, $\vec{G}_{j+1} = \vec{G}_j/b_j$, $b_{j+1} = b(\vec{G}_{j+1})$ inductively.

Stop when $\vec{G}_j$ is reduced to a vertex. Then set $r = j - 1$, $T = \{b_1, b_2, \ldots, b_r\}$.

EXAMPLE 1   Consider the directed graph $\vec{G}$ of Fig. 1, the ordering of the edge-set being given by the labelling. We have $\iota(\vec{G}) = 0$, $\varepsilon(\vec{G}) = 1$; the only active edge of $\vec{G}$ is 1. The directed cycles of $\vec{G}$ in lexicographical order are $\gamma_1 = 123 < \gamma_2 = 1246 < \gamma_3 = 12478 < \gamma_4 = 1457$. We have $z_1 = 3$, $z_2 = 6$, $z_3 = 8$, $z_4 = 7$. In $\vec{G}/3$ the edge 4 becomes active. In $\vec{G}/6$ the edge 7 becomes active. In $\vec{G}/8$ no new active edge appears. Hence $k = 3$, $b_1 = 8$, $\vec{G}_2 = \vec{G}/8$. Applying the algorithm to $\vec{G}_2$, etc., we get successively $b_2 = 6$, $b_3 = 3$, $b_4 = 2$ $(r = 4)$: $T = 2368$.

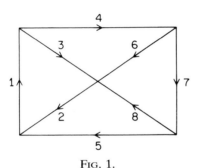

FIG. 1.

THEOREM 1   *The correspondence $T = f(\vec{G})$ given by Algorithm 1 has the required properties. Furthermore, this correspondence preserves the activities of edges: the internally (respectively the externally) active edges with respect to the orientation $\vec{G}$ and with respect to the spanning tree T are the same.*

The key property in the proof of Theorem 1 is that for $i = 2, \ldots, k$ we have $\gamma_{i-1} \setminus \{z_{i-1}\} \subseteq \gamma_i$. To show that the correspondence has the right multiplicity we have to define a reciprocal correspondence:

Algorithm 2

Let G be a (connected) undirected graph with a total ordering of its edge-set. Let T be a spanning tree of G with $\iota(T) = 0$.

(1) Let a be the greatest active non-loop edge of G with respect to T.

Set $a_1 = a$ and let z_1 be the greatest edge of the unique cycle contained in $T \cup \{a_1\}$.

(2) Suppose a_i, z_i have been constructed, $i \geqslant 1$. If in G/z_i there is at least one new active edge (with respect to $T \backslash \{z_i\}$) which is not a loop greater than z_i, let a_{i+1} be the smallest one. Let z_{i+1} be the greatest edge of the unique cycle contained in $T \cup \{a_{i+1}\}$. Then set $i = i + 1$ and go to (2).

Otherwise set $k = i$, $b = z_i$ and go to (3).

(3) If a is undirected, then direct it arbitrarily. Direct b consistently with the direction induced by a on the unique cycle contained in $T \cup \{a_1, a_2, \ldots, a_k\} \backslash \{z_1, z_2, \ldots, z_{k-1}\}$. Direct the edges e parallel to b: in the direction of b if e is inactive; arbitrarily if e is active.

Form $G' = G/b$, $T' = T \backslash \{b\}$. Eliminate loops. If all edges have been directed then end (G is reduced to one vertex). Otherwise go to (1) with $G = G'$, $T = T'$.

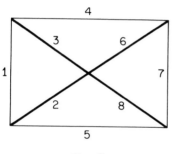

FIG. 2.

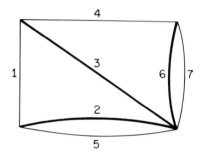

FIG. 3.

EXAMPLE 2   Consider the undirected graph of Fig. 2, the ordering of the edge-set being given by the labelling. Let T be the spanning tree 2368. We have $\iota(t) = 0$, $\varepsilon(T) = 1$; the only active edge is 1. We have $a = a_1 = 1$, $z_1 = 3$. In $G/3$, 4 becomes active, hence $a_2 = 4$, $z_2 = 6$. In $G/6$, 7 becomes active, hence $a_3 = 7$, $z_3 = 8$. In $G/8$, no new active edge appears, hence $b = z_3 = 8$. We direct 1 arbitrarily, then 8 consistently with the direction induced by 1 on the unique cycle 12478 contained in $T \backslash 36 + 147$. We form $G/8$ (Fig. 3) and apply again the procedure. We have $a = a_1 = 1$, $z_1 = 3$, $a_2 = 4$, $z_2 = 6$. In $G/6$ the only new active edge 7 is a loop 6. Hence $b = z_2 = 6$. We direct 6 and 7, etc.

THEOREM 2   *Algorithms 1 and 2 are reciprocal in the following sense:*
Let G be an undirected graph, with a total ordering of its edge-set. Let $\vec{G}$ be an orientation of G. Let T be the spanning tree given by Algorithm 1

*applied to $\vec{G}$. Then $\vec{G}$ is the directed graph given by Algorithm 2 applied to
T when the active edges are directed as they are directed in $\vec{G}$.*

*Conversely let T be a spanning tree of G. Let $\vec{G}$ be the orientation of G
given by Algorithm 2 applied to T for some choice of the directions of the
active edges. Then T is the spanning tree given by Algorithm 1 applied to
$\vec{G}$.*

Theorem 2 is illustrated by Examples 1 and 2. We note that for any
orientation associated with T by Algorithm 2, for $i = 1, 2, \ldots, k$, γ_i is the
unique cycle contained in $T \cup \{a_1, a_2, \ldots, a_i\} \backslash \{z_1, z_2, \ldots, z_{i-1}\}$, and a_i is
the smallest element of $\gamma_i \backslash \gamma_{i-1}$.

In Example 2 the direction of the only active edge 1 forces the
orientation of the whole graph (in this case up to the reversal of all
directions we get the graph of Fig. 1). In general let $E(a)$ be the set of
edges whose direction are forced by the direction of an active edge a in
Algorithm 2. The sets $E(a)$ partition the edge-set of G. The $2^{\varepsilon(T)}$
orientations of G associated with T (recall that $\iota(T) = 0$) are obtained
from any one of them by reversing for each subset A of active edges all
edge directions in the $E(a)$ for $a \in A$.

EXAMPLE 3   We give some further examples of the correspondence. The
graph G is the undirected graph considered in Examples 1 and 2 (see Fig.
1 or Fig. 2), with the ordering of the edge-set given by the labelling.

The Tutte polynomial of G is

$$t(G; x, y) = x^4 + y^4 + 4x^3 + 4x^2y + 4xy^2 + 4y^3 + 6x^2 + 9xy + 6y^2 + 3x + 3y.$$

We have $t_{0,1} = 3$. The three spanning trees T of G with $\iota(T) = 0$,
$\varepsilon(T) = 1$, namely 1257, 2367 and 2368, are shown in Fig. 4, together with
a corresponding orientation. In this case the correspondence between
spanning trees and orientations is one to two, that is one to one up to the
reversing of all edge directions.

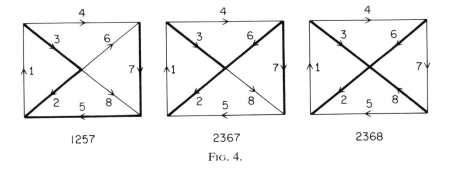

| 1257 | 2367 | 2368 |

FIG. 4.

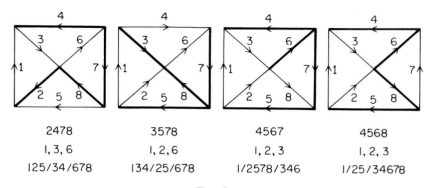

2478	3578	4567	4568
I, 3, 6	I, 2, 6	I, 2, 3	I, 2, 3
I25/34/678	I34/25/678	I/2578/346	I/25/34678

FIG. 5.

We have $t_{0,3} = 3$. The four spanning trees T of G with $\iota(T) = 0$, $\varepsilon(T) = 3$ are shown in Fig. 5, together with one of the $8 = 2^3$ corresponding orientations. The first spanning tree is 2478; its active edges are 1, 3, 6; the associated partition of E is $E(1) = 125$, $E(3) = 34$, $E(6) = 678$. The seven other orientations associated with $T = 2478$ are obtained by reversing all directions of edges in 125, 34, 678, 12345, 125678, 34678, 12345678 respectively.

References

1. M. Las Vergnas, The Tutte polynomial of a morphism of matroids. II. Activities of orientations. In *Progress in Combinatorial Optimization: Proceedings of the Waterloo Silver Jubilee Combinatorial Conference 1982* (W. R. Pulleyblank, ed.), Academic Press, New York (to appear).
2. M. Las Vergnas, Bases and orientations in graphs and matroids. Manuscript in preparation.
3. R. Stanley, Acyclic orientations of graphs. *Discrete Math.* **5** (1973), 171–178.
4. W. T. Tutte, A contribution to the theory of chromatic polynomials. *Canad. J. Math.* **6** (1954), 80–91.

21

EQUI-MATCHABLE GRAPHS

M. Lesk, M. D. Plummer and W. R. Pulleyblank

ABSTRACT We characterize those graphs in which every maximal matching is maximum. Using our characterization we show that membership in this class can be polynomially determined.

1. Introduction

A matching M in a graph G is *maximal* if it is not a proper subset of any larger matching. A matching M is *maximum* if it has the largest cardinality among all matchings in G. In 1974, Grübaum [5] posed the problem of characterizing those graphs in which every maximal matching is maximum. We will call these graphs *equi-matchable*. In the same year, Meng [11] obtained the following characterization of equi-matchable graphs. A set of lines L in G is a *line cover* if every point of G is incident with a line of L.

THEOREM (Meng)   *A graph G is equi-matchable if and only if $|M| + |L| \geq |V(G)|$ holds for some minimum maximal matching M and for some minimum line cover L in G.*

This theorem, which in fact can be easily deduced from a theorem of Gallai [3] and of Norman and Rabin [12], does not provide a very satisfactory characterization of equi-matchable graphs for two reasons. First, it provides no information concerning the structure of such graphs; second, it does not provide a "good" characterization of equi-matchable graphs, from the point of view of complexity. (See Garey and Johnson [4] for definitions of the terms P, NP, co-NP, NP-complete and an excellent exposition of the associated theory.)

A *good* (or NP-) characterization of a class of graphs is the following. For each graph in the class we provide a certain amount of additional information, the amount of which is polynomially bounded in the length of the description of the graph, such that a person can verify in polynomial time that the graph belongs to the class. (In some cases, no additional information need be provided.) For example, the definition of equi-matchable graphs provides *a priori* a good characterization of those

GRAPH THEORY AND COMBINATORICS
ISBN 0-12-111760-X

graphs which are *not* equi-matchable. For if we have a non equi-matchable graph G, then by providing as additional information two maximal matchings M_1 and M_2 of different cardinalities, we easily enable someone to verify in polynomial time that M_1 and M_2 are indeed maximal matchings of different cardinalities and hence that G is not equi-matchable. Note that we are not at all concerned by the amount of time that it takes us to find M_1 and M_2. The point of a good characterization is simply that such additional information should exist, and the properties attributed to it should be polynomially checkable.

Meng's theorem does not provide a good characterization of those graphs which are equi-matchable (modulo the current state of knowledge) because if we wish to show that T is equi-matchable using this theorem we must exhibit both a minimum line cover L in G and a minimum maximal matching M. Yannakakis and Gavril [15] have shown that it is *NP*-complete to determine whether or not, for a fixed k, a graph has a maximal matching containing k or fewer lines. Consequently, unless $NP = \text{co-}NP$ (a situation which most researchers consider to be extremely unlikely), no good characterization can exist of a minimum maximal matching. In other words, there is no "small" amount of extra information which could be presented, along with G and a maximal matching M, that would enable someone to verify polynomially that M is indeed a minimum maximal matching. Consequently, by merely providing G, M and L, and some polynomial amount of extra information, it is not possible to enable a person to verify polynomially that G is equi-matchable using Meng's result.

A second characterization of equi-matchable graphs by Lewin [8] suffers from a defect similar to that of Meng. Lewin defines *an odd blocked snake* with respect to a matching M of G as a simple path P in $G - V(M)$ of odd length having distinct end-points which are non-adjacent in G and such that any point of $G - V(M)$ adjacent to an end-point of P belongs to P. He then shows the following:

THEOREM (Lewin) *A graph is equi-matchable if and only if for every matching M, there exists no odd blocked snake with respect to M.*

Once again we have a good characterization of non-equi-matchable graphs. (Just exhibit a matching M and an odd blocked snake with respect to M.) However, it does not seem to provide any efficient means for showing that a graph is equi-matchable, as it seems no easier to verify equi-matchability using this theorem than it does using the original definition.

Relating this somewhat informal preceding exposition to the language of complexity theory, there exists a good characterization of those graphs belonging to a set C if and only if the recognition problem for C belongs to the class *NP*. There exists a polynomially bounded algorithm for determining whether or not a given graph G belongs to C if and only if the recognition problem for C is in the class *P*. The two outstanding questions in complexity theory are the following:

(i) Is $P = NP$? If the answer is "yes" (most people suspect it is "no"), then this would imply that whenever there exists a good characterization of a class of graphs, there exists also a polynomially bounded algorithm for determining whether or not a given graph belongs to the class.

(ii) Is $NP = $ co-*NP*? If the answer is "yes" (and again, most people suspect it is "no"), then this would imply that whenever there exists a good characterization of membership in a class of graphs, there also exists a good characterization of non-membership in the class.

For the case of equi-matchable graphs, we provide in this paper the following: a good characterization of those graphs which *are* equi-matchable and a polynomially bounded algorithm, which uses our characterization, to determine *whether or not* a given graph is equi-matchable.

Our approach is as follows. We will use the Gallai–Edmonds theorem [**10**, p. 52] on maximum matchings and a result by Sumner [**14**] on so-called "randomly matchable" graphs to help reduce the problem to the bipartite case. Then we will present a good characterization of the bipartite graphs which are equi-matchable.

If $S \subset V(G)$ we will denote by $\langle S \rangle$ the subgraph induced by S, and by $\Gamma(S)$ the set of points adjacent to at least one point of S. (Of course, we may have $\Gamma(S) \cap S \neq \emptyset$. For any other concepts used, but not defined, in this paper, the reader is referred to Harary [**6**], Garey and Johnson [**4**] and Lovász [**10**].

2. Randomly matchable graphs and factor-critical graphs

A matching in a graph G is *perfect* if it covers all points in G. A graph G is *randomly matchable* if every matching is a subset of a perfect matching. Sumner [**14**] has characterized these graphs. We first present this result, but with a new proof.

THEOREM 1 *A connected graph is randomly matchable if and only if* $G = K(n, n)$ *or* $G = K(2n)$.

PROOF   Clearly $K(n, n)$ and $K(2n)$ are randomly matchable.

Conversely, suppose G is a connected randomly matchable graph. It is then easy to see that, in fact, G must be 2-connected. Now suppose G is bipartite, say $G = (U, W)$, but not complete. So let $u \in U$, $w \in W$ be non-adjacent. Since G is connected there is an odd path P joining u and w. Put the first, third, ..., and last lines of P into a matching and extend this matching to a perfect matching M for G. Then the symmetric difference $M \oplus P$ is a matching for G which leaves only points u and w exposed. Hence $M \oplus P$ cannot extend to a perfect matching, a contradiction. Hence $G = K(n, n)$ for some $n \geqslant 1$.

Now we consider the non-bipartite case.

Claim 1   If G is any 2-connected non-bipartite graph, every point lies on an odd cycle. For let u be any point and let C be any odd cycle. If $u \in V(C)$ we are done, so suppose $u \notin V(C)$. Then by Meneg's theorem there exist two openly disjoint paths from u to two different points on C and these two paths, together with one of the two parts of C intercepted, form an odd cycle.

Claim 2   If G is any 2-connected non-bipartite graph, then every pair of points is joined by an odd path.

Let u and v be any two points of G. If $uv \in E(G)$, there is nothing to prove, so suppose $uv \notin E(G)$. By Claim 1 we know there is an odd cycle C containing v. If $u \in V(C)$ again we are done, so suppose $u \notin V(C)$. Then there is a path P from u to cycle C. If P meets C in a point $\neq v$ we are done, so assume all such paths P meet C at v. Thus v is a cutpoint for G, a contradiction, and Claim 2 is proved.

Suppose G is not bipartite. Let u and v be any pair of non-adjacent points in G. By Claim 2, there is an odd path P joining u and v. Now proceed as in the bipartite case to get the same contradiction. Thus $G = K(2n)$, for some $n \geqslant 1$.   □

The following results from the *Gallai–Edmonds theorem* [**10**, p. 52] will be useful.

Let G be any graph and denote by D the set of those points which are left uncovered by at least one maximum matching of G. Let A denote the points of $V(G) - D$ which are neighbours of at least one point in D. Finally, let $C = V(G) - (D \cup A)$. Then:

(a) Every component of the graph $\langle D \rangle$ is *factor-critical*; i.e. if $\langle D_i \rangle$ is such a component and v is any point of $\langle D_i \rangle$, then $\langle D_i - v \rangle$ is a graph with a perfect matching.

(b) If M is any matching of G, then M is maximum if and only if (i) for each $\langle D_i \rangle$, at most one line of M joins D_i to A; (ii) for each $a \in A$, there is a line of M joining a to some $\langle D_i \rangle$; (iii) for each $\langle D_i \rangle$, M leaves at most one point of $\langle D_i \rangle$ uncovered; and (iv) M perfectly matches C to itself.

The following properties of an equi-matchable graph with no perfect matching are now immediate.

LEMMA 1 *Let G be a connected equi-matchable graph with no perfect matching, having Gallai–Edmonds decomposition (D, A, C). Then:*

(a) *$C = \phi$, and*
(b) *A is an independent set in G.* $\square$

The case in which G has no perfect matching, but $A = \phi$, must be treated separately. In this case every component of G is factor-critical and so it is sufficient to characterize those factor-critical graphs which are equi-matchable.

First, we prove a lemma which applies to any graph. This result was first proved by Lovász [**9**, Theorem 8.2] as a corollary to some more general results. For the sake of completeness we include a direct proof.

LEMMA 2 *Let G be a graph and suppose $X \subseteq V(G)$. Then exactly one of the following holds:*

(1) *G has a matching which covers X or*
(2) *there exist disjoint odd subsets $S_1, S_2, \ldots, S_k$ of X such that no line of G joins S_i to S_j for $i \neq j$ and $|N(S_1, \ldots, S_k)| < k$, where $N(S_1, \ldots, S_k) = \Gamma(S_1 \cup \ldots \cup S_k) - (S_1 \cup \ldots \cup S_k)$.*

PROOF If (2) holds then it is easy to see that every matching of G leaves some point of X exposed.

Conversely, suppose every matching of G leaves some $x \in X$ exposed. Among all matchings of G choose one, M, which leaves the minimum possible number of points of X exposed. Now delete all exposed points of $V(G) - X$ and let G' be the resulting graph. Then M is a maximum matching of G' which is not perfect; so, letting (D', A', C') be the Gallai–Edmonds decomposition of G', we must have $D' \neq \varnothing$.

First note that $D' \subseteq X$. For if not, we could find a maximum matching M' of G' leaving some $w \in D' - X$ exposed and M' would then be a matching of G leaving fewer points of X exposed than does M, contradicting the choice of M.

Secondly, no $v \in D'$ can be adjacent in G to any $w \in V(G) - V(G')$, for suppose such a v and w existed. Take a maximum matching M'' of G'

which leaves v exposed. Now $|M| = |M''|$ since both are maximum in G' and hence both contain perfect matchings of C' and matchings of all of A' into D'. Thus each covers the same number of points in X. But then $M'' \cup \{vw\}$ is a matching in G covering one more point of X than does M, thus contradicting the choice of M.

Thus if $\langle D_1' \rangle, \langle D_2' \rangle, \ldots, \langle D_k' \rangle$ are the components of $\langle D' \rangle$, then $N(D_1', D_2', \ldots, D_k')$ (taken *in G*) equals A'. But by the Gallai–Edmonds theorem applied to G', $|A'| < k$, so letting $S_i = D_i'$, $i = 1, \ldots, k$, we have shown that (2) holds. □

Now we are prepared to give a good characterization of factor-critical equi-matchable graphs.

THEOREM 2   *Let G be a connected factor-critical graph. Then the following are equivalent*:

(1) *G is equi-matchable.*
(2) *For every pair u, v of independent points and for every matching M of $G - u - v$, there is a third point w adjacent to u or to v which is not covered by M.*
(3) *For every pair u, v of independent points in G there exist disjoint sets $S_1, S_2, \ldots, S_k \subseteq \Gamma(u) \cup \Gamma(v)$ such that*
    (i) $|S_i|$ *is odd for* $1 \le i \le k$;
    (ii) *no line joins S_i to S_j for $i \ne j$; and*
    (iii) $|N(S_1, S_2, \ldots, S_k)| < k$, *where*

$$N(S_1, \ldots, S_k) = \Gamma\left(\bigcup_{i=1}^{k} S_i\right) - \bigcup_{i=1}^{k} S_i - \{u, v\}$$

(*that is, $N(S_1, \ldots, S_k)$ is the set of points in $G - u - v$ each of which does not lie in any S_i, but each of which is adjacent to a point in some S_i*).

PROOF   First we show the equivalence of (1) and (2). If G is not equi-matchable, then there is a maximal matching M_{uv} in G leaving some two points u and v exposed and so u and v are non-adjacent. If any $w \in \Gamma(u) \cup \Gamma(v)$ is not covered by M_{uv}, the maximality of M_{uv} in G is contradicted. So all such points w are covered by M_{uv} and so this choice of points u and v together with matching M_{uv} contradicts (2).

Conversely, if there exist two points u and v in G and a matching M_{uv} in $G - u - v$ which contradict (2), then clearly M cannot be extended to a maximum cardinality matching of G, for any extension leaves both u and v exposed and a maximum matching of a factor-critical graph can leave only one point exposed.

The equivalence of (2) and (3) follows by letting X and G in Lemma 2 be replaced by $\Gamma(u) \cup \Gamma(v)$ and $G - u - v$ respectively. $\square$

We note that Theorem 2 does give us a polynomial method for determining whether or not a factor-critical graph is equi-matchable. Consider each pair u, v of non-adjacent points in turn and, for each pair, let $X' = \Gamma_G(u) \cup \Gamma_G(v)$ and $G' = G - u - v$. Now to each line of G' assign a weight equal to the number of its end-points lying in X'. (Clearly, all weights in G' will be 0, 1, or 2.) Note that the sum of the weights of the lines of a matching in G' will equal the number of points of X' covered by the matching. Then find a maximum weight (not necessarily perfect) matching M^* of G' using Edmonds' *weighted* matching algorithm [2]. Then M^* will leave exposed the minimum possible number of points of X'. If, for some u and v, the matching M^* leaves no points of X' exposed, then property (2) of Theorem 2 does not hold, so G is not equi-matchable. (Moreover, extending M^* to any maximal matching M_G of G yields a maximal matching of G which is not of maximum cardinality.)

On the other hand, if for each pair u and v, the corresponding M^* leaves some point of X' exposed, then by (2) of Theorem 2, G is equi-matchable. Moreover, the matching algorithm automatically produces the sets S_i required in part (3) of Theorem 2. (Alternatively, we can apply the Gallai–Edmonds decomposition theory to the subgraph of G' obtained by deleting all points of $V(G') - X'$ not covered by M^* in a manner analogous to that used in the proof of Lemma 2.)

3. Reduction to the bipartite case

In this section we consider those graphs which have a non-trivial Gallai–Edmonds decomposition, i.e. they are not factor-critical and do not contain a perfect matching. For a connected graph this is equivalent to saying that $A \neq \varnothing$.

THEOREM 3 *Let G be a connected equi-matchable graph without a perfect matching, let (D, A, C) be its Gallai–Edmonds decomposition and suppose $A \neq \varnothing$. Let $\langle D_i \rangle$ denote any component of $\langle D \rangle$ with $|D_i| \geq 3$.*

(1) *Then $\langle D_i \rangle$ must be one of the following types of graphs (see Fig. 1):*
  I. *$\langle D_i \rangle = K(2m - 1)$ for some $m \geq 2$ and every point of $\langle D_i \rangle$ is joined to exactly one common point $a \in A$.*
  II. *$\langle D_i \rangle$ contains a cutpoint d_i of G (call it the "hook" of D_i) which is the only point of D_i adjacent to a point of A.*

I.

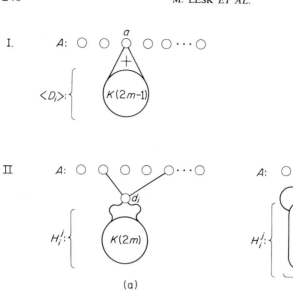

II

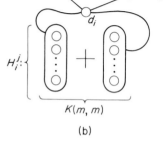

(a)

(b)

III

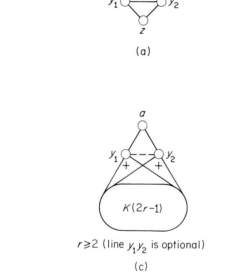

(a)

(b)

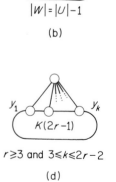

$r \geq 2$ (line $y_1 y_2$ is optional)

(c)

$r \geq 3$ and $3 \leq k \leq 2r - 2$

(d)

Fig. 1.

Let $H_i^1, \ldots, H_i^r$, $r \geq 1$, be the components of $\langle D_i - d_i \rangle$. Consider any one of these, H_i^j. There are two possibilities: (a) $H_i^j = K(2m)$ for some $m \geq 1$ and at least two lines join d_i to H_i^j, or (b) $H_i^j = K(m, m)$ for some $m \geq 1$ and if (U, W) is the bipartition of $K(m, m) = H_i^j$, at least one line joins d_i to a point of U and at least one line joins d_i to a point of W.

III. At least two points of D_i are adjacent to points of A and at least one point of $\langle D_i \rangle$ is adjacent to no point of A. In this case there is a point $a \in A$ such that a separates $\langle D_i \rangle$ from the rest of G.

Here we have four subcases. If $\langle D_i \rangle$ contains exactly two points y_1 and y_2 of attachment to a, then $\langle D_i \rangle$ must be one of the following three types: (a) $\langle D_i \rangle$ is a $K(3)$; (b) $\langle D_i - y_1 - y_2 \rangle$ is a complete bigraph $K(r, r-1)$ where $r \geq 2$, and if (U, W) is the bipartition of $\langle D_i - y_1 - y_2 \rangle$ where $|U| = r$, then y_1 and y_2 are both adjacent to all points of U and to each other; (c) $\langle D_i - y_1 - y_2 \rangle$ is a $K(2r-1)$, $r \geq 2$, y_1 and y_2 are both adjacent to all points of $\langle D_i - y_1 - y_2 \rangle$ and y_1 and y_2 may or may not be adjacent to each other. The fourth subcase may be stated as follows: (d) if $\langle D_i \rangle$ has between 3 and $|D| - 1$ points of attachment to a, then $\langle D_i \rangle$ is a $K(2r-1)$ for some $r \geq 3$.

(2) Suppose we delete all type II and type III components of $\langle D \rangle$ from G and contract all type I components to single points. Then there is a matching of the resulting (bipartite) graph $\hat{G}_1$ which covers all points of A and $\hat{G}_1$ is equi-matchable.

(Note that (2) implies that the number of singleton and type I components of $\langle D \rangle$ is at least $|A|$ and moreover each point of A is adjacent to such a component. Condition (1) implies that all components of $\langle D \rangle$ which are not singletons are equi-matchable without perfect matchings. However, there are equi-matchable factor-critical graphs, for example the cycle on seven points, which cannot serve as a $\langle D_i \rangle$ in this decomposition.)

PROOF Let us first note that there cannot exist two independent lines joining D_i to A for, if there were, they could be extended to a maximum matching of G of a form violating conclusion (b-(i)) of the Gallai–Edmonds theorem.

Suppose all points of D_i are adjacent to a common point $a \in A$. Let $G_i = \langle V(D_i) \cup \{a\} \rangle$. We will show G_i is a $K(2m)$ for some $m \geq 2$ by first showing that G_i is randomly matchable.

Let M_i be any matching in G_i, let M be a maximum matching of G with $M_i \subseteq M$ and let e be the line of M covering a.

Suppose e matches a to a point not in G_i. Then since M covers all points of D_i except one, say v_i, the set $([M \cap E(G_i)] - \{e\}) \cup \{av_i\}$ is a perfect matching of G_i which contains M_i.

On the other hand, if e matches a to a point in G_i (i.e. to a point in D_i) then, by the Gallai–Edmonds theorem, $M \cap E(G_i)$ is a perfect matching in G_i.

Thus in either case M_i extends to a perfect matching of G_i and hence, by Theorem 1, G_i is either a $K(2m)$ or a $K(m, m)$ for some $m \geq 2$. But if $G_i = K(m, m)$, $\langle D_i \rangle$ is a $K(m, m-1)$ which is not factor-critical, contradicting part (a) of the Gallai–Edmonds theorem. So $G_i = K(2m)$; i.e. $\langle D_i \rangle$ is of type I.

Now suppose $\langle D_i \rangle$ contains a unique point d_i adjacent to a point of A. Let M_i be any matching in $\langle D_i - d_i \rangle$ and let $e = ad_i$ be any line joining d_i to A. Let $M'_i = M_i \cup \{e\}$. Then M'_i extends to a maximum matching M of G which must, therefore, perfectly match $\langle D_i - d_i \rangle$. So again by Theorem 1, each component of $\langle D_i - d_i \rangle = K(2m)$ or $= K(m, m)$ for some $m \geq 1$.

Since each $\langle D_i \rangle$ is factor-critical, it is 2-line-connected, and hence there are at least two lines joining d_i to each component H_i^j of $\langle D_i - d_i \rangle$. If H_i^j is a $K(2m)$ then we have II(a). In the case that H_i^j is a $K(m, m)$ and the bipartition of H_i^j is (U, W) with $|U| = |W| = m$, suppose d_i is not adjacent to any point of W. Then if $u \in U$, $H_i^j - u$ has no perfect matching, contradicting the fact that $\langle D_i \rangle$ is factor-critical. So d_i is joined to at least one point of W and by symmetry also to a point of U. Therefore $\langle D_i \rangle$ is of the form of II(b).

Now suppose $\langle D_i \rangle$ is not of type I or II, i.e. at least two points y_1 and y_2 of $\langle D_i \rangle$ are adjacent to a point $a \in A$ and at least one point is not adjacent to any point of A. Note that a separates $\langle D_i \rangle$ from the rest of G.

If there is only one point $z \in D_i - y_1 - y_2$, then $\langle D_i \rangle$ is a factor-critical graph on three points and hence is a $K(3)$. Since $\langle D_i \rangle$ is not type I, points z and a are not adjacent. Thus we have case III(a).

So suppose $\langle D_i \rangle$ contains at least three points other than y_1 and y_2. Consider $\langle D_i - y_1 \rangle$. Let E_1 be any matching in $\langle D_i - y_1 \rangle$. Then $\{ay_1\} \cup E_1$ extends to a maximum matching M of G covering y_1. Hence, since $\langle D_i - y_1 \rangle$ contains a perfect matching, $M \cap E(\langle D_i - y_1 \rangle)$ must be a perfect matching of $\langle D_i - y_1 \rangle$. So $\langle D_i - y_1 \rangle$ is randomly matchable and hence is a $K(m, m)$ or a $K(2m)$ by Theorem 1. Since $|D_i| \geq 5$, we have $m \geq 2$.

First suppose that $\langle D_i - y_1 \rangle$ is a $K(m, m)$ having bipartition (U, W') with $y_2 \in W'$. Note that $|U| \geq 2$. Now by a symmetric argument $\langle D_i - y_2 \rangle$ is randomly matchable and since it contains the independent set U, $\langle D_i - y_2 \rangle$ is also a $K(m, m)$. Hence y_1 is adjacent to all of U but none of $W = W' - \{y_2\}$. However, y_1 and y_2 are adjacent for otherwise $\langle D_i \rangle$ would be a bipartite factor-critical graph, which is impossible.

Next we claim that a is joined to no point of $\langle D_i \rangle$ other than y_1 and y_2, for if there exists $y_3 \in D_i - y_1 - y_2$ and a line ay_3 then $\langle D_i - y_3 \rangle$ contains a $K(3)$, e.g. with points y_1, y_2 and u, where u is any point of $U - \{y_3\}$. Hence, since $\langle D_i - y_3 \rangle$ must be randomly matchable, $\langle D_i - y_3 \rangle = K(2m)$. But this is a contradiction since one of the independent sets U and W must be contained in $D_i - y_3$. Thus a is adjacent to D_i only at y_1 and y_2 as claimed.

Thus we have the configuration of case III(b) with $r \geq 2$.

Now suppose $\langle D_i - y_1 \rangle$ is a $K(2m)$, as is $\langle D_i - y_2 \rangle$.

First suppose there are only two lines joining a and $\langle D_i \rangle$, namely ay_1 and ay_2. If $m = 2$ we have case III(a), so suppose $m \geq 3$. Thus $\langle D_i - y_1 - y_2 \rangle$ is a $K(2m - 1)$ and all points of $\langle D_i - y_1 - y_2 \rangle$ are joined to both y_1 and y_2. Thus we have configuration III(c) where line $y_1 y_2$ may or may not be present.

Finally suppose there is a line ay_3 where $y_3 \in D_i - \{y_1, y_2\}$. Recall both $\langle D_i - y_1 \rangle$ and $\langle D_i - y_2 \rangle$ are $K(2m)$s and so $\langle D_i - y_3 \rangle$ must also be a $K(2m)$. Hence y_1 and y_2 are adjacent and $\langle D_i \rangle$ is a $K(2m + 1)$ as claimed. Thus, where $y_1, y_2, \ldots, y_k$ are the points of $\langle D_i \rangle$ adjacent to a, we have the configuration of III(d). This completes the proof of (1).

Now let $\langle D_1 \rangle, \ldots, \langle D_k \rangle$ be those components of $\langle D \rangle$ which are neither singletons nor type I. For each $1 \leq i \leq k$, let M_i be a matching in $\langle D_i \rangle$ which covers all hooks of type II components and all points of D_i which are joined to A in case $\langle D_i \rangle$ is of type III. (It is routine to verify that such M_is exist.) Then $M_0 = M_1 \cup \ldots \cup M_k$ is a matching in G which therefore extends to a maximum matching M of G. Thus M must cover A and $M \cap E(\hat{G}_I)$ is a matching of $\hat{G}_I$ which covers all points of A. Now we show that $\hat{G}_I$ is equi-matchable. Let N_1 be any matching in G_I. For each type II or III component $\langle D_i \rangle$ in $\langle D \rangle$ choose a matching M_i in $\langle D_i \rangle$ which covers all points of $\langle D_i \rangle$ adjacent to A. Let N_2 be the union of these matchings M_i. Then $N_1 \cup N_2$ induces a matching in G which extends to a maximum matching M_0 of G which must match all of A into those components of $\langle D \rangle$ which are not of type II or of type III. Consequently, if $\hat{N}_I$ is the matching of $\hat{G}_I$ induced by M_0, then $\hat{N}_I$ covers all of A in $\hat{G}_I$ and hence is a maximum matching in $\hat{G}_I$ containing N_1. So $\hat{G}_I$ is equi-matchable. $\square$

We require some preliminary results before proving the converse of Theorem 3. A *point cover* for a graph G is a set of points collectively incident with all the lines of G. Such a set of minimum cardinality is called a *minimum point cover*.

LEMMA 3 *If $G = (U \cup W, E)$ is a connected bipartite graph with $|U| \leq |W|$ and in which each line extends to a maximum matching of G, then* (a) *if*

$|U| < |W|$, *U is the only minimum point cover for G, while* (b) *if* $|U| = |W|$, *U and W are the only minimum point covers for G.*

PROOF   Let *P* be a minimum point cover for *G* and suppose $U \cap P \neq \varnothing$ $\neq W \cap P$. Let $e = uw$ be a line joining $U \cap P$ and $W \cap P$. (Such an *e* must exist since *G* is connected.) Let *M* be a maximum matching in *G* containing *e*. Since *P* must cover *M* and since *P* contains both end-points of line *e*, we have $|P| \geq |M| + 1$. But this contradicts the König minimax theorem for bipartite graphs (i.e. that $|P| = |M|$).

Thus either $U \cap P = \varnothing$ or $W \cap P = \varnothing$ and, again by connectivity, either $P = U$ or $P = W$. Since *P* is a *minimum* cover the proof is complete. $\square$

COROLLARY 3(a)   *If* $G = (U, W)$ *is a connected bipartite graph with* $|U| \leq$ $|W|$ *in which each line extends to a maximum matching, then every maximum matching covers U.*

PROOF   Follows immediately from the preceding lemma and König's minimax theorem. $\square$

We now prove the converse to Theorem 3.

THEOREM 4   *Let G be a connected graph without a perfect matching, which is not factor-critical and which has Gallai–Edmonds decomposition* (D, A, C). *Suppose*

(1) $C = \varnothing$.
(2) *A is an independent set.*
(3) *All components of* $\langle D \rangle$ *are singletons or of types I, II or III as described in Theorem 3.*

Let $\hat{G}$ *be the bigraph obtained from G by shrinking all components of* $\langle D \rangle$ *to singletons and let* $\hat{G}_I$ *be obtained from* $\hat{G}$ *be deleting all points corresponding to type II or III components of* $\langle D \rangle$. *Suppose*:

(4) $\hat{G}_I$ *is equi-matchable and* $|A| \leq \frac{1}{2}|V(\hat{G}_I)|$.

Then G is equi-matchable.

Before we prove this theorem let us observe that there are connected graphs which satisfy hypotheses (1)–(3) which are not equi-matchable. See Fig. 2 for such a graph and note that the three-element matching shown does not extend to a maximum matching. (Note that $\hat{G}_I$ is not equi-matchable either.)

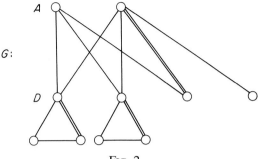

Fɪɢ. 2.

PʀᴏᴏF   Suppose that (1)–(4) hold. By hypothesis, if (A, H) represents the bipartition of $\hat{G}_I$, then $|A| \leq |H|$. Thus by Corollary 3(a) every maximum matching of $\hat{G}_I$ covers A.

Now let M be a matching in G. In particular let $M = \{e_1, \ldots, e_k, f_{k+1}, \ldots, f_r\} \cup M_D$ where each f_i joins A to a singleton or a type I component in $\langle D \rangle$, e_i joins A to a component of $\langle D \rangle$ of type II or III, and M_D is a matching in $\langle D \rangle$.

Since $\hat{G}_I$ is equi-matchable, we can find an $\hat{M}_I$, a maximum matching of $\hat{G}_I$ containing all the f_is. Then $\hat{M}_I$ covers A. Let M_I be the matching in G corresponding to $\hat{M}_I$. For any $a \in A$ which is matched by $\hat{M}_I$ to a point on $\hat{G}_I$ corresponding to a type I component $\langle D_s \rangle$, we can, with no loss of generality, assume that the line of M_I incident with a has as its other end-point a point of $\langle D_s \rangle$ not covered by M_D. For $i = 1, \ldots, k$, let g_i be the line of M_I adjacent to line e_i. So $M' = M_I - \{g_1, \ldots, g_k\} \cup \{e_1, \ldots, e_k\}$ is a matching in G covering A which uses only lines of G. Because of the structure of the type I, II and III components, we can now extend M' to a matching M_0 of G which contains both M' and M_D and which covers all but at most one point of each component $\langle D_s \rangle$ of $\langle D \rangle$. By the Gallai–Edmonds theorem, M_0 is a maximum matching of G containing M. $\square$

3. Good characterizations of equi-matchable bipartite graphs

In this section we characterize bipartite equi-matchable graphs in a manner analogous to the characterization of factor-critical graphs given in Theorem 2. We denote the bipartite graphs with point partition $U \cup W$ by $G = (U, W)$.

Tʜᴇᴏʀᴇᴍ 5   *A connected bipartite graph $G = (U, W)$ with $|U| \leq |W|$, is equi-matchable if and only if, for all $u \in U$, there exists a non-empty $X \subseteq \Gamma(u)$ such that $|\Gamma(X)| \leq |X|$.*

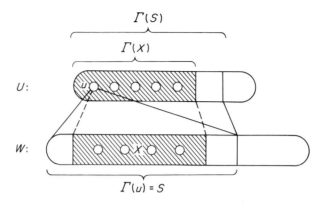

FIG. 3.

PROOF First suppose $G = (U, W)$ is equi-matchable. Then, by Corollary 3(a), all maximum matchings in G cover U. Now suppose there is a $u \in U$ such that for every $X \subseteq \Gamma(u) = S$, $|\Gamma(X)| > |X|$. Consider the subgraph $G_u = \langle \Gamma(u) = S, \Gamma(S) - \{u\}\rangle$. (Cf. Fig. 3.)

For all non-empty $X \subseteq S$, we have $|\Gamma_{G_u}(X)| = |\Gamma_G(X)| - 1 \geq |X|$. Thus by Hall's theorem there is a matching M of all of S into $\Gamma(S) - u$. But since G is equi-matchable M extends to a maximum matching M' of G which must cover U and hence u. But this is a contradiction since M already covers $\Gamma(u) = S$, but not u.

Conversely, let M be an arbitrary maximal matching in G. We must show it to be maximum. If M covers U, we are done, so suppose there is a point $u \in U$ not covered by M. Denote the lines of M by $u_i w_i$, $i = 1, \ldots, k, u_i \in U$, $w_i \in W$. Since G is connected u is not an isolate and hence $\Gamma(u) \subseteq \{w_1, \ldots, w_k\}$ by the maximality of M.

Now let X by any subset of $\Gamma(u)$, say $X = \{w_{i_1}, \ldots, w_{i_s}\}$. Then since $\{u_{i_1}, \ldots, u_{i_s}, u\} \subseteq \Gamma(X)$, we have $|X| = s < s + 1 \leq |\Gamma(X)|$, contradicting the hypothesis and completing the proof of the theorem. $\square$

The preceding theorem, together with the polynomial matching algorithm for general graphs due to Edmonds [1], allows us to show that the class of connected equi-matchable graphs is polynomially recognizable. That is, not only do we have good (i.e. polynomially verifiable) characterizations of both those graphs which are equi-matchable and those which are not, but moreover, given an arbitrary graph G, we can in polynomial time determine whether or not G is equi-matchable.

First note that if G has a perfect matching, by Sumner's result we need only exhibit whether or not $G = K(2m)$ or $G = K(m, m)$ for some m.

So suppose G has no perfect matching. Edmonds' matching algorithm can then be used to produce the Gallai–Edmonds decomposition (D, A, C) of G. If $A = \varnothing$, then G is factor-critical and we proceed in the manner discussed following Theorem 2. Otherwise $A \neq \varnothing$ and if G fails to satisfy one of the properties listed as hypotheses (1)–(3) of Theorem 4, then G is not equi-matchable. So suppose G satisfies these hypotheses. One can then form the bipartite graph $\hat{G}_I$. Denote the bipartition of $\hat{G}_I$ by (A, H). If $\hat{G}_I$ does not have a matching which covers all $a \in A$, then by Theorem 3, G is not equi-matchable. Otherwise, for each point $a \in A$ we consider the subgraph $\hat{G}_I(a) = \langle \Gamma(a) = S_a, \Gamma(S_a) - \{a\} \rangle$. For each of these $|A|$ subgraphs run any (polynomial) bipartite matching algorithm which either obtains a matching covering all points for $\Gamma(S_a) - a$ or else finds an $X \subseteq \Gamma(S_a) - \{a\}$ such that $|X| > |\Gamma(X)|$.

If any $\hat{G}_I(a)$ has a matching M_a covering S_a then any maximal matching M of $\hat{G}_I$ that contains M_a will leave a exposed. Hence $\hat{G}_I$ is not equi-matchable. (Moreover, by the "easy" direction of Hall's theorem, for all $X \subseteq \Gamma(a)$ we will have $|\Gamma(X)| > |X|$.) Therefore, by Theorem 3, neither is G.

Conversely, suppose that no $\hat{G}_I(a)$ has a matching M_a covering S_a. Then for each $a \in A$ the algorithm finds $X \subset \Gamma(S_a) - a$ such that $|X| > |\Gamma(X)|$. By Theorem 5, therefore, $\hat{G}_I$ is equi-matchable; so by Theorem 4 G is equi-matchable.

The total amount of work performed by this process is bounded as follows. First we apply the maximum matching algorithm to determine whether G has a perfect matching and, if not, to obtain the Gallai–Edmonds decomposition. If G has a perfect matching then it can be verified in time $O(|V|^2)$ whether or not G is a $K(2m)$ or a $K(m, m)$. If we have the Gallai–Edmonds decomposition (D, A, C) of G with $A = \varnothing$, i.e. G is factor-critical, then we require $O(|V|^2)$ applications of the weighted matching algorithm to determine whether or not G is equi-matchable, as described following Theorem 2. If we have the Gallai–Edmonds decomposition with $A \neq \varnothing$, then we can verify in time $O(|V|^2)$ whether or not D and A have the requisite structure, and $C = \varnothing$. Then we form the graph $\hat{G}_I$ and apply a maximum (bipartite) matching algorithm $O(|A|)$ times to $\hat{G}_I$.

Thus if we use versions of the matching algorithm which run in $O(|V|^3)$, as described, for example, in Lawler [7], then the time for this process of determining whether or not a graph is equi-matchable is $O(|V|^3)$ if G has a perfect matching, $O(|V|^5)$ if G is factor-critical and $O(|V|^4)$ if G has no perfect matching, but is not factor-critical.

Acknowledgements

We are grateful to J.-C. Bermond, O. Favaron, M. Habib and J.-C. Fouquet for careful reading of an earlier version of this paper and for bringing to our attention an important case we had not considered. The first author worked with Université Pierre et Marie Curie, Paris 6 and with the Centre National D'Etudes des Telecommunications, France whose support is gratefully acknowledged. Much of the work of the second and third authors was done while visiting the Institüt für Ökonometrie und Operations Research, Universität Bonn, and they are grateful for support from the Sonderforschungsbereich 21 (DFG) during that visit. The second author is also grateful for support from the Vanderbilt Research Council and the third author for support from the National Science and Engineering Research Council of Canada.

References

1. J. Edmonds, Paths, trees and flowers. *Canad. J. Math.* **17** (1965), 449–467.
2. J. Edmonds, Maximum matching and a polyhedron with (0, 1) vertices. *J. Res. Nat. Bur. Stand.* **69B** (1965), 125–130.
3. T. Gallai, Über extreme Punkt- und Kantenmengen, *Ann. Univ. Sci. Budapest, R. Eötvös Sect. Math.* **2** (1959), 133–138.
4. M. R. Garey and D. S. Johnson, *Computers and Intractability: A Guide to the Theory of NP-completeness*, Freeman, San Francisco, 1979.
5. B. Grünbaum, Matchings in polytopal graphs. *Networks* **4** (1974), 175–190.
6. F. Harary, *Graph Theory*, Addison-Wesley, Reading, Mass., 1969.
7. E. L. Lawler, *Combinatorial Optimization: Networks and Matroids*, Holt, Rinehart and Winston, New York, 1976.
8. M. Lewin, Matching-perfect and cover-perfect graphs. *Israel J. Math.* **18** (1974), 345–347.
9. L. Lovász, Subgraphs with prescribed valencies. *J. Combinat. Theory* **8** (1970), 391–416.
10. L. Lovász, *Combinatorial Problems and Exercises*, North Holland, Amsterdam, 1979.
11. D. H.-C. Meng, Matchings and coverings for graphs. PhD thesis, Michigan State University, East Lansing, 1974.
12. R. Z. Norman and M. O. Rabin, An algorithm for a minimum cover of a graph. *Proc. Amer. Math. Soc.* **10** (1959), 315–319.
13. M. D. Plummer, On n-extendable graphs. *Discrete Math.* **31** (1980), 201–210.
14. D. P. Sumner, Randomly matchable graphs. *J. Graph Theory* **3** (1979), 183–186.
15. M. Yannakakis and F. Gavril, Edge dominating sets in graphs. *SIAM J. Appl. Math.* **38** (1980), 364–372.

22

UNISECANTS OF FAMILIES OF SETS

Richard Rado FRS

ABSTRACT   The paper examines the problem of whether a family of sets contains a unisecant. A certain condition on the indexing set is shown to be necessary and sufficient for the existence of a unisecant.

1. Introduction

A unisecant of the family $(A_\nu : \nu \in I)$ of sets is, by definition, a set X satisfying

$$|A_\nu \cap X| = 1 \quad \text{for} \quad \nu \in I.$$

E. C. Milner raised the problem of deciding whether a given family of sets possesses a unisecant. In this note we give a solution of this problem. It must, however, be admitted that this solution is not quite satisfactory in so far as the existence of a unisecant is made to depend on the existence of a certain partition of the index set I. It would be desirable to have a criterion which does not involve the existential quantifier.

2. Notation

We consider a fixed family $(A_\nu : \nu \in I)$. For $M \subseteq I$ put

$$A_M = \bigcup (\nu \in M) A_\nu$$

and, in the case $M \neq \varnothing$,

$$A_{[M]} = \bigcap (\nu \in M) A_\nu.$$

The expression $\bigcup' (\nu \in M) A_\nu$ denotes the set $\bigcup (\nu \in M) A_\nu$ and at the same time expresses the fact that

$$A_\mu \cap A_\nu = \varnothing \quad \text{for} \quad \{\mu, \nu\}_{\neq} \subseteq M.$$

The atoms of the family are the sets of the form

$$A_{M,I} = A_{[M]} \backslash A_{I \backslash M},$$

defined for $\varnothing \neq M \subseteq I$.

GRAPH THEORY AND COMBINATORICS
ISBN 0-12-111760-X

3. The main theorem

THEOREM 1   *The family $(A_\nu : \nu \in I)$ of sets possesses a unisecant if and only if there are a set L and sets I_λ for $\lambda \in L$ such that*

$$I = \bigcup{}' (\lambda \in L) I_\lambda ;$$
$$I_\lambda \neq \varnothing \quad and \quad A_{I_\lambda, I} \neq \varnothing \quad for \quad \lambda \in L.$$

PROOF   (a) Let

$$I = \bigcup{}' (\lambda \in L) I_\lambda; \quad A_{I_\lambda, I} \neq \varnothing \quad \text{for} \quad \lambda \in L. \tag{1}$$

Then there is

$$x_\lambda \in A_{I_\lambda, I} \quad \text{for} \quad \lambda \in L. \tag{2}$$

Put

$$X = \{x_\lambda : \lambda \in L\}. \tag{3}$$

Let $\nu \in I$. Then, by (1),

$$\nu \in I_\lambda \quad \text{for some} \quad \lambda \in L. \tag{4}$$

Also,

$$x_\lambda \in A_{I_\lambda, I} [\text{by (2)}] \subseteq A_\nu \ [\text{by (4)}] \tag{5}$$

and, by (3), $x_\lambda \in A_\nu \cap X$.

Now suppose that $\lambda_0, \lambda_1 \in L$ and

$$x_{\lambda_0}, x_{\lambda_1} \in A_\nu \cap X. \tag{6}$$

Let $t \in \{0, 1\}$. Then, by (5),

$$x_{\lambda_t} \in A_{I_{\lambda_t}, I}. \tag{7}$$

If

$$\nu \notin I_{\lambda_t}, \tag{8}$$

then

$$x_{\lambda_t} \in A_\nu \cap A_{I_{\lambda_t}, I} \ [\text{by (6), (7)}]$$
$$\subseteq A_\nu \cap (A_I \setminus A_\nu) \ [\text{by (8)}]$$
$$= \varnothing,$$

which is a contradiction. Hence $\nu \in I_{\lambda_t}$; $\nu \in I_{\lambda_0} \cap I_{\lambda_1}$ and, by (1), $\lambda_0 = \lambda_1$. Hence $|A_\nu \cap X| = 1$ and X is a unisecant of the family.

(b) Let $|A_\nu \cap X| = 1$ for $\nu \in I$. Then there is x_ν such that

$$A_\nu \cap X = \{x_\nu\} \quad \text{for} \quad \nu \in I, \tag{9}$$

and there is $L \subseteq I$ such that we have, using an obvious notation,

$$\{x_\nu : \nu \in I\} = \{x_\lambda : \lambda \in L\}_{\neq}. \tag{10}$$

Put, for $\lambda \in L$,

$$I_\lambda = \{\nu \in I : x_\lambda \in A_\nu\}. \tag{11}$$

Then, by (9), we have $\lambda \in I_\lambda$ so that $I_\lambda \neq \varnothing$ for $\lambda \in L$. Also, by (11),

$$x_\lambda \in A_{[I_\lambda]} \backslash A_{I \backslash I_\lambda} = A_{I_\lambda, I} \quad \text{for} \quad \lambda \in L. \tag{12}$$

Now consider any $\nu \in I$. By (10) we have $x_\nu = x_\lambda$ for some $\lambda \in L$. Then $x_\nu = x_\lambda \in A_{I_\lambda, I}$ by (12); $x_\lambda = x_\nu \in A_\nu$ by (9); and $\nu \in I_\lambda \subseteq I_L$ by (11). Since ν was arbitrary, we conclude that $I \subseteq I_L$, i.e. $I_L = I$. Finally, assume that

$$\{\lambda_0, \lambda_1\}_{\neq} \subseteq L, \qquad \nu_0 \in I_{\lambda_0} \cap I_{\lambda_1}. \tag{13}$$

Then, by (10), $x_{\lambda_0} \neq x_{\lambda_1}$. Let $t \in \{0, 1\}$. Then

$$x_{\lambda_t} \in A_{I_{\lambda_t}, I} \cap X \,[\text{by } (12)] \subseteq A_{\nu_0} \cap X \,[\text{by } (13)].$$

Hence, by (10),

$$1 = |A_{\nu_0} \cap X| \geqslant |\{x_{\lambda_0}, x_{\lambda_1}\}| = 2,$$

which is a contradiction. Therefore $\{\lambda_0, \lambda_1\}_{\neq} \subseteq L$ implies $I_{\lambda_0} \cap I_{\lambda_1} = \varnothing$, and Theorem 1 is proved. □

4. An obvious extension of the notion of a unisecant

Definition *Let there be given a family*

$$(A_\nu : \nu \in I) \tag{14}$$

of sets and a family $(a_\nu : \nu \in I)$ *of cardinal numbers. A* $(a_\nu : \nu \in I)$ *secant of* (14) *is, by definition, a set* X *such that* $|X \cap A_\nu| = a_\nu$ *for* $\nu \in I$.

Theorem 2 *The family* (14) *possesses a* $(a_\nu : \nu \in I)$ *secant if and only if for every* $\nu \in I$ *there is a partition*

$$A_\nu = \bigcup{}' (\lambda \in L_\nu) A_{\nu\lambda} \tag{15}$$

such that $|L_\nu| = a_\nu$ *and the family*

$$(A_{\nu\lambda} : \nu \in I; \lambda \in L_\nu) \tag{16}$$

possesses a unisecant.

PROOF　(a) Let X be a $(a_\nu: \nu \in I)$ secant of (14). Then, for every $\nu \in I$ we have $|X \cap A_\nu| = a_\nu$ and, clearly, we can write (15), where $|L_\nu| = a_\nu$ and $|X \cap A_{\nu\lambda}| = 1$ for $\lambda \in L_\nu$. Thus X is a unisecant of (16).

　　(b) Let X be a unisecant of (16). Let (15) hold and $|L_\nu| = a_\nu$ for $\nu \in I$. Then $|X \cap A_{\nu\lambda}| = 1$ for $\lambda \in L_\nu$ and $(X \cap A_{\nu\lambda}) \cap (X \cap A_{\nu\mu}) = \varnothing$ for $\lambda \neq \mu$. Then X is a $(a_\nu: \nu \in I)$ secant of (14) and Theorem 2 follows.　□

23

PLANAR PERMUTATIONS DEFINED BY TWO INTERSECTING JORDAN CURVES

P. Rosenstiehl

ABSTRACT   A planar permutation of size $2M$ is defined by two oriented Jordan curves intersecting in the plane in $2M$ points, one point being the root. When the points are labelled from 1 to $2M$ in the order in which they occur on one of the curves, beginning at the root, the permutation appears on the other curve. For the sake of solving different problems, each planar permutation of size $2M$ can be represented by two particular sets of $2M$ nested parentheses, by a particular kind of plane graph with $2M$ edges, or by a way of folding a ring of $2M$ stamps.

1. Introduction

Let A and B designate two oriented Jordan curves with $2M$ intersection points, among which one is called the root. Starting at the root, the intersection points are given *ranks* 1 to $2M$ in the order in which they occur on the curve A, in the given orientation. Along the curve B, starting similarly at the root and following the orientation, the intersection points are given *locations* 1 to $2M$. Let the point with location l on B have the rank $\rho(l)$ on A, and let the point of rank r on A have the location $\lambda(r)$ on B. The mappings ρ and λ so defined are inverse permutations of the set $E = \{1, 2, \ldots, 2M\}$, and will be called *planar permutations*. Notice that $\rho(1) = \lambda(1) = 1$. Figure 1 shows a particular example.

It will be convenient to say that a pair of Jordan curves is *rooted* if a *root pattern* has been defined for them, namely the choice of one intersection point, of an orientation on both curves, and the choice of the curve which carries the rank labels. In Fig. 1, the root pattern is represented by a couple of arrows, the first one being tangential to the curve of the ranks, the second one tangential to the curve of the locations.

DEFINITION 1   *A rooted pair of Jordan curves determines a unique permutation ρ as described above. Permutations which can be obtained in this way will be called* planar *permutations.*

These planar permutations arose in the context of sorting geographical

GRAPH THEORY AND COMBINATORICS
ISBN 0-12-111760-X

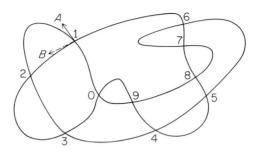

Fig. 1. Two rooted Jordan curves defining $\rho = 1\ 2\ 3\ 0\ 9\ 4\ 5\ 8\ 7\ 6$. The plain arrow represents the first arrow, pointing to the ranks; the dashed arrow represents the second arrow, pointing to the locations.

map data. A conjecture of Masao Iri, which we proved in [2], asserts that a planar permutation can be sorted in linear time.

We shall use below two other ways of representating planar permuations, for which we require additional definitions, which we give now. A *nested set of pairs in E* is a set S of disjoint pairs of distinct elements of E, such that, for any $\{a, b\}$, $\{c, d\} \in S$, we never have $a < c < b < d$. A pair $\{a, b\} \in S$ is a *short pair* if no other pair of S has its elements between a and b. It is obvious that a non-empty set S has at least one short pair.

In a previous paper [3], a one-to-one mapping was displayed between the families of q Jordan curves intersecting in $2M$ points, and even plane graphs with $2M$ edges and a bicycle space (intersection of cycle and cocycle spaces) of dimension $q - 1$. Here, we are concerned with the case $q = 2$, that is, graphs with a *unique bicycle* – assumed to be non-empty – which has to be the whole set of edges. As seen in Fig. 2, Jordan curves A and B define regions colourable into two colours Z and Z'. Let G be the incidence graph of the regions of colour Z, and G' the incidence graph for Z' as shown in Fig. 2. Graph G' is the dual of G. Let us define the *checker graph* (G, G') of (A, B) to be the superposition of the two plane graphs G and G', with their dual edges crossing at the intersection points of A and B. Each face of (G, G') is a quadrangle with two opposite vertices being intersection points of A and B, and the two others being points in two regions, one of each colour. If we draw inside each quadrangular face of (G, G') a diagonal line joining the two vertices which are intersection points of A and B, and stick them all together, we generate the pair of curves (A, B) that we started with. The pair (A, B) is called the *set of diagonals of* (G, G').

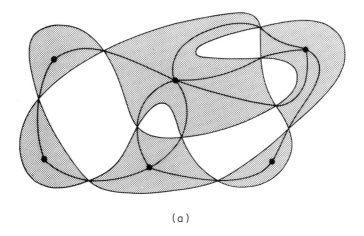

(a)

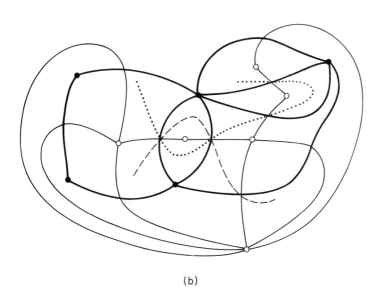

(b)

Fig. 2. Correspondence between a pair of Jordan curves and a pair of dual plane graphs with a unique bicycle. (a) Intersecting Jordan curves divide the plane into two-colourable regions. Graph G is the incidence graph of the regions of colour Z (stippled areas are colour Z; plain areas are colour Z'). (b) The diagonals of graph G (and of its dual G') are the Jordan curves which defined G and G'.

2. Operations

Definition 1 relies on topological objects. Let us give now a combinatorial definition of planar permutations on the set $E = \{1, 2, \ldots, 2M\}$.

DEFINITION 2   *A planar permutation of size $2M$ is a cyclic permutation ρ on E such that*

(a) $\rho(1) = 1$;

(b) *the set of pairs $\{\rho(i), \rho(i+1)\}$ for $i = 1, 3, \ldots, 2M - 1$, and the set of pairs $\{\rho(i), \rho(i+1)\}$ for $i = 2, 4, \ldots, 2M$ are both nested.*

It is easily seen from Fig. 3 that Definitions 1 and 2 are equivalent. We have ordered the points according to their rank on a straight line, which can be closed underneath by an arc to form a Jordan curve A. One set of pairs of Definition 2 is represented by nested arcs above the line, and the other set by nested arcs below it. The permutation being cyclic, the property of these arcs to be nested is equivalent to the property for B to be a Jordan curve. The root pattern at point 1 is clear. It follows that a planar permutation ρ, by Definition 2, generates a rooted pair of Jordan curves, which by Definition 1 generates the permutation ρ we started

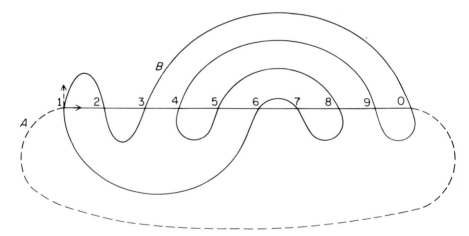

FIG. 3. Representation by two nested sets of arcs of the two nested sets of pairs of the planar permutation $\rho = 1\ 2\ 3\ 0\ 9\ 4\ 5\ 8\ 7\ 6$. (First set above the line; second set below the line.) In parenthesis code this representation is

$$\rho = \frac{(\ )\ ((\ ((\ ))\ ))}{((\ )(\ ))\ ()\ ()}.$$

with. Therefore, two non-isomorphic pairs of Jordan curves cannot be rooted to represent the same planar permutation.

Let us define three operations which transform a planar permutation ρ of size $2M$ into any other corresponding to the same pair of Jordan curves.

OPERATION INVERSE   This is the usual inverse ρ^{-1} of ρ defined as follows: if $\rho(r) = l$, then $\rho^{-1}(l) = r$. As explained in the introduction, ρ^{-1} is obtained by interchanging the two arrows in the root pattern of ρ.

OPERATION OBVERSE ρ^* is defined by $\rho^*(1) = \rho(1)$ and $\rho^*(k) = \rho(2M+2-k)$ for $k = 2, 3, \ldots, 2M$, which is obviously the planar permutation obtained by reversing the direction of the second arrow in the root pattern of ρ.

NOTE   The operation of reversing the first arrow of ρ is obtained by a combination of the above two operations, namely by $((\rho^{-1})^*)^{-1}$.

OPERATION SHIFT ρ^+ is defined by $\rho^+(k) = \rho(h) - 1$, where $h = \rho^{-1}(2) + k - 1$ (mod $2M$), which is obviously the planar permutation obtained by moving the root pattern of ρ from the point of rank 1 to the point of rank 2, that is by shifting forward on its curve the first arrow up to the next point, and placing the second arrow at this new root such that the orientation of its curve is not changed.

For the example of Fig. 3, we have

$$\rho = 1\ 2\ 3\ 0\ 9\ 4\ 5\ 8\ 7\ 6,$$
$$\rho^{-1} = 1\ 2\ 3\ 6\ 7\ 0\ 9\ 8\ 5\ 4,$$
$$\rho^* = 1\ 0\ 9\ 2\ 3\ 8\ 7\ 4\ 5\ 6,$$
$$\rho^+ = 1\ 2\ 9\ 8\ 3\ 4\ 7\ 6\ 5\ 0.$$

THEOREM 1   *The set of planar permutations of size $2M$ is partitioned by the equivalence relations generated by operations inverse, obverse and shift, the classes being represented by non-isomorphic pairs of Jordan curves intersecting in $2M$ points.*

The first part of the theorem is proved by checking that, given a planar permutation ρ, corresponding to a pair (A, B) of Jordan curves with a root pattern, all possible root patterns of (A, B) can be generated from ρ, by compositions of the three operations. The second part of the theorem is a straightforward consequence of the equivalence of Definitions 1 and 2.

3. Automorphisms

Planar permutations of size $2M$ may help in the problem of automorphism of a pair of Jordan curves. Dealing with a pair of Jordan curves
(A, B) amounts to dealing with the corresponding checker graph (G, G').
This suggests a third definition of planar permutations. For that purpose,
let us define a *root pattern* of a plane graph to be the choice of one of its
edges, at the midpoint of which are two perpendicular arrows making
angles of 45° with the edge (see Fig. 4). A *rooted plane graph* is a
connected plane graph with a root pattern.

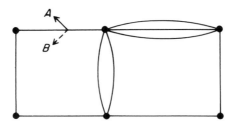

Fig. 4. Root pattern of a plane graph.

DEFINITION 3   *A rooted plane graph with $2M$ edges and a unique bicycle
determines a unique planar permutation of size $2M$.*

It is clear that a graph with a unique bicycle is an even graph. So,
Definition 3 appears easily equivalent to Definition 1. First, the diagonals
of an even plane graph are all Jordan curves. Second, a connected plane
graph G with a unique bicycle has a pair of diagonals (A, B). Third, the
root pattern of G, or of (G, G') is equivalent to a root pattern of their
pair of diagonals (A, B).

The theorem below deals with automorphisms of (G, G'), that is
automorphisms which map G on G and G' on G', or G on G' and G' on
G, the last case requiring that (A, B) defines as many regions of colour Z
as ones of colour Z'. It is a direct consequence of Definition 3.

THEOREM 2   *Any automorphism ϕ of (G, G'), where G is a connected
plane graph with a unique bicycle, corresponds to two ways of rooting
(G, G'), with root patterns P_1 and P_2 generating the same planar permutation; ϕ is a mapping of G onto G' if and only if, for one of the root patterns,
the two arrows are on the same side of the root-edge of G, and for the other
pattern on opposite sides.*

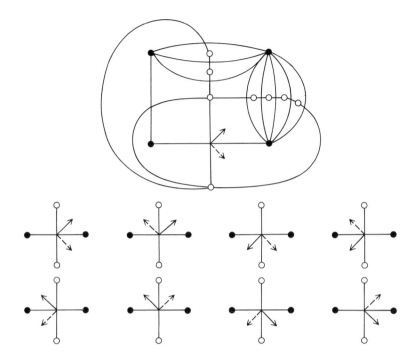

Fɪɢ. 5. For graphs G and G' without non-trivial automorphism (of G onto G, or of G onto G') the eight root patterns of each of their 10 edges generate 80 distinct planar permutations.

It is clear that the mapping of P_1 onto P_2 maps the root edges e_1, e_1' of P_1 onto the root edges e_2, e_2' of P_2, either in the order e_2, e_2' or e_2', e_2. The second case corresponds to an automorphism of G onto its dual G', and the positions of the arrows of P_1 and P_2 are as described in the theorem.

As an example, let us consider the graph G of Fig. 5, which has no non-trivial automorphism, neither with itself nor with its dual. It follows that, each edge being able to be rooted in eight ways, and G having 10 edges, (G, G') generates 80 different planar permutations. On the other hand, in the case of Fig. 3, we find that

$$(((((\rho^{-1})^*)^{-1})^*)^{-1})^+ = (\rho^*)^{-1}.$$

In that case an automorphism of G onto G' can be deduced explicitly from the mapping of the two root patterns defined by the two members of the equality above.

3. Testing and generating

Using Definition 2, we shall generate, as long as $M > 1$, a planar permutation of size $2M - 2$ from one of size $2M$. Consider the first set of nested pairs. Since $M > 1$, one pair is a short pair. Let us call it $\{k, k + 1\}$ and remove it from the first set, which is still nested.

In the second set, k and $k + 1$ appear in two distinct pairs, say $\{k, \alpha\}$ and $\{k + 1, \beta\}$. These may have three kinds of positions, as displayed in Fig. 6 in the parenthesis code, with the first set above the line and the second set below it. Let us remove these two pairs from the second set and replace them by the new pair $\{\alpha, \beta\}$. The second set remains nested in the three cases since no pair of the set could have one element between k and $k + 1$. Notice that the planar permutation of size $2M - 2$ obtained has two consecutive ranks missing which may be other than $2M - 1$ and $2M$; so ranks over the missing ranks have to be decreased by two in order to get the canonical form of a planar permutation. One advantage of representing a planar permutation in the parenthesis code is that this renumbering does not have to be carried out explicitly. For this reason, we shall describe operations on the planar permutations in terms of this code.

If we start with a valid pair of nested sets of parentheses, and apply this procedure many times, we necessarily reach the "trivial" planar permutation shown in Fig. 7(a). Figure 7(b) shows the corresponding rooted pair of Jordan curves and Fig. 7(c) the corresponding rooted graph G. (A further application of the procedure would make the configuration vanish completely!) Thus we have a test of validity of an alleged planar permutation. It can be carried out in linear time. Conversely, let us define two constructive operations as follows.

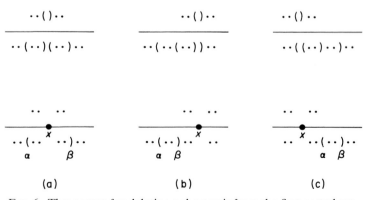

Fɪɢ. 6. Three cases for deleting a short pair from the first nested set.

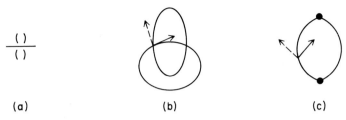

(a)   (b)   (c)

FIG. 7. The three representations of the trivial planar permutation: (a) by the parenthesis code; (b) by a rooted pair of Jordan curves; (c) by a rooted plane graph with a unique bicycle.

Take a point x on the line, midway between two consecutive parentheses, or at either end of the line. We shall say that a matching pair of parentheses below the line "grips" x if one is on the left of x and the other one on the right of x, and there is no other pair with this property between them. It can happen that there is no such pair.

OPERATION P Let α and β be the ranks for a pair of parentheses which grip x. Operation P will consist of inserting two consecutive matching parentheses above the line at the point x, that is, between the parentheses on either side of x, and inserting in the corresponding positions below the line two parentheses the other way round. Thus operation P reverses the operation that we performed above in Fig. 6(a).

OPERATION Q Let F be the set of parentheses below the line that lie between the pair which grips x. If there is no pair which grips x, let F be all the parentheses below the line. In F find a matching pair of parentheses not contained within another matching pair in F. Let their ranks be α and β. Operation Q will consist of inserting two consecutive matching parentheses above the line at the point x, and inserting in the corresponding positions below the line two parentheses oriented so as to match the parentheses with ranks α and β, after the reversal of whichever of these latter is closer to x. This is illustrated by Fig. 6(b) and (c).

In any given example, and for any choice of x, one at least of these two operations is possible. Since operations P and Q reverse the procedure by which any planar permutation is reduced to the trivial one, we have the following theorem.

THEOREM 3   *Any planar permutation can be generated from the trivial planar permutation by a sequence of operations of types P and Q.*

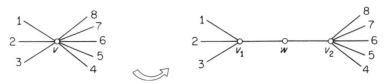

FIG. 8. Odd stretching.

The two operations P and Q can be interpreted in terms of their effect on the graph G.

The operation P corresponds to an operation on G, which we shall call *odd stretching* of a vertex v. This is achieved by dividing the edges incident to v into two odd sets of consecutive edges, and splitting v into two vertices v_1 and v_2, each adjacent to a new vertex w of degree 2. This is illustrated in Fig. 8. The relation between odd stretching and operation P is shown in Fig. 9.

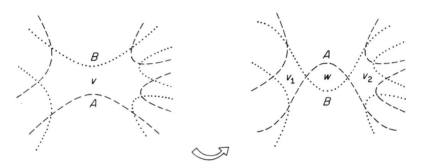

FIG. 9. Relation between odd stretching and operation P.

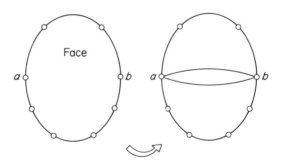

FIG. 10. Double linking.

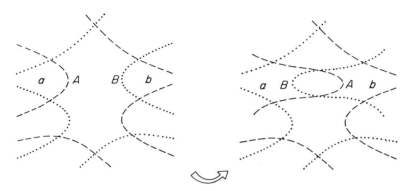

FIG. 11. Relation between double linking and operation Q.

The operation Q corresponds to an operation on G, which we shall call *double linking* across a face of G. Let a and b be two vertices on the boundary of a face r, one from each of the two vertex sets that make up the bipartition of G. Double linking consists of adding two extra edges joining a and b, thus dividing r into three regions. This is illustrated in Fig. 10. The relation between double linking and operation Q is shown in Fig. 11.

The fact that at least one of the operations P and Q can be performed gives the following theorem.

THEOREM 4 *A connected plane graph with a unique bicycle has a region bounded by two edges, or a vertex of degree 2, and possibly both.*

The generation of all planar permutations by operations P and Q, starting with the trivial permutation, implies the next theorem.

THEOREM 5 *Any connected plane graph with a unique bicycle may be generated from the digon by carrying out a sequence of double linking or odd stretching operations.*

It is easily verified that the operations odd stretching and double linking preserve the equality between the number of vertices in the two vertex sets that make up the bipartition of G. Hence the following theorem:

THEOREM 6 *A connected planar graph with a unique bicycle is vertex-colourable in two colours, and has the same number of vertices of each colour.*

4. About folding a ring of stamps

Imagine a closed ring of $2M$ postage stamps. In how many ways can this ring of stamps be folded down on to a single stamp? (We ignore the thickness of the stamps.) This appears to be an open problem, although the solution to the corresponding problem for an open strip of stamps has been known for some time [1]. Figure 12(a) shows a way of folding a ring of eight stamps. By rounding off the angles in this figure, corresponding to the perforations between the stamps – an operation which does not alter the figure in any significant way – we obtain Fig. 12(b), which corresponds

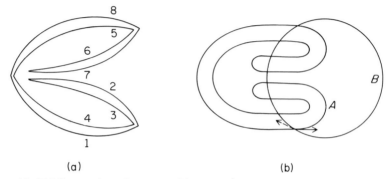

(a)   (b)

Fig. 12. Folding a ring of stamps: (a) a way of folding a ring of eight stamps, stamp 1 being placed at the bottom of the pile; (b) the corresponding rooted pair of Jordan curves.

Fig. 13. Representation of a medieval labyrinth: *l'église paroissiale de Saint-Quentin.*

to a pair of intersecting Jordan curves, with a distinguished root pattern. This correspondence is quite general, and it follows that the problem of enumerating the ways of folding a ring of stamps is equivalent to that of enumerating planar permutations or enumerating rooted planar graphs with one bicycle.

The reader who is interested in this aspect of the problem may like to contemplate a generalization of the postage stamp problem that is exemplified by Fig. 13! [5].

Acknowledgement

We wish to thank R. C. Read for his suggestions and improvements.

References

1. J. E. Koehler, Folding a strip of stamps. *J. Combinat. Theory* **5** (1968), 135–152.
2. P. Rosenstiehl and R. E. Tarjan, Geographical sorting in linear time. To appear.
3. P. Rosenstiehl, Bicycles et diagonales des graphes planaires. *Cahiers du CERO, Bruxelles* **17** (1975), 365–382.
4. P. Rosenstiehl, Solution algébrique du problème de Gauss sur la permutation des points d'intersection d'une ou plusieurs courbes fermées du plan. *Compt. Rend. Acad. Sci. Paris A* **283** (1976), 551–553.
5. P. Rosenstiehl, Le dodécadédale ou l'éloge de l'heuristique. *Critique, Paris* (1982), 785–796.

A CLASS OF SIZE RAMSEY PROBLEMS INVOLVING STARS

J. Sheehan, R. J. Faudree and C. C. Rousseau

ABSTRACT In this paper we study the size Ramsey number of a pair of graphs, one of which is a star and the other a complete bipartite graph, with two vertices in one of the classes. We determine exactly the size Ramsey number for a rather large family of such pairs. Our results lead to several conjectures.

1. Introduction

Let F, G and H be finite, simple and undirected graphs. The number of vertices and edges of a graph F are denoted by $p(F)$ and $q(F)$ respectively. A graph $F \to (G, H)$ if every 2-colouring (say red and blue) of the edges of F produces either a "red" G or a blue "H". The *Ramsey number*, $r_p(G, H)$, of G and H, the *size Ramsey number*, $r_q(G, H)$, of G and H and the *restricted size Ramsey number*, $r_q^*(G, H)$, of G and H are defined respectively by $r_p(G, H) = r(G, H) = \min\{p(F): F \to (G, H)\}$, $r_q(G, H) = \min\{q(F): F \to (G, H)\}$ and $r_q^*(G, H) = \min\{q(F): F \to (G, H), \ p(F) = r_p(G, H)\}$. As usual, $r_i(G)$ is the abbreviation for $r_i(G, G)$ $(i = p, q)$.

Size Ramsey numbers were introduced in [7] and a survey (written in 1979) of some of the few results concerning this function can be found in [5]. Obviously

$$r_q(G) \leqslant \binom{r(G)}{2}.$$

When G is a complete graph, equality holds. On the other hand, for stars just the opposite holds. Trivially $r(K_{1,n}) = 2n - 1$, n even and $r(K_{1,n}) = 2n$, n odd, whereas $r_q(K_{1,n}) = 2n - 1$. Therefore

$$r_q(K_{1,n}) \bigg/ \binom{r(K_{1,n})}{2} \to 0. \tag{1}$$

Recently, Beck [1] has answered a question which was asked in [7]. Erdös, Faudree, Rousseau and Schelp raised the question whether (1) holds when $K_{1,n}$ is replaced by P_n, the path of length $n - 1$ $(n \geqslant 1)$. Since

GRAPH THEORY AND COMBINATORICS
ISBN 0-12-111760-X

(see [10])

$$r(P_n) = n + \left[\frac{n}{2}\right] - 1,$$

this determines the answer to the question. In fact Beck proves that there exists a graph F with $q(F) \leq 900n$ such that $F \to (P_n, P_n)$.

The exact determination of $r_q(P_n)$ is probably very difficult, whereas the determination of $r_p(P_n)$ is not so difficult. In this paper, see Section 3, we give examples where this situation is reversed. We prove that (i) for $n \geq 5$, $r_q(K_{1,n}, K_{2,2}) = 4n$, and (ii) if $m \geq 9$ and n is sufficiently large, $r_q(K_{1,n}, K_{2,m}) = 2m + 4(n-1)$. In contrast the numbers $r_p(K_{1,n}, K_{2,m})$ are, in the main, unknown.

In Section 4 some more recent results for the function r_q involving stars are surveyed and we consider some unresolved questions arising from our study.

2. Terminology and notation

For the most part, our terminology and notation will conform with that used in [4]. For example, if G is a graph $V(G)$ and $E(G)$ denote respectively its set of vertices and its set of edges. The degree of a vertex v in G is denoted by $d(v)$. We denote by $\delta(G)$ and $\Delta(G)$ the minimum and maximum degrees, respectively, of vertices of G. For any set $S \subseteq V(G)$, the neighbour set of G is denoted by $N(S)$. However, there are certain special conventions which we shall follow. Let G be a graph with $p = p(G)$. By a 2-colouring of G we mean a partition $E(G) = (R, B)$. Equivalently we ascribe to each edge of G a colour, either red or blue. This 2-colouring defines two edge-induced spanning subgraphs and we use $\langle R \rangle$ and $\langle B \rangle$ as symbols for these graphs. Let F and G be graphs without isolated vertices. The *Ramsey number* $r(F, G)$ of F and G, in this notation, is the smallest value of p such that in every possible 2-colouring (R, B) of the complete graph K_p either $\langle R \rangle$ contains (a subgraph isomorphic to) F or $\langle B \rangle$ contains G.

Finally, if (X, Y) is a partition of $V(G)$, $[X, Y]$ denotes the spanning subgraph of G with $E([X, Y]) = \{xy : x \in X, y \in Y\} \cap E(G)$.

3. Size Ramsey numbers versus Ramsey numbers

Before considering $r_q(K_{1,n}, K_{2,m})$, the next theorem and its corollary summarize the analogous known results for $r_p(K_{1,n}, K_{2,m})$.

THEOREM 1 [**12, 14**] *Let n, λ be positive integers. Then*

$$r_p(K_{1,n}, K_{2,\lambda+1}) \leq 1 + n + \tfrac{1}{2}(\lambda + 1 + ((\lambda - 1)^2 + 4\lambda n)^{1/2})$$

with equality if and only if there exists an $(n + k, k, \lambda)$-graph. (A regular graph G of degree k with $p(G) = n + k$ is an $(n + k, k, \lambda)$-graph if any two distinct vertices of G have exactly λ vertices mutually adjacent to both of them.)

PROOF Suppose there exists a colouring (R, B) of $K_m (m \geq 1)$ such that $\langle R \rangle \not\supseteq K_{1,n}$ and $\langle B \rangle \not\supseteq K_{2,\lambda+1}$. Write $G = \langle B \rangle$. Suppose $V(G) = \{v_1, v_2, \ldots, v_m\}$. Since $\langle R \rangle \not\supseteq K_{1,n}$, $\delta(G) \geq m - n$. We have

$$m\binom{\delta}{2} \leq \sum_{i=1}^{m} \binom{d(v_i)}{2} \leq \lambda \binom{m}{2}. \tag{2}$$

The first inequality coming from convexity and the second since $G \not\supseteq K_{2,\lambda+1}$. The theorem follows immediately from (2). $\square$

COROLLARY 2 [**14**] *If a (v, k, λ)-design with a polarity exists then*

$$r(K_{2,\lambda+1}, K_{1,v-k+1}) \in \{v + 1, v + 2\}. \quad \square$$

(For the relevant definitions and more on this subject we refer the reader to [**2**, p. 372].)

The known results for $r_q(K_{1,n}, K_{2,m})$ are contained in Theorems 4 and 5. The proofs of these theorems are the result of earlier joint work [**15**]. We first require a lemma (a proof of a special case of which was first given by Woodall [**16**]).

LEMMA 3 *Let m, s be integers $(m, s \geq 1)$. Let $A_1, A_2, \ldots, A_s$ be a family of finite sets and let $p_1, p_2, \ldots, p_s$ be integers satisfying $p_i \geq m - 1$, $i = 1, 2, \ldots, s$. If, for $t = 1, 2, \ldots, s$,*

$$|A_t| \geq \sum_{i=1}^{t} p_i - (t - 1)(m - 1),$$

then there exists a family of sets $X_1, X_2, \ldots, X_s$ such that $X_i \subseteq A_i$, $|X_i| = p_i$, and $|X_i \cap X_j| \leq m - 1$ $(i \neq j)$, for $i, j = 1, 2, \ldots, s$.

PROOF We proceed by induction on s. When $s = 1$ there is nothing to prove. Assume that the lemma is true for $s = 1, 2, \ldots, k - 1$ $(k \geq 2)$ and for any choice of integers $p_1, p_2, \ldots, p_s$ such that $p_i \geq m - 1$, $i = 1, 2, \ldots, s$. Now suppose $s = k$. Select a subset X_1 of A_1 with $|X_1| = p_1$ and then select a subset C of X_1 with $|C| = m - 1$. Let $D = X_1 - C$ and, for

$t = 1, 2, \ldots, k - 1$, write $B_t = A_{t+1} - D$ and $q_t = p_{t+1}$. Then

$$|B_t| \geq \sum_{i=1}^{t} q_i - (t - 1)(m - 1), \qquad t = 1, \ldots, k - 1.$$

By induction, there exist subsets $X_i \subseteq B_{i-1} \subseteq A_i$ with $|X_i| = q_{i-1} = p_i$ and $|X_i \cap X_j| \leq m - 1$ $(i \neq j)$, for $i, j = 2, \ldots, k$. These sets, together with X_1, form the required family. $\square$

THEOREM 4   Let $m \geq 9$. Then, if n is sufficiently large, $r_q(K_{1,n}, K_{2,m}) = 4n + 2m - 4$.

PROOF   Let $k = 2n + m - 2$. We first prove that $K_{2,k} \to (K_{1,n}, K_{2,m})$. Suppose, to the contrary, that (R, B) is a 2-colouring of $K_{2,k}$ such that $\langle R \rangle \not\supseteq K_{1,n}$ and $\langle B \rangle \not\supseteq K_{2,m}$. Let (X, Y) be the bipartition of $K_{2,k}$ with $|X| = 2$, $|Y| = k$, $[X, Y] \cong K_{2,k}$. Since $\langle B \rangle \not\supseteq K_{2,m}$ in $\langle B \rangle$, at most $m - 1$ vertices of Y have degree 2. Hence $|B| \leq 2(m - 1) + (2n - 1)$. In $\langle R \rangle$ neither vertex of X has degree greater than $n - 1$ and so $|R| \leq 2(n - 1)$. These inequalities lead to a contradiction since $|R| + |B| = 2k$.

Now we prove that $r_q(K_{1,n}, K_{2,m}) \geq 4n + 2m - 4$ when m is fixed $(m \geq 9)$ and n is sufficiently large. Throughout the remainder of this discussion the sufficiently large character of n will be tacitly assumed. Suppose, contrary to the proposition, that there exists a graph G such that $G \to (K_{1,n}, K_{2,m})$ and $q(G) \leq 4n + 2m - 5$. We shall make a series of observations concerning G, culminating with the conclusion that no such graph exists. Without loss of generality, we assume that G is minimal, i.e. that for every $e \in E(G)$, $G - e \nrightarrow (K_{1,n}, K_{2,m})$. Let

$$\mathcal{H} = \{v \in V(G): d(v) \geq n\}$$

and $\mathcal{L} = V(G) \backslash \mathcal{H}$. We think of $\mathcal{H}$ and $\mathcal{L}$ as the "high" and "low" vertices respectively. The minimality of G implies that $E(\langle \mathcal{L} \rangle) = \varnothing$.

Let $|\mathcal{H}| = s$. Clearly $s \geq 2$ for otherwise $G \not\supseteq K_{2,m}$. Also, since $\mathcal{H}$ contributes at least sn to the degree sum and $q(G) \leq 4n + 2m - 5$, $s \leq 8$. However, using the lemma, we can show that $s = 2$. Let G^* be the graph obtained from G by adding edges to $E(\mathcal{H})$ so that $\langle \mathcal{H} \rangle \cong K_s$. We now only consider G^*. Colour all of the edges of $\langle \mathcal{H} \rangle$ red. We may assume that $[\mathcal{H}, \mathcal{L}] \to (K_{1,n-s+1}, K_{2,m})$ (otherwise clearly $G \nrightarrow (K_{1,n}, K_{2,m})$). Let $\mathcal{H} = \{x_1, x_2, \ldots, x_s\}$ and, for $i = 1, 2, \ldots, s$, define $A_i = N(x_i) \cap \mathcal{L}$, $d_i = |A_i|$. Without loss of generality, we may assume that $d_1 \leq d_2 \leq \ldots \leq d_s$ and that $d_i \geq n - s + m - 1$, $i = 1, 2, \ldots, s$. (If any x_i has $d_i < n - s + m - 1$, we may safely colour its incident edges and then delete it from further considerations.) The statement that $[\mathcal{H}, \mathcal{L}] \to (K_{1,n-s+1}, K_{2,m})$ is equivalent to saying that it is not possible to select subsets $X_i \subseteq A_i$ such that $|X_i| = p_i = d_i - (n-s)$ and $|X_i \cap X_j| \leqslant m - 1$ $(i \neq j)$, $i, j = 1, 2, \ldots, s$. Hence, for some t, $2 \leqslant t \leqslant s$, the hypothesis of the lemma is not satisfied, i.e.

$$d_t < \sum_{i=1}^{t} (d_i - (n-s)) - (t-1)(m-1). \tag{3}$$

Since m, s and t are bounded, (3) implies that

$$\sum_{i=1}^{t-1} d_i \geqslant nt + O(1). \tag{4}$$

However, since $q(G) \leqslant 4n + 2m - 5$ and $d_1 \leqslant d_2 \leqslant \ldots \leqslant d_s$,

$$\sum_{i=1}^{t-1} d_i \leqslant \frac{(t-1)}{s} \sum_{i=1}^{s} d_i \leqslant \left(\frac{t-1}{s}\right) 4n + O(1). \tag{5}$$

Hence, from (4) and (5), $s^2 \leqslant 4(s-1)$. Hence $s \leqslant 2$. Hence $s = 2$.

Finally, suppose $s = 2$. If $x_1 x_2 \in E(G)$, colour this edge blue and observe that, whether or not $x_1 x_2 \in E(G)$, the assumption that $G \to (K_{1,n}, K_{2,m})$ means that in every 2-colouring of $[\mathcal{H}, \mathcal{L}]$ either $\langle R \rangle \supseteq K_{1,n}$ or $\langle B \rangle \supseteq K_{2,m}$. We argue using the lemma as above, this time setting $p_i = d_i - (n-1)$, $i = 1, 2$. We conclude that

$$d_2 < \sum_{i=1}^{2} (d_i - (n-1)) - (m-1)$$

and so $d_1 \geqslant 2n + m - 2$. But then $q(G) \geqslant d_1 + d_2 \geqslant 4n + 2m - 4$, whereas, by assumption, $q(G) \leqslant 4n + 2m - 5$. Hence the assumed graph G does not exist and so $r_q(K_{1,n}, K_{2,m}) \geqslant 4n + 2m - 4$. $\square$

THEOREM 5   For $n \geqslant 3$, $r_q(K_{1,n}, K_{2,2}) = 4n$.

PROOF We do not give the proof of this theorem here. The proof proceeds using exactly the same techniques as in the proof of Theorem 4 but the various inequalities are handled much more delicately.

COMMENT Our knowledge (in the context of Theorems 4 and 5) of the function r_q is more complete than for r_p. We know, for example, only that $r_p(K_{1,n}, K_{2,2}) \leqslant 1 + n + [\sqrt{n}]$ with equality for $n = q^2$ or $n = q^2 + 1$ when q is a prime power. Of course this is not typical and indeed in the next sections this will become transparently clear.

4. Size Ramsey results involving stars

None of the following results, due to Faudree and Sheehan (see [8]), involving r_q are difficult to prove. On the other hand, all of the corresponding results for r_p are trivial to prove.

THEOREM 6 [8]   Let $\alpha = k(n-1)+1$ $(k, n \geqslant 2)$. Then

$$
r_q^*(K_{1,k}, K_n) = \begin{cases} \binom{\alpha}{2} - \binom{k}{2}, & k \geqslant n \text{ or } k \text{ odd,} \\ \binom{\alpha}{2} - \dfrac{k(n-1)}{2}, & \text{otherwise} \quad \square \end{cases}
$$

THEOREM 7 [8]

$$
r_q(K_{1,2}, K_n) = r_q^*(K_{1,2}, K_n) = 2(n-1)^2 \qquad (n \geqslant 2). \quad \square
$$

THEOREM 8 [8]   For $t \geqslant k \geqslant 2$ and n sufficiently large,

$$
r_q(K_{1,k}, K_t + \bar{K}_n) = r_q^*(K_{1,k}, K_t + \bar{K}_n) = \binom{\beta}{2} + \beta(n+k-1),
$$

where $\beta = k(t-1)+1$. $\square$

In [8] it is conjectured that

CONJECTURE 9 $r_q(K_{1,k}, K_n) = r_q^*(K_{1,k}, K_n)$. $\square$

COMMENT   The determination of $r_p(K_{1,k}, K_n)$ is trivial. The result follows from a nice, but essentially simple, observation by Chvátal [6] that if T is any tree with m vertices then $r(T, K_n) = (m-1)(n-1)+1$. However, the determination of $r_q(K_{1,k}, K_n)$ seems to be very difficult. In [9] a table of size Ramsey numbers is determined for small order graphs. Easily the most difficult number to determine was $r_q(K_{1,3}, K_3)$, which turned out to be 18, consistent with Conjecture 9. Moreover, our attempts to prove this conjecture have met with negligible success.

We conclude with another example where results for r_q seem more difficult than for r_p. Lawrence [13] has proved

THEOREM 10

$$
r(K_{1,n}, C_m) = \begin{cases} 2n+1, & \text{if } m \text{ is odd, and } m \leqslant 2n+1, \\ m, & \text{if } m \geqslant 2n. \end{cases}
$$

(*The case of even* m, $m < 2n$, *is undecided.*) $\square$

We have obtained very few results for $r_q(K_{1,n}, C_m)$. However, some interesting remarks can be made. Let $F \twoheadrightarrow (G, H)$ mean that $F \to (G, H)$ but $F - e \not\to (G, H)$ $(e \in E(F))$. Then, as a simple consequence of well known results, we deduce

PROPOSITION 11   *Suppose $m = 2s$, s an integer $(s \geq 1)$. Then*

(i) $K_{s,2n} \twoheadrightarrow (K_{1,n}, C_m)$ $(m \leq 2n)$;
(ii) $K_{s,s} \to (K_{1,n}, C_m)$ but $K_{s,s} \not\twoheadrightarrow (K_{1,n}, C_m)$ $(m \geq 4n - 2)$.

PROOF   (i) We can prove that $K_{s,2n} \to (K_{1,n}, C_m)$ as a direct consequence of a result of Jackson [**11**]. The minimality is easily checked.

(ii) Again, a proof that $K_{s,s} \to (K_{1,n}, C_m)$ is a consequence of a result of Moon and Moser [**3**, p. 135]. The non-minimality is a consequence of the same result. $\square$

Suppose m is an even integer $(m \geq 3)$. Let n be any integer such that $n \mid m$ $(1 \leq 2n < m)$. Let $G(m, n)$ be the graph defined as follows: $V(G(m, n)) = \{1, 2, \ldots, m\}$ and $ij \in E(G(m, n))$ if and only if $i - j \in \{\pm(km/n - 1): k = 0, 1, \ldots, n - 1\}$ (all numbers here being modulo m). Then

CONJECTURE 12 $G(m, n) \twoheadrightarrow (K_{1,n}, C_m)$.

REMARK   Conjecture 12 is true when $n = 2$. The sort of techniques we use to prove this in the case $n = 2$ should give a proof in the more general case. $G(m, n)$ is regular of degree $2n$. It is bipartite if m is even, n divides m, and m/n is even. If the conjecture proves to be true, then it follows that, for $n \mid m$, m even, $r_q(K_{1,n}, C_m) \leq mn$. Even for the restricted size Ramsey number, subject to Conjecture 12 being true, we can prove only

PROPOSITION 13   *Suppose m is even, $n \mid m$, m/n even $(2 \leq 2n < m)$. Then*

$$\frac{mn}{2} \leq r_q^*(K_{1,n}, C_m) \leq mn.$$

PROOF   The upper bound is subject to the proviso above. The lower bound is a consequence of the following observations: (i) $r_p(K_{1,n}, C_m) = m$; (ii) if $p(G) = m$ and $G \to (K_{1,n}, C_m)$ then in any 2-colouring of G there is either a red $K_{1,n}$ or a blue Hamiltonian circuit; (iii) as a consequence of (ii), $\delta(G) \geq n$. $\square$

Let $G(m)$, m not necessarily even, $m \geq 3$, be the graph defined as follows: $V(G(m)) = \{1, 2, \ldots, m\}$, $ij \in E(G(m))$ if and only if $i - j \in \{\pm 1, \pm 2\}$ (all numbers here being modulo m). Then we can prove that $G(m) \twoheadrightarrow (K_{1,2}, C_m)$ $(m \geq 4, n \neq 5, 8)$ so $r_q(K_{1,2}, C_m) \leq 2n$. It is fairly easy to improve the lower bound in Proposition 13 in the case $n = 2$, but even so the bounds are hopelessly wide. In fact we can prove

$$\frac{5m}{3} \leq r_q^*(K_{1,2}, C_m) \leq 2m.$$

We conjecture

CONJECTURE 14 $r_q^*(K_{1,2}, C_m) = 2m$.   □

We have proved that $r_q(K_{2,2}) = 15$. Erdös posed the problem of determining $r_q(K_{2,n})$. Clearly $K_{3,6n-5} \twoheadrightarrow (K_{2,n}, K_{2,n})$, so our final, very wild conjecture is

CONJECTURE 15 $r_q(K_{2,n}) = 18n - 15$.

References

1. J. Beck, On size Ramsey numbers of paths, trees and circuits. 1. *J. Graph Theory* (in press).
2. L. W. Beineke and R. J. Wilson, *Selected Topics in Graph Theory*, Academic Press, London, 1978.
3. B. Bollobás, *Extremal Graph Theory*, Academic Press, London, 1978.
4. J. A. Bondy and U. S. R. Murty, *Graph Theory with Applications*, American Elsevier, New York, and Macmillan, London, 1976.
5. S. A. Burr, A survey of noncomplete Ramsey theory for graphs. *Ann. N. Y. Acad. Sci.* **328** (1979), 58–75.
6. V. Chvátal, Tree-complete graph Ramsey numbers. *J. Graph Theory* **1** (1977), 93.
7. P. Erdös, R. J. Faudree, C. C. Rousseau and R. H. Schelp, The size Ramsey number. *Period. Math. Hung.* **9** (1978), 145–161.
8. R. J. Faudree and J. Sheehan, Size Ramsey numbers involving stars. *Discrete Math.* **46** (1983), 151–157.
9. R. J. Faudree and J. Sheehan, Size Ramsey numbers for small-order graphs. *J. Graph Theory* **7** (1983), 53–55.
10. L. Geréncser and A. Gyárfás, On Ramsey-type problems, *Ann. Univ. Sci. Budapest. Eötvös Sect. Math.* **10** (1967), 167–170.
11. B. Jackson, Cycles in bipartite graphs. *J. Combinat. Theory B* **30** (1981), 332–342.
12. S. L. Lawrence, Cycle-star Ramsey numbers. *Notices Amer. Math. Soc.* **20** (1973), Abstract A-420.

13. S. L. Lawrence, Bipartite Ramsey theory. *Notices Amer. Math. Soc.* **20** (1973), Abstract A-562.
14. T. D. Parsons, Ramsey graphs and block designs. *J. Combinat. Theory A* **20** (1976), 12–19.
15. C. C. Rousseau and J. Sheehan, Size Ramsey numbers for bipartite graphs. *Notices Amer. Math. Soc.* **25** (1978), Abstract A-36.
16. D. R. Woodall, private communication.

25

WORD RAMSEY THEOREMS

Imre Simon

ABSTRACT   The paper contains several Ramsey-type results concerning words over a finite alphabet. Among others, we shall discuss
homomorphic colourings of words, and factorizations with bounded
gap size.

1. Introduction

In this paper we survey a number of results which grew out of a
fundamental application of Ramsey's theorem to words over a finite
alphabet. The results describe certain regularities which are unavoidable
in long enough words and can be viewed as a part of Ramsey theory, as
described in Graham, Rothschild and Spencer [4].

A remarkable aspect of this set of theorems is that they were discovered for their applications in finite automaton theory and in the theory
of finitely generated semi-groups. Indeed, they were proved as "combinatorial lemmas" in those subjects, as in the case of Ramsey's theorem
itself. The applications we refer to are outside the scope of this paper;
some of them can be found in [2, 5, 6, 9, 10, 13, 14].

The original application of Ramsey's theorem to words as well as the
realization of the importance of unavoidable regularities to the theories
we mentioned are due to M. P. Schützenberger [12]. A modern descen-
dent of his pioneering work is the book *Combinatorics on Words*, by M.
Lothaire [8], where further results and applications can also be found.

2. Ramsey's theorem and words

We shall consider finite and infinite words over finite alphabets. More
precisely, let A be a finite non-empty set: A will be called the *alphabet*
and its elements will be called the *letters*. A *word over A* will be a finite
sequence of letters, or, equivalently, a function $[1, n] \to A$, for some
$n \in \omega$. The set of all words is the free monoid A^* generated by A. The
empty word is denoted by 1 and the set of non-empty words over A will
be denoted by A^+. We shall consider an *infinite word over A* as a function
$s: \omega \to A$; accordingly, the set of infinite words will be A^ω.

GRAPH THEORY AND COMBINATORICS
ISBN 0-12-111760-X

Let s be a word in A^* (infinite word in A^ω) and let $\mathbf{t} = (t_1, t_2, \ldots, t_n)$ be an n-tuple of non-empty words in A^+. We say that $\mathbf{t}$ is a *factorized segment* of s if $s = ut_1t_2 \ldots t_nv$ for some u and v in A^* (u in A^* and v in A^ω). The *gap size* of $\mathbf{t}$ is given by $\max\{|t_i| \mid 1 \le i \le n\}$, where $|x|$ denotes the length of a word x. We shall also consider the case $n = \omega$; then $\mathbf{t} = (t_1, t_2, \ldots)$ is a factorized segment of s in A^ω if $s = ut_1t_2 \ldots$ for some u in A^*.

A *colouring* of a set S is a function $f\colon S \to E$; if card $E = k$ then f is a k-colouring of S; if $f(S) \subseteq E$ is finite, then f is a *finite colouring* of S.

Let $f\colon A^* \to E$ be a colouring of A^*. We say that $\mathbf{t}$ is an *nth power modulo f* if $f(t_1) = f(t_2) = \ldots = f(t_n)$. The common value of the $f(t_i)$s is denoted by $f(\mathbf{t})$. If further $f(\mathbf{t}) = f(t_it_{i+1} \ldots t_j)$, for every $1 \le i \le j \le n$, then we say that $\mathbf{t}$ is *monochromatic* of colour $f(\mathbf{t})$ or that f is constant on $\mathbf{t}$ with value $f(\mathbf{t})$.

The main application of Ramsey's theorem to infinite words is given below:

THEOREM 1   *For every finite colouring of A^* and for every infinite word s in A^ω, s has an infinite monochromatic factorized segment.*

PROOF   From the word $s\colon \omega \to A$ and from the colouring $f\colon A^* \to E$ we define a colouring $f'\colon [\omega]^2 \to E$ as follows: for every $0 \le i < j < \omega$,

$$f'(i, j) = f(s_is_{i+1} \ldots s_{j-1}),$$

where s_k denotes the kth letter $s(k)$ of s. By Ramsey's theorem, there exists an infinite subset B of ω, such that $[B]^2$ is monochromatic. Let $b_1 < b_2 < \ldots$ be an enumeration of the elements of B and let $t_i = s_{b_i}s_{b_i+1} \ldots s_{b_{i+1}-1}$. Then $\mathbf{t} = (t_1, t_2, \ldots)$ is an infinite monochromatic factorized segment of s.   □

We give now the finite version of Theorem 1, whose proof is similar to the proof just given.

THEOREM 2   *There exists a minimal number $p = p(n, k)$, such that for any k-colouring of A^* every word of length at least p has a monochromatic factorized segment of n components.*   □

The weakening of Theorem 2 for nth powers modulo f has a special interest because in that case it is easy to determine the exact value of p': see [12] or [8, Theorem 4.1.1 and Example 4.1.2].

THEOREM 3   *Let $p' = n^k$. Then for any k-colouring f of A^* every word of*

length at least p' in A has an nth power modulo f. The above value of p' is
best possible.*

3. Homomorphic colourings of words

A very important special case of Theorem 2 is obtained when the
colouring $f: A^* \to M$ is a morphism of monoids. In this case, if $\mathbf{t} =
(t_1, t_2, \ldots, t_n)$ is a monochromatic factorized segment of a word s, with
$n > 1$, then $f(\mathbf{t})$ has to be an idempotent in M. Indeed, we have

$$f(t_1) = f(t_2) = f(t_1 t_2) = e;$$

hence

$$e = f(t_1) = f(t_1 t_2) = f(t_1) f(t_2) = e^2.$$

Thus, the following holds:

COROLLARY 4   *There exists a minimal number $q = q(n, k)$, such that for
any morphism $f: A^* \to M$, with M a finite monoid of k elements, every
word of length at least q has a factorized segment $\mathbf{t} = (t_1, t_2, \ldots, t_n)$ on
which f is constant and $f(\mathbf{t})$ is idempotent.*   □

Before proceeding we note that the property of Corollary 4, for $n = 1$,
characterizes finite monoids. Indeed, the following has been proved [14]:

THEOREM 5   *Let M be a finitely generated monoid such that for some n
every sequence $(m_1, m_2, \ldots, m_n)$ of n elements of M has an idempotent
segment (i.e. there exist $1 \leq i \leq j \leq n$, such that $m_i m_{i+1} \ldots m_j$ is idempotent). Then M is finite.*   □

Now, we investigate the function $q(n, k)$, defined in Corollary 4. Using
two different ideas one gets two incomparable upper bounds, while the
lower bound which follows is given by the example used to prove
Theorem 3.

THEOREM 6   *For every n, $k \geq 1$,*

$$n^k \leq q(n, k) \leq \min\{2^{nk}, (nk)^k\}.$$ □

In the case $n = 1$ one can get a sharper result:

THEOREM 7   *For every $k \geq 1$,*

$$2^{k/2} \leq q(1, k) \leq 2^k.$$ □

4. Factorizations with bounded gap size

Theorem 1, presented in Section 2, guarantees, for every infinite word in A^ω and for every finite colouring $f: A^* \to E$ of A^*, the existence of a monochromatic factorized segment with infinitely many components. An important fact, first discovered by T. C. Brown [1], and later developed by Brown [2], Justin [6] and Jacob [5] is that for some classes of colourings every infinite word in A^ω has arbitrarily long monochromatic factorized segments of bounded gap size! Next, we give a generalization of those results.

Let $[\omega]^2$ denote the family of 2-subsets of ω. A colouring f of $[\omega]^2$ is *monotonic* if, for every $a \leq b \leq c < c' \leq b' \leq a'$,

$$f(c, c') = f(a, a') \quad \text{implies} \quad f(b, b') = f(a, a').$$

We will need the gap size of (finite) subsets of ω. Let P be an n-subset of ω and let $p_1 < p_2 < \ldots < p_n$ be the elements of P in increasing order. The *gap size* of P is

$$\max\{p_{i+1} - p_i \mid 1 \leq i < n\}.$$

THEOREM 8   *For every monotonic finite colouring of $[\omega]^2$ there exists an integer g such that, for every n, ω has an n-subset X of gap size g, for which $[X]^2$ is monochromatic.*

Perhaps it is worth rephrasing Theorem 8 in terms of graphs: *For every monotonic colouring of the edges of the complete graph G with vertex set ω there exists an integer g such that, for every n, G has a monochromatic complete subgraph K_n whose vertices have gap size g.*

PROOF OF THEOREM 8   Let $f: [\omega]^2 \to E$ be a monotonic finite colouring of $[\omega]^2$ and let $m = |E|$. For every $0 \leq i < i' < \omega$ we define a subset $F(i, i')$ of E and a natural $r(i, i')$ given by

$$F(i, i') = \{f(j, j') \mid 0 \leq i \leq j < j' \leq i'\},$$
$$r(i, i') = |F(i, i')|.$$

Thus, $r(i, i')$ is the number of colours of the 2-subsets $\{j, j'\}$ "contained" in $\{i, i'\}$.

We begin by noting two properties of the function r:

(i) For every $0 \leq i \leq k < k' \leq i'$, $r(k, k') \leq r(i, i')$.
(ii) For every $0 \leq i \leq j \leq k < k' \leq j' \leq i'$, if $r(k, k') = r(i, i')$ then $f(i, i') = f(j, j') = f(k, k')$.

Indeed, (i) follows from the fact that $F(k, k') \subseteq F(i, i')$. To see (ii), we note that $F(k, k') \subseteq F(i, i')$ and if $r(k, k') = r(i, i')$ then $F(k, k') = F(i, i')$. Then, there exist l and l' such that $0 \leq k \leq l < l' \leq k'$ and $f(l, l') = f(i, i')$. Since f is monotonic, it follows that $f(i, i') = f(j, j') = f(k, k')$.

Next, we define a natural $h(p)$, for each $1 \leq p \leq m$, given by

$$h(p) = \max\{i' - i \mid 0 \leq i < i' < \omega \quad \text{and} \quad r(i, i') = p\}.$$

Here we consider max $A = \omega$, for any infinite subset A of ω. For convenience, we define $h(0) = 0$.

Clearly, $h(p) = \omega$ for some p, since $r(i, i')$ is defined for every $0 \leq i < i' < \omega$ and p is bounded by $m = |E|$. Then let q be the smallest integer in $[1, m]$, such that $h(q) = \omega$, and let

$$g = 1 + \max\{h(p) \mid 0 \leq p < q\}.$$

We claim that for every n, ω has an n-subset X of gap size g, for which $[X]^2$ is monochromatic. Indeed, by construction, there exist $0 \leq i < i' < \omega$, such that $r(i, i') = q$ and $i' - i \geq (n-1)g$. Consider now $x_j = i + (j-1)g$, for $1 \leq j \leq n$, and let $X = \{x_j\}$. Clearly, X is an n-subset of ω with gap size g.

We claim now that for every $1 \leq j < j' \leq n$, $f(x_j, x_{j'}) = f(i, i')$; that is, $[X]^2$ is monochromatic. We begin by showing that, for every $1 \leq j < n$, $r(x_j, x_{j+1}) = q$. Indeed, by property (i),

$$r(x_j, x_{j+1}) \leq r(i, i') = q,$$

and since $x_{j+1} - x_j = g > h(p)$, for every p, $1 \leq p < q$, it follows that $r(x_j, x_{j+1}) \neq p$. Thus, $r(x_j, x_{j+1}) = q = r(i, i')$, for every $1 \leq j < n$. By property (ii), $f(x_j, x_{j'}) = f(i, i')$, for every $1 \leq j < j' \leq n$. This completes the proof of Theorem 8. $\square$

It is natural to ask at this point whether Theorem 8 holds for arbitrary colourings of $[\omega]^2$? The answer is no!

EXAMPLE 9   Consider the following 2-colouring ϕ of $\omega - 0$, given by colouring the first element red, the second one blue, the next two elements red, the next two ones blue, the next three ones red and so on. Formally,

$$\phi(t) = \begin{cases} 1 & \text{if } t \in [m^2 - m + 1, m^2] \text{ for some } m \geq 1, \\ 2 & \text{otherwise.} \end{cases}$$

Consider now the 2-colouring f of $[\omega]^2$ given by

$$f(i, j) = \begin{cases} 1 & \text{if } \phi(|i - j|) = 1, \\ 2 & \text{otherwise.} \end{cases}$$

Now, for every n, if P is an n-subset of ω with gap size g and for which $[P]^2$ is monochromatic (with respect to f), then $n \leqslant g^2 + g$. Indeed, let $p_1 < p_2 < \ldots < p_n$ be the elements of P. The gap size being g, we have that

$$n - 1 \leqslant p_n - p_1 \leqslant (n-1)g,$$

and that if $I = [x, x']$ is an interval of g elements contained in $[p_1, p_n]$ then I contains some p_j. Assuming then that $n > g^2 + g$, it follows that there exists a p_j in $[p_1 + g^2 - g + 1, p_1 + g^2]$ and a p_k in $[p_1 + g^2 + 1, p_1 + g^2 + g]$. Hence, $1 = f(p_1, p_j) \neq f(p_1, p_k) = 2$. Thus, for no integer g are there arbitrarily long factorized segments of gap size g on which f is constant. $\square$

A particular case of Theorem 8 is an infinite version of a result of Jacob [5] on words (see [7, p. 290]). A *rank* r on A^* is a colouring $r: A^* \to \omega$, such that, for every u and v in A^*,

$$r(uv) \leqslant \inf\{r(u), r(v)\}.$$

Since we assume throughout that A is a finite alphabet, it follows that a rank is a finite colouring of A^*.

COROLLARY 10   *For every rank r on A^* and for every infinite word s in A^ω, there exists an integer g, such that, for every n, s has a factorized segment of n components, of gap size g, on which r is constant.*

PROOF   From the word $s: \omega \to A$ and from the rank $r: A^* \to \omega$ we define a finite colouring $f: [\omega]^2 \to \omega$ as follows: for every $0 \leqslant i < j < \omega$,

$$f(i, j) = r(s_i s_{i+1} \ldots s_{j-1}),$$

where s_k denotes the kth letter $s(k)$ of s. It is easy to show that f is monotonic. Thus, the result follows from Theorem 8. $\square$

We list now two interesting consequences of Theorem 8, due, respectively, to Brown [1] and Justin [6].

COROLLARY 11   *Let $f: \omega \to E$ be a finite colouring of ω. There exists an integer g in ω such that for every n, ω has a monochromatic n-subset of gap size at most g.*

Corollary 11 is easily interpreted in terms of words in E^ω. It suffices to take the infinite word $s = s_0 s_1 \ldots$ over E, such that $s_i = f(i)$. Then Corollary 11 guarantees that for any m the word s has a factorized segment of the form

$$\mathbf{p} = (e, p_1, e, p_2, \ldots, e, p_m, e)$$

with gap size at most g, with e in E and p_i in E^*.

PROOF OF COROLLARY 11   Let $E = \{e_1, e_2, \ldots, e_n\}$ be an enumeration of E. From the colouring f of ω we define a colouring

$$f': [\omega]^2 \to E$$

as follows: for every $0 \leqslant i < j < \omega$,

$$f'(i, j) = \min\{k \mid e_k = f(l) \quad \text{and} \quad i \leqslant l < j\}.$$

Clearly, f' is a monotonic colouring. On the other hand, for every $(n + 1)$-subset X of ω of gap size g, for which $[X]^2$ is monochromatic (with respect to f'), there exists a monochromatic (with respect to f) n-subset Y of ω of gap size at most $2g - 1$. Indeed, let $x_1 < x_2 < \ldots < x_n < x_{n+1}$ be an enumeration of X, and let us suppose that $f'(x_1, x_{n+1}) = k$. Then, for each $1 \leqslant j \leqslant n$, $f'(x_j, x_{j+1}) = k$ and there exists $y_j, x_j \leqslant y_j < x_{j+1}$, such that $f(y_j) = e_k$; that is, $Y = \{y_1, \ldots, y_n\}$ is a monochromatic (with respect to f) n-subset of ω. Since

$$y_{j+1} - y_j < x_{j+2} - x_j = 2g,$$

it follows that Y has gap size at most $2g - 1$. Thus, Corollary 11 follows from Theorem 8.   □

A consequence of Corollary 11 is the following important result; proved in [2].

COROLLARY 12   Let $f\colon A^* \to M$ be a morphism from A^* to the finite monoid M and let s be an infinite word in A^ω. There exist an integer g in ω and an idempotent e in M, such that, for any n, s has a factorized segment $\mathbf{t}$ of n components with gap size at most g on which f is constant with value $f(\mathbf{t}) = e$.   □

Corollary 12 has an intriguing aspect. Let $f\colon A^* \to M$ be a morphism and let $s\colon \omega \to A$ be an infinite word on A. We can associate to these data a colouring $f'\colon [\omega]^2 \to M$ as follows: for every $0 \leqslant i < j < \omega$,

$$f'(i, j) = f(s_i s_{i+1} \ldots s_{j-1}),$$

where s_k is the kth letter $s(k)$ of s. Even though in general f' is not a monotonic colouring of $[\omega]^2$, Corollary 12 guarantees that the conclusion of Theorem 8 holds for f'. This suggests the existence of weaker conditions under which Theorem 8 can be proved.

Theorem 8 and Corollaries 10–12 admit finite versions which follow from them by König's lemma, and which can also be proved by induction. For instance, the finite version of Corollary 12 is due to Justin [6]; a proof (for words) can be found in [7]. Let $g(A)$ denote the gap size of a subset A of ω.

THEOREM 13   *For every k and for every function $v: \omega \to \omega$ there exists a minimal integer $n = n(k, v)$, such that for every k-colouring of $[1, n]$, there is a monochromatic subset A of $[1, n]$, with at least $v(g(A))$ elements.*   □

Brown [3] noted that Theorem 13 is similar to the celebrated version of Ramsey's theorem which has been shown [11] not to be provable in Peano's first-order arithmetic. That result is:

For every k and r and for every function $v: \omega \to \omega$ there exists a minimal integer $m = m(k, r, v)$, such that, for every k-colouring of $[1, m]^r$, there is a subset A of $[1, m]$, with at least $v(\mathrm{sm}(A))$ elements, for which $[A]^r$ is monochromatic.

Here, $\mathrm{sm}(A)$ is the smallest element of $A \subseteq \omega$, and Theorem 13 is obtained by substituting $\mathrm{sm}(A)$ by $g(A)$, for $r = 1$. Observe that Example 9 shows that for $r = 2$, $\mathrm{sm}(A)$ can not be substituted by $g(A)$!

References

1. T. C. Brown, On van der Waerden's theorem on arithmetic progressions. *Notices Amer. Math. Soc.* **16** (1969), 245.
2. T. C. Brown, An interesting combinatorial method in the theory of locally finite semigroups. *Pacific J. Math.* **36** (1971), 285–289.
3. T. C. Brown, On van der Waerden's theorem and the theorem of Paris and Harrington. *J. Combinat. Theory A* **30** (1981), 108–111.
4. R. L. Graham, B. L. Rothschild and J. H. Spencer, *Ramsey Theory*, John Wiley and Sons, New York, 1980.
5. G. Jacob, La finitude des representations lineaires de semi-groupes est decidable. *J. Algebra* **52** (1978), 437–459.
6. J. Justin, Généralisation du théorème de van der Waerden sur les semi-groupes répétitifs. *J. Combinat. Theory A* **12** (1972), 357–367.
7. G. Lallement, *Semigroups and Combinatorial Applications*, John Wiley and Sons, New York, 1979.
8. M. Lothaire, *Combinatorics on Words*, Addison-Wesley, Reading, Mass., 1983.

9. A. Mandel and I. Simon, On finite semigroups of matrices. *Theor. Comput. Sci.* **5** (1977), 101–112.

10. R. McNaughton, Algebraic decision procedures for local testability. *Math. Syst. Theory* **8** (1974), 60–76.

11. J. Paris and L. Harrington, A mathematical incompleteness in Peano arithmetic. In *Handbook of Mathematical Logic* (J. Barwise, ed.), North Holland, Amsterdam, 1977, pp. 1133–1142.

12. M. P. Schützenberger, *Quelques problèmes combinatoires de la théorie des automates* (rédigé par J. F. Perrot), Université de Paris, Paris, 1966.

13. I. Simon, Limited subsets of a free monoid. In *Proceedings: 19th Annual Symposium on Foundations of Computer Science*, IEEE, Piscataway, N.J., 1978, pp. 143–150.

14. I. Simon, Conditions de finitude pour des semigroupes. *Compt. Rend. Acad. Sci. Paris A*, **290** (1980), 1081–1082.

26

UNIT DISTANCES IN THE EUCLIDEAN PLANE

J. Spencer, E. Szemerédi and W. Trotter, Jr

ABSTRACT   Consider a set of n points in the Euclidean plane and let $U(n)$ denote the maximum number of pairs of points which can be at unit distance. Erdös observed that $U(n) > n^{1+c/(\log \log n)}$ and conjectured that $U(n)$ has an upper bound of the same form. Progress on this conjecture has been slow, and to date the best known upper bound, $U(n) < n^{1.499}$, is due to Beck and Spencer. In this paper, we modify techniques first used to solve extremal problems involving configurations of points and lines in the plane to obtain the improved bound: $U(n) < cn^{4/3}$.

1. Introduction

Combinatorial problems in discrete geometry have been investigated extensively by numerous researchers. The reader is encouraged to consult Erdös's survey papers [2–4] and Moser's annual summary [6] for a compilation of results and an extensive bibliography. Here, we concentrate on one of the oldest and most tantalizing problems in this area: What is the maximum number $U(n)$ of unit distances determined by n points in the Euclidean plane!

Consideration of the lattice points shows that $U(n) > n^{1+c/(\log \log n)}$ and Erdös conjectures that this inequality is essentially best possible. However, progress on upper bounds has been slow. In [5], Józsa and Szemerédi showed that $U(n) = o(n^{3/2})$. This result was improved by Beck and Spencer [1], who showed that $U(n) < n^{1.499}$. In this paper, we will show that $U(n) < cn^{4/3}$. In order to establish this result, we will employ techniques similar to those developed by Szemerédi and Trotter in [7] and [8]. The results in these papers provide inequalities involving incidences between points and lines. We will need a number of modifications in order to obtain results for points and circles.

2. The Covering Lemma

Assume that a pair of perpendicular lines has been chosen and used to establish a coordinatization of the plane. The following lemma is proved

GRAPH THEORY AND COMBINATORICS
ISBN 0-12-111760-X

in [7]:

LEMMA 1   *Let r_1 and r_2 satisfy $r_2 \geqslant 256r_1$ and let P be a set of n points in the plane. Then there exists a family Q of squares so that:*

  (i) *The sides of the squares in Q are parallel to the coordinate axes.*
  (ii) *Each square in Q contains at least r_1 but no more than r_2 points from P.*
  (iii) *No point in the plane belongs to the interior of two or more squares from Q.*
  (iv) *At least $n/16$ of the points in P are covered by the squares in Q.* □

In this paper, we will require a slight modification of this covering lemma. Instead of covering points with squares, we will use rectangles for which the ratio between the width and the height is some fixed constant. It is easy to see that the lemma remains valid with this modification since a linear transformation can be used to interchange squares with rectangles of the desired shape.

3. The principal theorem

In this section, we prove the existence of an absolute constant c for which $U(n) < cn^{4/3}$. The result will follow as an easy corollary to a somewhat more technical result. Consider two squares resting on the x-axis. Each side has length 10^{-6} and the centers of the squares are exactly 1 unit apart. We then denote by $U(n, t)$ the maximum number of unit distances which can occur between a set X of n points in one square and a set Y of t points in the other. For convenience, we consider the points in X as belonging to the leftmost square.

The approximation of $U(n, t)$ is relatively simple if one of the parameters is quite large in comparison with the other since the optimal configuration (up to a constant factor) places all the points in one set on a unit circle determined by one point in the other set. However, when $\sqrt{n} \leqslant t \leqslant n^2$, the problem is much more difficult. Here, we prove the following inequality:

THEOREM 1   *There exists an absolute constant c so that $U(n, t) < cn^{2/3}t^{2/3}$ whenever $\sqrt{n} \leqslant t \leqslant n^2$.*

PROOF   Let $M_0 = 1$. Then define $M_{i+1} = 10M_i^{10}$ for each $i \geqslant 0$. We show that this inequality is valid when $c = M_{10^{12}}$.

The argument is by contradiction. We suppose the result is false and choose a counterexample with $n + t$ as small as possible. For this config-uration, $U(n, t) \geq cn^{2/3}t^{2/3}$. The remaining part of the proof is subdivided into several parts. First, we introduce some appropriate labels. Second, we present several counting lemmas. Third, we describe in a series of steps a method for producing a subconfiguration satisfying several key properties. Finally, we count the number of times arcs cross and show that it is more than the total number of pairs of arcs.

Label the points in X as $x_1, x_2, \ldots, x_n$ and the points in Y as $y_1, y_2, \ldots, y_t$. For each $i = 1, 2, \ldots, n$, we let d_i count the number of unit distances from x_i to Y. Similarly, for each $j = 1, 2, \ldots, t$, we let e_j count the number of unit distances from y_j to X. For simplicity, we write $U = U(n, t)$. Note that

$$\sum_{i=1}^{n} d_i = \sum_{j=1}^{t} e_j = U.$$

Now a pair of points in one of our two sets can belong to at most one unit circle whose center is a point in the other set. Thus,

$$\sum_{i=1}^{n} \binom{d_i}{2} \leq \binom{t}{2}.$$

We conclude that

$$t^2 \geq \frac{1}{10} \sum_{i=1}^{n} d_i^2 \geq \frac{1}{10n} \left(\sum_{i=1}^{n} d_i \right)^2 \geq \left(\frac{c^2}{10} \right) n^{1/3} t^{4/3}.$$

From this, it follows that (being generous) $c^2 \sqrt{n} < t$. Since the argument is dual, we know that $c^2 \sqrt{t} < n$, i.e. $t < n^2/c^4$. The important observation to make here is that since c is extremely large, we can apply the inductive hypothesis to subsets of size n/M_i and t/M_j respectively, where M_i and M_j are large – but relatively small in comparison to c; for example, when i, $j \leq 1000$. That is, we can be assured that $\sqrt{(n/M_i)} \leq t/M_i \leq (n/M_i)^2$ and that the number of unit distances determined by these subsets is less than $c(n/M_i)^{2/3}(t/M_j)^{2/3}$.

Another easy conclusion that can be drawn from the inequality

$$\binom{t}{2} \geq \sum_{i=1}^{n} \binom{d_i}{2}$$

is that $U(n, t) \geq 10t^2$. Dually, $U(n, t) \geq 10n^2$. Note that these inequalities hold for all n, t with $\sqrt{n} \leq t \leq n^2$. Furthermore, we observe that the function $U(n, t)$ increases monotonically in both coordinates.

Next, we present several claims concerning unit distances between subsets of X and Y. In each case, the proof is an elementary counting argument. Furthermore, we make no attempt to obtain the best possible inequalities. Instead, we present the claims in simple form to facilitate their use later in our proof. Let d denote the average value of d_i, i.e.

$$d = \frac{1}{n} \sum_{i=1}^{n} d_i.$$

CLAIM 1 Let $i \leq 100$ and let A and B be subsets of X and Y respectively. Suppose there are at least U/M_i unit distances in $A \cup B$. Then there are at least n/M_{i+1} points in A each having at least d/M_{i+1} unit distances in B.

PROOF Let A_0 denote the subset of A consisting of those points with at least d/M_{i+1} unit distances in B. Suppose that $|A_0| < n/M_{i+1}$. Then the number of unit distances in $A_0 \cup B$ is less than $c(n/M_{i+1})^{2/3} t^{2/3} < U/2M_i$. This inequality follows from the observation that $M_{i+1} = 10M_i^{10} > 2^{3/2}M_i^{3/2}$. However, the number of unit distances in $(A - A_0) \cup B$ is less than $nd/M_{i+1} < U/2M_i$. These inequalities imply that there are less than U/M_i unit distances in $A \cup B$. The contradiction completes the proof. □

CLAIM 2 Let $i \leq 100$. Suppose that A and B are subsets of X and Y respectively with $|A| \geq n/M_i$. Suppose further that each point in A has at least d/M_i unit distances in B. Then let $K = 16M_i^2$ and let $B = B_1 \cup B_2 \cup \ldots \cup B_K$ be an arbitrary partition into subsets of equal size. Then there exists a pair (α, β) with $1 \leq \alpha < \beta \leq K$ and a subset $A(\alpha, \beta)$ of A with $|A(\alpha, \beta)| \geq N/M_{i+1}$ so that each point in $A(\alpha, \beta)$ has at least d/M_{i+1} unit distances in B_α and at least d/M_{i+1} unit distances in B_β.

PROOF For each $\alpha = 1, 2, \ldots, K$, let A_α denote those points in A which have at least d/M_{i+1} unit distances in B_β if and only if $\alpha = \beta$. Then each point in A_α has at least $d/M_i - (K-1)d/M_{i+1} > d/2M_i$ unit distances in B_α.

Now suppose there is some $\alpha \leq K$ with $|A_\alpha| = xn$ and $x \geq 8M_i/K^2$. Then we may conclude that $xnd/2M_i = |A_\alpha| \, d/2M_i < c \, |A_\alpha|^{2/3}(t/K)^{2/3} \leq Ux^{2/3}/K^{2/3} = ndx^{2/3}/K^{2/3}$, and thus $x < 8M_i/K^2$. The contradiction shows that $x < 8M_i/K^2$ for all α. Therefore, $A_0 = A_1 \cup A_2 \cup \ldots \cup A_k$ contains at most $K(8M_i/K^2)n$ points. However, $K(8M_i/K^2)n = n/2M_i$. Thus $|A - A_0| \geq n/2M_i$.

Since

$$(n/2M_i) \bigg/ \binom{K}{2} > n/M_{i+1},$$

the existence of the desired set $A(\alpha, \beta)$ follows by the box principle and the observation that, for each point a from A, there must be some B_α so that a has at least d/M_{i+1} unit distances in B_α. $\square$

CLAIM 3   Let $i \leqslant 100$. Suppose that A and B are subsets of X and Y respectively with $|A| \geqslant n/M_i$. Suppose further that each point in A has at least d/M_i unit distances in B. Then let K be an arbitrary positive integer and let $A_1, A_2, A_3, \ldots, A_K$ be disjoint subsets of A whose union contain at least half the points in A. If no point in B has unit distances in two or more of the A_αs, then there is at least one α for which $|A_\alpha| \geqslant n/M_{i+2}$.

PROOF   Suppose to the contrary that $|A_\alpha| < n/M_{i+2}$ for every α. Let $A_0 = A_1 \cup A_2 \cup \ldots \cup A_K$. Then we can form new partitions of A_0 by merging subcollections of the A_αs. It is easy to see that there exists a partition of this type $A_0 = A_1' \cup \ldots \cup A_{K'}'$, where $n/2M_{i+1} \leqslant |A_\alpha'| \leqslant 2n/M_{i+1}$ for each α. Then for each $\alpha = 1, 2, \ldots, K'$, let B_α denote the subset of B with at least one unit distance in A_α'. Let B_α' denote those points in B_α with two or more unit distances in A_α'.

The number of unit distances in $A_\alpha' \cup B_\alpha$ is at least $nd/2M_iM_{i+1}$ and the number of unit distances in $A_\alpha' \cup (B_\alpha - B_\alpha')$ is at most $|B_\alpha - B_\alpha'| \leqslant t < nd/4M_iM_{i+1}$. On the other hand, the number of unit distances in $A_\alpha' \cup B_\alpha'$ is less than $c(2n/M_{i+1})^{2/3} |B_\alpha'|^{2/3}$. Thus, we must have

$$\frac{nd}{4M_iM_{i+1}} < c\left(\frac{2n}{M_{i+1}}\right)^{2/3} |B_\alpha'|^{2/3}.$$

This requires $|B_\alpha'| > t/M_{i+1}^{1/2}$. Since this inequality holds for every $\alpha = 1, 2, \ldots, K'$, we conclude that $K' < M_{i+1}^{1/2}$. However, since each A_α' contains at most $2n/M_{i+1}$ points and their union contains at least $n/2M_i$ points, we conclude that $K' \geqslant M_{i+1}/4M_i > M_{i+1}^{1/2}$. The contradiction completes the proof. $\square$

The reader should note that these three claims have dual forms in which the roles played by the two squares and the sets X and Y are reversed. Now we are ready to use these claims to obtain a subconfiguration satisfying several essential properties.

In these arguments, we will be subdividing rectangles by horizontal or vertical lines. When the subdivision is by horizontal lines we call the pieces *strips* and use the term *slices* when the subdivision is by vertical lines. In each case the horizontal dimension of a rectangle is called its *width* and the vertical dimension its *height*.

Step 1   Apply Claim 1 and choose a subset X_1 of X so that $|X_1| \geqslant n/M_1$ and each point in X_1 has at least d/M_1 unit distances in Y.

Step 2   Apply Claim 2 with $i = 1$. Divide Y into K subsets of size t/K

by dividing the right most square into K strips. The height of the strips may vary, but each is to contain t/K points from Y.

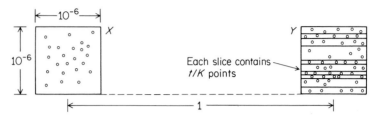

We obtain two strips on the right each having at most t points and a subset X_2 of X containing at least n/M_2 points each having at least n/M_2 unit distances in each of these two strips.

Step 3   Apply Claim 2 again to each of the two strips obtained in Step 2. We obtain four strips each of size at most t and a subset X_4 of X containing at least n/M_4 points each having at least d/M_4 unit distances in each of the four strips.

Step 4   Label the four strips from bottom to top as S_1, S_2, S_3, S_4 and let s_1, s_2, s_3, s_4 denote their respective heights. Without loss of generality, we assume that $s_2 \leqslant s_3$. Otherwise, turn the entire plane upside down. Delete from consideration all points in Y except those in strips S_2 and S_4. Note that these two strips each contain at least t/M_5 of the points in Y, and that the gap between them is at least s_2.

Step 5   Try to find a strip of thickness s_2 which covers at least n/M_{10} of the points in X_4. If this can be done, go to Step 8.

Step 6   In this case, we are unable to cover n/M_{10} of the points in X_4 with a strip of height s_2. Now consider the n/M_4 points in X_4 and the set of at least t/M_4 points in S_2. In these two sets there are at least $nd/M_4M_4 > U/M_5$ unit distances. So it follows that there is a subset Y_6 of S_2 with $|Y_6| \geqslant t/M_6$ so that each point in Y_6 has at least e/M_6 unit distances in X_4.

Step 7   Apply Claim 2 to obtain four strips in the left most rectangle each containing at least n/M_{10} points from X and a subset Y_{10} of S_2 of size at least t/M_{10} so that each point in Y_{10} has at least e/M_{10} unit distances in each of these four strips. Label these strips from bottom to top as R_1, R_2, R_3, R_4 and let r_1, r_2, r_3 and r_4 denote their respective heights. Again, we may assume that $r_2 \leqslant r_3$.

Step 8   From either Step 5 or Step 7 we have a strip of height z_1 in one square and two strips in the other square of heights z_1 and z_2 respectively with the two strips separated by a gap of at least z_1. Furthermore, each of these three strips contains at least an M_{10}th of the points in the entire square. The points in the solitary strip have at least d/M_{10} (e/M_{10} if the solitary strip is in the rightmost square) unit distances

in each of the two strips in the other square. For convenience we may
assume that the solitary strip belongs to the leftmost square.

Next we partition the solitary strip vertically into slices whose width is
$z_1/10$. Label the slices from left to right as $A_1, A_2, \ldots, A_K$. Since no
point in either of the strips in the right square has unit distances in two or
more of these subrectangles unless they are consecutive, we may apply
Claim 3 to the larger of $A_1 \cup A_3 \cup A_5 \cup \ldots$ and $A_2 \cup A_4 \cup A_6 \cup \ldots$.

Step 9 So now we have a set A of size at least n/M_{12} so that all points
in A are contained in a rectangle whose height is z_1 and whose width is
$z_1/10$. Every point in A has at least d/M_{12} unit distances in each of two
sets B_1 and B_2 which are contained respectively in strips of thickness z_1
and z_2. Furthermore, these strips are separated by a gap of at least z_1. It
is clear that all points in B_1 which have a unit distance in A can be
covered with a rectangle of height z_1 and width $3z_1/10$. Since we do not
know the height of the rectangle containing B_2 and it could be much
larger than z_1, we make no such claim for B_2. Now choose a strip S of the
rectangle containing A of height $z_1/1000 = z$ so that S contains at least
n/M_{13} of the points in A. Then apply Claim 1 to choose a subset B_1' of B_1
of size t/M_{15} with each point in B_1' having at least e/M_{15} unit distances in
S. By the box principle, we can then choose a strip in B_1 containing at
least t/M_{16} points each having at least e/M_{15} unit distances in S with the
height of the strip in B_1 also at most $z_1/1000$. Applying Claims 1 and 3
again and rotating the configuration if necessary, we have the following
configuration:

Each point in A has at least d/M_{18} unit distances in B_1 and B_2.

Next, we divide A vertically into slices of width z^2. Note that no point of B_1 can be adjacent to points from non-consecutive slices:

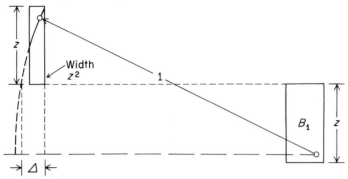

Note also that $\Delta < 2z^2$.

By Claim 3, we conclude after relabeling that we have a set A in a rectangle whose height is z and whose width is z^2 with $|A| \geq n/M_{20}$. Furthermore, we have two strips B_1 and B_2 with the thickness of B_1 also z and the gap between B_1 and B_2 at least $100z$. Also, each point in A has at least d/M_{20} unit distances in B_1 and in B_2.

Next, we present the essential idea for counting crossings between arcs of the unit circles determined by points in this subconfiguration. Let A' be a subrectangle of A. We say A' is *similar* to A if both rectangles have the same ratio between their height and their width and if the height of A' is at most half the height of A.

THREE SIDES LEMMA   *Let A' be a subrectangle of A with A' similar to A. Then let p be any of the points in A' and consider any two unit circles determined by a point in B_1 and a point in B_2. Then these circles intersect at least three of the sides of A'. Furthermore, the circle determined by the point in B_1 cannot intersect both vertical sides of A'.*

We illustrate the statement of this lemma with the following diagram:

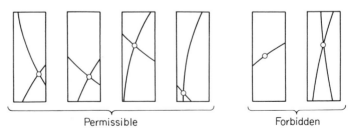

Permissible   Forbidden

PROOF   To prove the lemma, we proceed as follows. We can easily dismiss the first of the two forbidden configurations. Let u and v be as shown below:

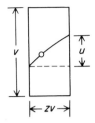

Since the arc hits the vertical sides, we must have $zv < 2u^2 \leq 2v^2$. But this implies that $z < 2v$, which is a contradiction.

For the second configuration, we consider the angle between the tangent lines to the arcs at the point p. For this configuration to arise, it is clear that this angle is less than $10z$. On the other hand, since B_1 has thickness z and the gap between B_1 and B_2 is at least $100z$, we know the angle between these tangent lines is at least $50z$. Thus, neither forbidden configuration can occur.   □

The next lemma is quite elementary but involves the systematic examination of several cases. For the sake of brevity, we provide only a sketch of the argument and leave the details to the reader.

INTERSECTION LEMMA   *Let A' be similar to A. Divide A' into four similar rectangles by perpendicular bisectors to the sides. Then let p_1 and p_2 be points in the same quarter of A'. For each of these points, choose a pair of unit circles passing through them as determined by points in B_1 and B_2. Then at least one of the arcs through p_1 crosses one or both of the arcs through p_2 at a point inside A'.*

PROOF   The arcs through p_1 partition the quarter into four regions. The result is immediate unless the region containing p_2 touches three of the four sides of the quarter rectangle. A dual statement holds for the regions determined by P_2. For example,

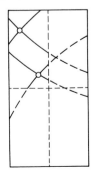

By analyzing the four quadrants individually, it is easy to see the desired crossing exists. $\square$

We are now ready to make the final computation. We apply the Covering Lemma in Section 2 to the points in A with rectangles similar to A. We use $r_1 = d/M_{23}$ and $r_2 = d/M_{22}$.

Now consider a rectangle R used in the covering. Choose a quarter rectangle R' of R which contains at least $r_1/4 \geq d/M_{24}$ points. Each of these points has at least d/M_{20} unit distances in B_1 and at least d/M_{20} unit distances in B_2. So for each point p in R, there are at least d/M_{21} points in B_1 for which p is the unique point of A in R at unit distance. A similar statement holds for B_2. There are at least d^2/M_{24} pairs of points from A which belong to R'. For each such pair there are at least d^2/M_{22} positions where arcs cross inside R. These crossings are wasted since they occur where there is no point in A. So inside each rectangle R, there are at least d^4/M_{25} wasted crossings. However, there are at least

$$\frac{n}{16 M_{20}} \frac{d}{M_{22}} > n/d$$

rectangles in the covering. Thus there are at least

$$nd^3/M_{25} > \frac{c^3 t^2}{M_{25}} > \binom{t}{2}$$

wasted crossings altogether. Clearly this is impossible. The contradiction completes the proof of Theorem 1. $\square$

It is easy to see that the following results may be obtained as easy corollaries from Theorem 1 using nothing more than the box principle.

COROLLARY 1   *There exists an absolute constant c so that whenever* $\sqrt{n} \leq t \leq n^2$ *and X and Y are sets of size n and t respectively, the number of pairs* $(x, y) \in X \times Y$ *which are at unit distance is less than* $cn^{2/3} t^{2/3}$. $\square$

COROLLARY 2   *There exists an absolute constant c so that* $U(n)$, *the maximum number of unit distances determined by n points in the plane, satisfies the inequality* $U(n) < cn^{4/3}$. $\square$

Acknowledgements

The authors gratefully acknowledge the assistance of Professor Paul Erdös through many stimulating conversations on the topics discussed in this paper and a host of other topics in combinatorial mathematics. The research of all three authors was supported in part by grants from the National Science Foundation.

References

1. J. Beck and J. Spencer, Unit distances. To appear.
2. P. Erdös, On some problems of elementary and combinatorial geometry. *Annali Mat., Ser. 4* **103** (1974), 99–108.
3. P. Erdös, Some applications of graph theory and combinatorial methods to number theory and geometry. In *Algebraic Methods in Graph Theory*, Coll. Math. Soc. T. Bolyai no. 25, Elsevier, Amsterdam, 1981, pp. 131–148.
4. P. Erdös, Problems and results in combinatorial geometry. *Proc. N.Y. Acad. Sci.* to appear.
5. S. Józsa and E. Szemerédi, The number of unit distances in the plane. In *Infinite and Finite Sets* (A. Hajnal et al., eds), North Holland, Amsterdam, 1975.
6. W. Moser, Research problems in discrete geometry. Mimeograph notes, 1981.
7. E. Szemerédi and W. Trotter, Jr, Extremal problems in discrete geometry. *Combinatorica* **3** (1983), 381–397.
8. E. Szemerédi and W. Trotter, Jr, A combinatorial distinction between the Euclidean and projective plane. *Europ. J. Combinat.* to appear.

CONNECTIVITY IN TOURNAMENTS

Carsten Thomassen

ABSTRACT We show that a tournament of large connectivity has a path connecting any two prescribed vertices such that the deletion of that path leaves a tournament of large connectivity. We also derive a result of this type involving local connectivity. We combine these results with results in [5] and determine almost completely for which natural numbers m there are paths of length m between any two vertices in any k-connected tournament of order n and we obtain a general result on spanning configurations in tournaments of large connectivity. In particular, any k independent edges are contained in a Hamiltonian cycle (in any prescribed order) in a tournament of large connectivity as conjectured in [6].

1. Introduction

In [4, 5] connections between connectivity and Hamiltonian paths and cycles in tournaments were studied. For example, it was shown in [5] that, for any three vertices in a strong tournament, there is a Hamiltonian path connecting two of them and that any 4-connected tournament is strongly Hamiltonian-connected (meaning that there is a Hamiltonian path with any prescribed first vertex and last vertex). These results were applied in various directions in [5–7].

In this paper we apply the results to paths or path systems connecting prescribed vertices and to more general configurations in tournaments with constraints on the connectivity. Since regular tournaments (or more generally, tournaments with small irregularity) and random tournaments have large connectivity (see [5] and [3] respectively), our results also apply to such tournaments, although probably much stronger assertions hold for those classes of tournaments.

We recall the following unsolved problem of Lovász (see [2]): Does there exist a function $g(k)$ (k being a natural number) such that, for any two vertices x and y in a $g(k)$-connected undirected graph G, there is a path P from x to y such that $G - V(P)$ is k-connected?

We investigate this question for tournaments. In particular, we show that $g(k) \leq k + 4$ if we replace "graph" by "tournament" in the above question and we apply this result to the path length distribution in k-connected tournaments: It is easy to describe, for each natural number

GRAPH THEORY AND COMBINATORICS
ISBN 0-12-111760-X

k and infinitely many natural numbers n, a k-connected tournament of order n and diameter almost n/k. On the other hand, we show that, for any two vertices x and y in a k-connected tournament of order n, there are xy paths of almost all lengths between n/k and $n-1$. We also obtain a sufficient condition, in terms of local connectivity, for the existence of disjoint paths with prescribed first vertices and last vertices in a tournament and show that these paths can even be chosen to be a spanning path system if the (global) connectivity of the tournament is large enough. In particular, if the connectivity of a tournament is large compared to the natural number k, then any k independent edges of the tournament are contained in a Hamiltonian cycle as conjectured in [6].

Our terminology and notation are the same as in [5, 6]. In particular, a tournament is *k-connected* if the removal of any set of fewer than k vertices leaves a strong (i.e. strongly connected) tournament. When the edge xy (from vertex x to vertex y) is present, we say that x *dominates* y. In addition, we denote by $\kappa(x, y)$ the maximum number of internally disjoint xy paths in a given tournament.

2. Deleting shortest paths and connecting prescribed vertices with paths of prescribed length in k-connected tournaments

PROPOSITION 2.1 *Let x and y be any vertices in a $(k+4)$-connected tournament T $(k \geq 1)$ and let P be shortest xy path. Then $T - V(P)$ is k-connected.*

PROOF Let $P: x_0 x_1 \ldots x_m$, where $x_0 = x$ and $x_m = y$. The minimality of P implies that x_i dominates x_j whenever $i \geq j+2$. Now suppose (*reductio ad absurdum*) that $T - V(P)$ is not k-connected, i.e. its vertex set has a decomposition into sets S, S_1, S_2 such that $|S| \leq k-1$, $|S_1| \geq 1$, $|S_2| \geq 1$ and each vertex of S_1 dominates each vertex of S_2. Since $T - S$ is 5-connected, it has five paths $P_1, P_2, \ldots, P_5$ from S_2 to S_1 which are pairwise disjoint or (if $|S_1| < 5$ or $|S_2| < 5$) have at most the initial or terminal vertex in common pair by pair. We choose P_q such that there is no path from S_2 to S_1 whose vertex set is a proper subset of $V(P_q)$ $(1 \leq q \leq 5)$. This implies that the intersection of P_q and the tournament induced by $V(P)$ is either a segment of P (possibly of length 0) or a path of the form $x_i x_j$ (with $i \geq j+2$) or $x_i x_{i+1} x_{i-1}$ or $x_i x_{i-2} x_{i-1}$. We let j be the largest number such that some P_q $(1 \leq q \leq 5)$ contains an edge of the form $u'x_j$, where u' is in S_2 and i the smallest number such that some P_q contains an edge of the form $x_i v'$, where v' is in S_1. Then it is easy to see that $j \geq i+4$ and thus $P - \{x_{i+1}, x_{i+2}, \ldots, x_{j-1}\}$ together with the path $x_i v' u' x_j$ contradicts the minimality of P. □

In the next section we extend the ideas in Proposition 2.1, but first we apply Proposition 2.1 by combining it with results in [5]. Corollary 2.2 below generalizes the result [5] that every 4-connected tournament is strongly Hamiltonian-connected.

COROLLARY 2.2 *Let x and y be any vertices in a $4(s+t)$-connected tournament (where s and t are non-negative integers not both zero). Then T contains a collection of $s+t$ internally disjoint paths, s of which are xy paths and t of which are yx paths and whose union contains all vertices of T.*

PROOF (by induction on $s+t$) The case $s+t=1$ is the result of [5, Corollary 5.7], so we assume that $s+t \geq 2$. Without loss of generality we can assume that $s \geq 1$. Let P be a shortest xy path of length at least 2. Then $P-\{x, y\}$ is a shortest path from the successor of x to the predecessor of y. By Proposition 2.1, $T-(V(P)\backslash\{x, y\})$ is $4(s+t-1)$-connected and we apply the induction hypothesis to complete the proof. □

It is easy to describe a k-connected (regular) tournament of order $2k+1$. Now we take two disjoint such tournaments T' and T'' and add k disjoint paths of length r from T' to T''. Then it is easy to add edges such that we obtain a k-connected tournament of order $n = (r-1)k + 4k + 2$ such that the distance from some vertex in T' to some vertex in T'' is $r+2 = (n-2)/k - 1$. (J. C. Bermond has pointed out that the largest distance can even be made $(n-2)/k+1$.) On the other hand, if x and y are any vertices of a k-connected tournament T of order n, then T has k internally disjoint xy paths and one of these has length at most $(n-2)/k + 1$. The next result shows that T has xy paths of almost all lengths between $n-1$ and $(n-2)/k+1$.

THEOREM 2.3 *If x and y are vertices of a tournament T of connectivity $k \geq 13$ and order n, and m is any natural number, $(n-3)/(k-5)+7 \leq m \leq n-1$, then T has an xy path of length m.*

PROOF Let T' be the tournament obtained from T by deleting the edge between x and y and adding the edge yx. Then T' is $(k-1)$-connected. Let P_1 be a shortest xy path in T' and let P_2 be a shortest xy path in $T'' = T' - (V(P_1)\backslash\{x, y\})$. Then P_1 has at most $(n-2)/(k-1)$ intermediate vertices and, since T'' is $(k-5)$-connected, P_2 has at most $(n-3)/(k-5)$ intermediate vertices. Furthermore, $T''' = T'' - V(P_2)$ and $T''-(V(P_2)\backslash\{x, y, x_2\})$ are both $(k-9)$-connected (where x_i is the successor of x on P_i for $i = 1, 2$). In particular, x_2 dominates a vertex z of T''' and is dominated by a vertex z' of T'''. Let z'' be a vertex of T''' distinct from z and z'. Since T'''' is 4-connected, it has a Hamiltonian path from z to z'' (and from z'' to z') and hence

$$T'''' = T - ([V(P_1) \cup V(P_2)] \backslash \{x_1, x_2\})$$

has a Hamiltonian path connecting x_1 and x_2.

By [5, Lemma 3.1], T'''' has paths of all possible lengths not less than 3 connecting x_1 and x_2. Consider any such path P. By [5, Theorem 5.2], T' has an xy path with vertex set $V(P) \cup V(P_1) \cup V(P_2)$ and thus T' has xy paths of all lengths between $n-1$ and $|V(P_1)| + |V(P_2)| - 1$. If $|V(P_1)| \leqslant 6$, then

$$|V(P_1)| + |V(P_2)| - 1 \leqslant \frac{n-3}{k-5} + 7$$

and the proof is complete in that case.

So assume that $P_1: y_0 y_1 \ldots y_m$ and $P_2: u_0 u_1 \ldots u_q$, where $y_0 = u_0 = x$, $y_m = u_q = y$, $y_1 = x_1$, $u_1 = x_2$ and $6 \leqslant m \leqslant q$. Since P_1 is a shortest xy path, y_i dominates u_j whenever $1 \leqslant j+1 < i \leqslant m$. If y_i dominates u_{i+1} for some i, $1 \leqslant i \leqslant m-2$, then clearly T contains xy paths of all lengths between $|V(P_2)| + 2$ and $n - |V(P_1)| + 2$ (by combining paths of T'''' between x_1 and x_2 with segments of P_1 and P_2 and possibly the edge $y_i u_{i+1}$). Since $k \geqslant 13$ and hence $n \geqslant 27$ we have $n - |V(P_1)| + 2 \geqslant |V(P_1)| + |V(P_2)| - 1$, so the theorem holds in that case. We can therefore assume that u_{i+1} dominates y_i for each i, $1 \leqslant i \leqslant m-2$. Similarly, we can assume that y_{i+1} dominates u_i when $1 \leqslant i \leqslant m-2$. If we now combine paths of T'''' between x_1 and x_2 with one of the paths

$$u_1 u_2 u_3 y_2 y_3 y_4 y_5 u_4 u_5 \ldots y \quad \text{or} \quad y_1 y_2 y_3 y_4 y_5 u_2 u_3 u_4 \ldots y,$$

then we get xy paths of all lengths between $|V(P_2)| + 6$ and $n - |V(P_1)| + 6$, and the proof is complete. $\square$

3. Disjoint paths connecting prescribed vertices

In this section we extend the ideas of Proposition 2.1 to obtain a sufficient condition for the existence of disjoint paths with prescribed initial and terminal vertices in terms of local connectivity of a tournament. The key result is the following:

PROPOSITION 3.1 *Let x, y, u, v be distinct vertices in a tournament T. Suppose $\kappa(u, v) \geqslant q + 2$ and that $P_1, P_2, \ldots, P_p$ are internally disjoint xy paths, which together induce a tournament with no xy path of length $\leqslant 3$. Then T has q internally disjoint uv paths, the union of which intersects at most $2q$ of the paths $P_1, P_2, \ldots, P_p$.*

PROOF   We can assume that $p \geqslant 2q + 1$. Let $Q_1, Q_2, \ldots, Q_q$ be a set of internally disjoint uv paths in $T - \{x, y\}$. We select paths $Q_1', Q_2', \ldots, Q_q'$ such that, for each $i = 1, 2, \ldots, q$, either $Q_i' = Q_i$ or else Q_i' is a segment of Q_i from u to a vertex z on one of the paths $P_1, P_2, \ldots, P_p$, say P_j, such that the segment of P_j from z to y intersects none of the paths $Q_1', Q_2', \ldots, Q_q'$ (except Q_i'). In addition, we assume that the total number of edges in $Q_1', Q_2', \ldots, Q_q'$ is minimum subject to these conditions. If P_j intersects $Q_1' \cup Q_2' \cup \ldots \cup Q_q'$ and z is the last vertex (on P_j) of $P_j \cap (Q_1' \cup Q_2' \cup \ldots \cup Q_q')$, then the minimality property of $Q_1' \cup Q_2' \cup \ldots \cup Q_q'$ implies that z is the last vertex of one of $Q_1', Q_2', \ldots, Q_q'$. In particular, $Q_1' \cup Q_2' \cup \ldots \cup Q_q'$ intersects at most q of the paths $P_1, P_2, \ldots, P_p$. By a similar argument, there is, for each $i = 1, 2, \ldots, q$, a segment Q_i'' of Q_i from u or from a vertex z' of some P_j, $j \in \{1, 2, \ldots, p\}$, to v such that the segment of P_j from x to z' intersects none of $Q_1'', Q_2'', \ldots, Q_q''$ (except Q_i'') and such that $Q_1'' \cup Q_2'' \cup \ldots \cup Q_q''$ intersects at most q of the paths $P_1, P_2, \ldots, P_q$. Now if $Q_i' \neq Q_i$ and $Q_i'' \neq Q_i$, then we extend $Q_i' \cup Q_i''$ to a uv path Q_i''' by adding two segments of $P_1 \cup P_2 \cup \ldots \cup P_p$, namely the path from the last vertex z of Q_i' to a predecessor y' of y and the path from a successor x' of x to the first vertex z' of Q_i'', and then by adding the edge $y'x'$ which exists since otherwise $xx'y'y$ would be a path of length 3 contradicting the assumption of Proposition 3.1. If $Q_i' = Q_i$ or $Q_i'' = Q_i$ we put $Q_i''' = Q_i$ and now the paths $Q_1''', Q_2''', \ldots, Q_q'''$ are internally disjoint and intersect together at most $2q$ of the paths $P_1, P_2, \ldots, P_p$.   □

THEOREM 3.2   *There exists, for each natural number k, a smallest natural number $f(k)$ such that the following holds: If $x_1, x_2, \ldots, x_k, y_1, y_2, \ldots, y_k$ are distinct vertices in a tournament T such that $\kappa(x_i, y_i) \geqslant f(k)$ for each $i = 1, 2, \ldots, k$, then T has k disjoint paths $P_1, P_2, \ldots, P_k$ such that P_i is an $x_i y_i$ path for each $i = 1, 2, \ldots, k$.*

PROOF   (by induction on k)   Clearly $f(1) = 1$ so we proceed to the induction step assuming $k \geqslant 2$. Suppose that $x_1, x_2, \ldots, x_k, y_1, y_2, \ldots, y_k$ are distinct vertices in a tournament T and $\kappa(x_i, y_i) \geqslant 2(k-1)f(k-1) + 1$ $(i = 1, 2, \ldots, k)$. We shall show that $T - \{x_2, x_3, \ldots, x_k, y_2, \ldots, y_k\}$ has an $x_1 y_1$ path P_1 such that in $T - V(P_1)$ we have $\kappa(x_i, y_i) \geqslant f(k-1)$ for $i = 2, 3, \ldots, k$. Then the result follows by induction. If $T - \{x_2, x_3, \ldots, x_k, y_2, \ldots, y_k\}$ has an $x_1 y_1$ path of length $\leqslant 3$, this can play the role of P_1, so assume that no such path exists. Let $Q_1, Q_2, \ldots, Q_{\kappa(x_1, y_1)}$ be internally disjoint $x_1 y_1$ paths. We shall show that one of these can play the role of P_1. For each $i = 2, 3, \ldots, k$ there are, by Proposition 3.1, internally disjoint $x_i y_i$ paths $P_{1,i}, P_{2,i}, \ldots, P_{f(k-1),i}$ which together intersect at most $2f(k-1)$ of the paths Q_i, $1 \leq i \leq \kappa(x_1, y_1)$. Hence there is at least one path Q_i which intersects none of $P_{j,i}$, $2 \leq i \leq k$, $1 \leq j \leq f(k-1)$. That path Q_i can play the role of P_1 and the proof is complete. $\square$

COROLLARY 3.3   If $x_1 y_1, x_2 y_2, \ldots, x_k y_k$ are pairwise independent edges in a tournament T such that $\kappa(u, v) \geq f(k)$ for any $u, v \in \{x_1, x_2, \ldots, x_k, y_1, y_2, \ldots, x_k\}$, then $x_1 y_1, x_2 y_2, \ldots, x_k y_k$ are contained in a cycle of T occurring in any prescribed order. $\square$

For undirected graphs an analogous result was proved by Woodall [8]: *Any set of k independent edges in a $(2k-1)$-connected graph are contained in a cycle of the graph.* Subsequently, Häggkvist and Thomassen [1] showed that this remains true if $(2k-1)$-connectedness is replaced by the weaker condition that any two vertices which are ends of two of the k edges under consideration are connected by $k+1$ internally disjoint paths.

COROLLARY 3.4   Let D be a digraph with vertices $z_1, z_2, \ldots, z_n$, none of which are isolated, and with at most k edges, some of which may be parallel. If T is a tournament with $A = \{x_1, x_2, \ldots, x_n\} \subseteq T$ such that $\kappa(u, v) \geq f(k)$ for all $u, v \in A$, then T contains a subdivision of D such that z_i corresponds to x_i for each $i = 1, 2, \ldots, n$.

PROOF   We replace successively each x_i by vertices $x_{i,1}, x_{i,2}, \ldots, x_{i,d_i}$ (where d_i is the degree of z_i in D) such that $x_{i,j}$ dominates precisely the same vertices as x_i does for $j = 1, 2, \ldots, d_i$. Then we extend the resulting oriented graph to a tournament T' by adding edges randomly. In T' we have $\kappa(u, v) \geq f(k)$ for any $u, v \in \{x_{i,j} \mid 1 \leq i \leq n, 1 \leq j \leq d_i\}$ and hence we can apply Theorem 3.2 to prove Corollary 3.4. $\square$

If we let D be the digraph with n vertices, such that any two vertices are joined by $k+1$ edges in both directions, we obtain the following result, which is of interest in connection with the aforementioned problem of Lovász.

COROLLARY 3.5   If A is a vertex set of n vertices in a tournament T satisfying $\kappa(x, y) \geq f(k(n-1)n)$ for any two x, y in A, then for any two vertices u and v in A, T has a uv path P such that the removal of all intermediate vertices of P leaves a tournament in which $\kappa(x, y) \geq k$ for any two x, y in A. $\square$

It is easy to describe for each natural number k a planar graph with four vertices x_1, x_2, x_3, x_4 on the outer cycle such that $\kappa(x_i, x_j) \geq k$ for $1 \leq i < j \leq 4$. This shows that Theorem 3.2 and Corollaries 3.4 and 3.5 cannot be extended to digraphs (not even symmetric digraphs) in general. We do not know if Corollary 3.3 is true for digraphs in general. Maybe all results of this section become true for digraphs in general if the condition on the local connectivity is replaced by a condition on the global connectivity. If this is done for tournaments we get stronger results, as shown in the next section.

4. Spanning configurations in tournaments of large connectivity

If we replace the condition on local connectivity in Theorem 3.1 by a condition on the (global) connectivity, the paths in Theorem 3.2 can be chosen such that they together span the tournament.

THEOREM 4.1   If $Z = \{x_1, x_2, \ldots, x_k, y_1, \ldots, y_k\}$ is a set of distinct vertices in an $h(k)$-connected tournament T where $h(k) = f(5k) + 12k + 9$, then T has k pairwise disjoint paths $P_1, P_2, \ldots, P_k$ such that P_i is an x_iy_i path for $i = 1, 2, \ldots, k$ and such that $V(T) = V(P_1) \cup V(P_2) \cup \ldots \cup V(P_k)$.

PROOF (by induction on k)   For $k = 1$ the theorem was proved in [5] so we proceed to the induction step.

We first select $10k$ distinct vertices $x_{i,j}$ and $y_{i,j}$ in $T - Z$ where $1 \leq i \leq k$, $1 \leq j \leq 5$ such that x_i (respectively y_i) dominates each $x_{i,1}$, $x_{i,2}$, $x_{i,3}$, $x_{i,4}$, $x_{i,5}$ (respectively is dominated by each $y_{i,1}$, $y_{i,2}$, $y_{i,3}$, $y_{i,4}$, $y_{i,5}$) for $i = 1, 2, \ldots, k$.

Then we select ten distinct vertices in $T - Z$ (distinct from the vertices previously considered), u_j and v_j ($1 \leq j \leq 5$), such that $x_{1,j}$ dominates v_j and is dominated by u_j for $j = 1, 2, \ldots, 5$. All these vertices exist because T is $(12k + 9)$-connected. By Theorem 3.2, $T - (Z \cup \{u_1, u_2, \ldots, u_5, v_1, v_2, \ldots, v_5\})$ has $5k$ disjoint paths $P_{i,j}$ such that $P_{i,j}$ is an $x_{i,j}y_{i,j}$ path for $1 \leq i \leq k$, $1 \leq j \leq 5$. If some x_iy_i path has length 3 we delete that path and complete the proof by induction. So we can assume that each $P_{i,j}$ has length at least 2 and that the last vertices of $P_{i,1}$, $P_{i,2} \ldots$ all dominate the first vertices of $P_{i,1}$, $P_{i,2} \ldots$ for $i = 1, 2, \ldots, k$.

Let T' be the tournament obtained from T by deleting Z and all paths $P_{i,1}$, $P_{i,2}, \ldots$ for $i = 1, 2, \ldots, k$. If T' is strong we put $T'' = T'$. Otherwise we define T'' as follows: We consider in $T - Z$ a path P from the terminal component of T' to the initial component of T' (for the definition of terminal and initial component, see [5, p. 144]). If P intersects some of the paths $P_{i,1}$, $P_{i,2}, \ldots$ ($1 \leq i \leq k$), then we let z_i (respectively z_i') be the

first (respectively last) vertex on P which is on one of $P_{i,1}, P_{i,2}, \ldots$.
Without loss of generality we can assume that z_i is on $P_{i,1}$ and that z_i' is
on $P_{i,1}$ or $P_{i,2}$. Since $V(P_{i,1}) \cup V(P_{i,2})$ induces a strong tournament, it has a
path from z_i to z_i' and we replace the segment of P from z_i to z_i' by that
path. In this way we transform P into a path P' from the terminal
component of T' to the initial component of T' such that P' does not
intersect any of $P_{1,3}, P_{2,3}, P_{3,3}, \ldots, P_{k,3}$. Now we let T'' be obtained from
T' by adding P' and all those $P_{i,1} \cup P_{i,2}$ for which P' intersects $P_{i,1}$. Then
T'' is strong.

Adding $x_{1,3}$, $x_{1,4}$ and $x_{1,5}$ to T'' results in a strong tournament which
has, by [5, Corollary 2.2], a Hamiltonian path P'' connecting two of $x_{1,3}$,
$x_{1,4}$, $x_{1,5}$. For each $i = 1, 2, \ldots, k$ we consider the subtournament T_i of T
induced by those paths $P_{i,1}, P_{i,2}, \ldots$ which are not in T'' together with P'' if
$i = 1$. Then T_i has, by [5, Theorem 5.2], a Hamiltonian path P_i from x_i to
y_i and the proof is complete. $\square$

Theorem 4.1 can be reformulated as follows:

COROLLARY 4.2   If $e_1, e_2, \ldots, e_k$ are pairwise independent edges in an
$h(k)$-connected tournament T, then T has a Hamiltonian cycle containing
$e_1, e_2, \ldots, e_k$ in that cyclic order. $\square$

This implies Conjecture 8.7 in [6]. In [6] it was also proved that, for any
set A of at most k edges in a $5k$-connected tournament T, $T - A$ has a
Hamiltonian cycle. More generally we have

COROLLARY 4.3   For any set A of at most k edges in a $h(2k)$-connected
tournament T and for any set B of at most k independent edges of $T - A$,
$T - A$ has a Hamiltonian cycle containing B.

We shall only sketch the proof: If uv is an edge of A such that B has no
edge of the form uv', then we select an edge of the form uv' such that v'
is not incident with any edge of A or B and we add that edge to B. Doing
this for every edge of A we extend B into B' and we then apply
Corollary 4.2 with B' instead of B.

THEOREM 4.4   Let D be a digraph with vertices $z_1, z_2, \ldots, z_n$, none of
which are isolated, and with at most k edges, some of which may be
parallel. If T is an $h(k)$-connected tournament with $A = \{x_1, x_2, \ldots, x_n\} \subseteq$
T, then T has a subdivision of D containing all vertices of T such that z_i
corresponds to x_i for $i = 1, 2, \ldots, n$.

The function $f(k)$ in Theorem 3.2 and hence also the function $h(k)$ in this section grow fast with k. Proposition 2.1 suggests that perhaps there exists a natural number $c > 0$ such that any ck-connected graph satisfies the conclusion of Theorem 4.1. If true, this would imply (using the same method as in [7]) that any k-connected tournament ($k \geq 3$) of order n can be covered by at most $cn^2/2k$ Hamiltonian cycles and that every regular tournament of order n can be covered by a collection of at most $\frac{3}{2}cn$ Hamiltonian cycles, since any regular tournament of order n has connectivity at least $\frac{1}{3}n$ by [5, Lemma 4.1]. These results would be best possible except for the values of the constant c. In [7] it was shown that every regular tournament of order n can be covered by a collection of at most $12n$ Hamiltonian cycles.

References

1. R. Häggkvist and C. Thomassen, Circuits through specified edges, *Discrete Math.* **41** (1982), 29–34.
2. W. Mader, Connectivity and edge-connectivity in finite graphs. In *Surveys in Combinatorics*, London Mathematical Society Lecture Note Series no. 38, London Mathematical Society, London, 1979, pp. 66–95.
3. J. W. Moon, *Topics on Tournaments*, Addison-Wesley, Reading, Mass., 1969.
4. C. Thomassen, On the number of Hamiltonian cycles in tournaments. *Discrete Math.* **31** (1980), 315–323.
5. C. Thomassen, Hamiltonian-connected tournaments. *J. Combinat. Theory B* **28** (1980), 142–163.
6. C. Thomassen, Edge-disjoint Hamiltonian paths and cycles in tournaments. *Proc. London Math. Soc.* **45** (1982), 151–168.
7. C. Thomassen, Hamiltonian cycles in regular tournaments. In *Cycles in Graphs* (B. Alspach, ed.), *Ann. Discrete Math.* (to appear).
8. D. R. Woodall, Circuits containing specified edges. *J. Combinat. Theory B* **22** (1977), 274–278.

PLANAR ENUMERATION

W. T. Tutte

ABSTRACT   This paper is concerned with problems of finding the number of combinatorially distinct rooted planar maps in some finite class, for example those with a given number of edges.

A planar map is the figure obtained by drawing a connected graph in the sphere. It is "rooted" by choosing an edge, called the root-edge, and specifying both a direction along this edge and a direction across it. The direction along the root-edge is from the "root-vertex" and along the root-edge to its other end. In the case of a loop the "other end" coincides with the root vertex. Similarly the direction across the root edge is from the "root-face" across the root edge to the face on the other side. That face coincides with the root-face in the case of an isthmus.

The enumeration of rooted maps is often easier than that of unrooted ones. Indeed if we wish to enumerate some class of unrooted maps it seems necessary to enumerate the corresponding rooted ones as the first step.

One rather ambitious project in the theory is that of enumerating rooted planar maps "with a vertex-partition". This means that the valency of the root-vertex is specified, and so is the number $n(i)$ of non-root vertices of each valency i.

We can attack this problem by means of a generating series f defined as follows:

$$f = \sum_M x^{n(M)} \prod_{i=1}^{\infty} y_i^{n(i)}. \tag{1}$$

Here M is an arbitrary rooted planar map and $n(M)$ is the valency of its root-face. The symbol $n(i)$ refers to M. For each M only a finite number of the $n(i)$ can be non-zero and the product in the formula reduces to a finite one. x and the y_i are independent variables.

It is convenient to allow M to be the "vertex-map", with one vertex, one face and no edges. This is counted as a rooted map by a special convention. Its contribution to f is the constant term 1. Otherwise M must be a "rooted map" as defined above. The loop-map has a single edge, which is a loop. It contributes to f the term xy_1. The "link-map", having a single edge with two distinct ends, contributes x^2.

GRAPH THEORY AND COMBINATORICS
ISBN 0-12-111760-X

In enumerative map theory the next step, after defining a generating series, is to find an equation that the series satisfies. This is done by graph-theoretical arguments. For example the function f of (1) is found to satisfy

$$f = 1 + x^2 f^2 + \sum_{j=1}^{\infty} x^{2-j} y_j [f]_{j-1}. \tag{2}$$

The symbol $[f]_m$ as used here means the truncation of f to x^m, that is the result of deleting from f all terms in which the index of x is less than m (see [7]).

We should now try to solve equation (2) for the power series f. The problem of enumerating rooted planar maps with a vertex-partition is simply that of finding the coefficients of the various products of variables in f.

It has to be admitted that equation (2) is as yet unsolved. And this is strange because solutions are known for so many of its special cases.

Consider especially the Eulerian case, in which we substitute zero for each y_i with an odd suffix. In other words we restrict our attention to "Eulerian" rooted maps, in which the valency of each vertex is even. Subject to this restriction the vertex-partition problem has a simple solution. The number of such rooted maps in which the root-vertex has valency $2t$ and the number of non-root vertices of valency $2j$ is s_j, denoted by $K(t; s_1, s_2, s_3, \ldots)$, is given by the following equation:

$$K(t; s_1, s_2, s_3, \ldots) = \frac{(n-1)!}{(n-s+1)!} \frac{(2t)!}{t!\,(t-1)!} \prod_{j=1}^{\infty} \frac{1}{s_j!} \left(\frac{(2j-1)!}{j!\,(j-1)!} \right)^{s_j} \tag{3}$$

Here

$$s = \sum s_j \quad \text{and} \quad n = t + \sum j s_j$$

Thus $s + 1$ is the number of vertices and n is the number of edges of each of the maps being enumerated.

Basically (3) is proved in [5]. In that paper we enumerate structures called "slicings". They are not the Eulerian planar maps but they are simply related to those maps. I call the relation "simple" but I have already made one mistake in applying it (in [8]). Formula (3) is the correct application, or at least a revised one.

We might hope for some formula analogous to (3) but applying to general rooted planar maps. Failing that we might hope for some plausible explanation of why the Eulerian case should be so much more tractable. So far both hopes have been disappointed.

There is another solved special case of (2). In it we replace the y_i by two variables y and z; we make the substitution $y_i = yz^i$. This special case is discussed in [7], where it is found convenient to make the further transformations

$$u = xz^{-1}, \qquad v = yz^2, \qquad w = z^2.$$

The generating series f then becomes a function

$$F = F(u, v, w)$$

of the three variables u, v and w. It is a power series. Its coefficient of $u^k v^i w^j$ is found to be the number of combinatorially distinct rooted planar maps with $i + 1$ faces and $j + 1$ vertices, and with a root-face of valency p. Equation (2) transforms into the following:

$$F - 1 - u^2 w F^2 = uv\left(\frac{uF - F(1, v, w)}{u - 1}\right). \tag{4}$$

The paper [7] gives a solution of (4) in terms of two parameters s and t related to v and w as follows:

$$v = \frac{s(s+2)}{4(s+t+1)^2}, \qquad w = \frac{t(t+2)}{4(s+t+1)^2}. \tag{5}$$

It is found that

$$F(1, v, w) = \frac{4(s+t+1)}{(s+2)(t+2)}, \tag{6}$$

after which we can get F in terms of u, s and t by solving (4) as a quadratic.

We can make a further simplification by putting $v = w$ and correspondingly $s = t$. It enumerates rooted planar maps with a given valency for the root-vertex and a given number of edges. In particular we find that the number of rooted planar maps with n edges is

$$\frac{2(2n)!\, 3^n}{n!\,(n+2)!}. \tag{7}$$

Going back to equation (2), let us note that the part of f that enumerates plane trees, that is rooted maps with only one face, can be found. For the number of such plane trees with a given root-vertex valency, and with a specified number $v(j)$ of non-root vertices of valency j can be inferred from the theory of [6]. (However, that paper deals explicitly only with the case in which the root-vertex is monovalent.)

There is one other special case of (2) that deserves mention. For it we substitute zero for each y_i other than y_3. This "cubic" case is solved by Mullin in [2]. One of his results is that the number of cubic planar maps with n non-root vertices and with a monovalent root-vertex is

$$\frac{3 \cdot 2^n}{(n+1)!} \left[\left(\frac{d}{dt}\right)^{n-1} t^{(3n-5)/2} \right]_{t=1}. \tag{8}$$

A systematic exposition of the enumerative theory of planar maps would come to a branch-point here. On the one hand it could proceed to an analogous theory of 2-connected rooted maps. On the other it could go on to the enumeration of λ-coloured general rooted maps.

In the realm of 2-connected maps there is still no solution of the general problem with a vertex-partition. Worse, there is not even as yet a solution of the Eulerian case. That ought to be one of the next problems to be attacked. The special case in which we fix only the root-vertex valency, the number of vertices and the number of faces is discussed in [1], albeit in dual form. The solution is simpler than in the general case, and explicit formulae for the coefficients are found. The cubic case, or its dual form dealing with triangulations, can also be solved.

There is a further theory of 3-connected maps [3]. It has been found helpful in the enumerative theory of convex polyhedra [4, 9].

As for the λ-coloured rooted maps, λ-coloured on the vertices that is, it would seem appropriate to modify the generating series (1) as follows:

$$f = \sum_M x^{n(M)} \prod_{i=1}^{\infty} y_i^{n(i)} P(M, \lambda). \tag{9}$$

Here $P(M, \lambda)$ is the number of ways of λ-colouring M. It can be expressed as a polynomial in λ and as such it is called the chromatic polynomial of M. Since it is a polynomial, we can consider it also for non-integral values of λ. For such values of λ, f no longer enumerates planar maps. We therefore call f a *chromatic sum* rather than an enumerating series.

In a further extension, apparently more ambitious, we insert into (9) a factor $t^{m(M)}$, where $m(M)$ is the valency of the root-face of M. Actually this extension is often forced upon us by the requirements of the theory, whether we wish it or not.

Alas there is as yet not even an equation for f taking account of vertex-partitions. There is an equation in the special case which takes account only of $m(M)$, $n(M)$, the numbers of faces and vertices and the colour-number λ. The equation is presented in [10], not merely for the chromatic polynomial but with the generalization known as the dichromate of M. It is presented indeed, but it is not solved. As yet significant progress has been made only in the special case in which M is 2-connected, and a triangulation. The results for chromatic sums of triangulations are summarized in [11] and [12].

References

1. W. G. Brown and W. T. Tutte, On the enumeration of rooted non-separable planar maps. *Canad. J. Math.* **16** (1964), 572–577.
2. R. C. Mullin, E. Nemeth and P. J. Schellenberg, The enumeration of almost cubic maps. In *Proceedings of the Louisiana Conference on Combinatorics, Graph Theory and Computing* (1970), 281–295.
3. R. C. Mullin and P. J. Schellenberg, The enumeration of c-nets via quadrangulations. *J. Combinat. Theory* **4** (1968), 259–276.
4. L. B. Richmond and N. C. Wormald, The asymptotic number of convex polyhedra. *Trans. Amer. Math. Soc.* **273** (1982), 721–735.
5. W. T. Tutte, A census of slicings. *Canad. J. Math.* **14** (1962), 708–722.
6. W. T. Tutte, The number of planted plane trees with a given partition. *Amer. Math. Monthly* **71** (1964), 272–277.
7. W. T. Tutte, On the enumeration of planar maps. *Bull. Amer. Math. Soc.* **74** (1968), 64–74.
8. W. T. Tutte, The use of numerical computations in the enumerative theory of planar maps (Jeffery-Williams Lectures 1968–1972). *Canad. Math. Congress* (1972), 73–89.
9. W. T. Tutte, On the enumeration of convex polyhedra. *J. Combinat. Theory B* **28** (1980), 105–126.
10. W. T. Tutte, Dichromatic sums for rooted planar maps. In *Proceedings of the 19th Symposium on Pure Mathematics*, American Mathematical Society, Providence, R.I., 1971, pp. 235–245.
11. W. T. Tutte, Chromatic solutions. *Canad. J. Math.* **34** (1982), 741–758.
12. W. T. Tutte, Chromatic solutions II. *Canad. J. Math.* **34** (1982), 952–960.

29

ON THE CHROMATIC NUMBER OF SPARSE RANDOM GRAPHS

W. Fernandez de la Vega

ABSTRACT   The following theorem, which was conjectured by Erdös and Spencer, is proved.
Let $c = c(n)$ be any function satisfying $c(n) \to \infty$ and $c(n) = o(n)$. The chromatic number of the random graph $G = G_{n,p}$ on n vertices where each edge is present with probability $p = c(n)/n$ "almost surely" satisfies the inequalities

$$\frac{c}{2 \log c} (1 + o(1)) \leqslant \chi(G) \leqslant \frac{c}{\log c} (1 + o(1)).$$

1. Introduction

Let $c \geqslant 0$ be fixed and let $G = G_{n,p}$, $p = c/n$, denote the random graph on n vertices with independent edges, each one present with probability p. The behaviour of the chromatic number of $G_{n,p}$ is discussed briefly in a problem of Erdös and Spencer [3, pp. 94, 96]. They imply that the inequalities

$$\frac{c}{2 \log c} (1 + o(1)) \leqslant \chi(G) \leqslant \frac{c}{\log c} (1 + o(1)), \tag{1}$$

where the o-terms are with respect to c, hold with a probability tending to 1 as $n \to \infty$. In other words, for each $\varepsilon > 0$, there exists $c_0 = c_0(\varepsilon)$ such that, for every fixed $c > c_0$, the probability that the chromatic number of $G = G_{n,p}$, $p = c/n$, verifies

$$\frac{c(1 - \varepsilon)}{2 \log c} \leqslant \chi(G) \leqslant \frac{c(1 + \varepsilon)}{\log c}$$

tends to 1 as n tends to infinity. We hear that a complete proof of this result can be found in a recent paper of C. McDiarmid.

We shall prove here a uniform version of (1), bringing a positive answer to a conjecture of Erdös and Spencer [3, Problem 9, p. 96].

THEOREM   Let $c = c(n)$ be any function satisfying $c(n) \to \infty$ and $c(n) = o(n)$ and let $G = G_{n,p}$, $p = c(n)/n$. Then, with a probability tending to 1 as

$n \to \infty$, the chromatic number of G satisfies

$$\frac{c}{2 \log c}(1 + o(1)) \leq \chi(G) \leq \frac{c}{\log c}(1 + o(1)).$$

Here we have to prove that, for every positive ε, there exists n_ε, which may depend on the function $c(n)$, such that the inequalities

$$P\left(\chi(G) \geq \frac{c}{2 \log c}(1 - \varepsilon)\right) \geq 1 - \varepsilon \qquad (2)$$

and

$$P\left(\chi(G) \leq \frac{c}{\log c}(1 + \varepsilon)\right) \geq 1 - \varepsilon \qquad (3)$$

hold for every $n \geq n_\varepsilon$.

The proof of (2) is almost immediate. The proof of (3) is rather lengthy and depends on a series of lemmas. We state these lemmas and prove the theorem in Section 2. The proofs of the lemmas are accomplished in Section 3. Previous results on the colouring problem for random graphs were obtained by Bollobás and Erdös [1], Fernandez de la Vega [4], Grimmet and McDiarmid [5] and Koršunov [6].

2. Proof of the theorem

In order to prove (3) we use, following the hint of Erdös and Spencer for the proof of the right-hand side of (1), the greedy algorithm for the determination of a colouring of G. The greedy algorithm picks vertices $P_1, P_2, \ldots$ from the vertex set, where P_{i+1} is the first vertex in the set $\{P: (P, P_j) \notin G, P \neq P_j, 1 \leq j \leq i\}$, until for some index $i = i_0$; this set is empty, so that a maximal independent set S_1 of cardinality i_0 has been obtained, which will be the first colour class. Then the vertices of S_1 are left apart and one starts again with the subgraph of G spanned by the remaining vertices, obtaining a second independent set S_2, and so on until the independent sets obtained cover the vertex set, defining a colouring of G using as many colours.

We shall require the following lemmas:

LEMMA 1   Let $G = G_{n,p}$, $p = c/n$. The cardinality of the first independent set S of G given by the greedy algorithm verifies, for every $c > 1$ and every $\varepsilon > 0$,

$$P(|S| \geq (1 - \varepsilon)nc^{-1} \log c) \geq 1 - \frac{c}{2n(1 - c^{-\varepsilon})^2}.$$

LEMMA 2   Let $G = G_{n,p}$, $p = c/n$. For every fixed c,

$$P(\chi(G) \leq 3 + 3c) \to 1 \quad as \quad n \to \infty,$$

i.e. there exists a function $N(\varepsilon, c)$ such that the above probability exceeds $1 - \varepsilon$ for $n \geq N(\varepsilon, c)$.

LEMMA 3   Let $X_0, X_1, \ldots$ be a non-decreasing sequence of random variables. Suppose that "almost surely" $X_0 \geq m_0$ and that there exist constants $h > 0$ and $p \leq 1/2$ such that

$$P(X_{i+1} - X_i \geq h \mid X_0, X_1, \ldots, X_i) \geq 1 - p$$

holds for every i and $X_i < m$. Let $T = T(m)$ be the first index for which $X_T \geq m$ holds and let us denote by $N = N(m)$ the number of indices $i \leq T$ for which the inequality $X_{i+1} - X_i < h$ holds. Then the expectation of N satisfies

$$EN \leq 2\left(1 + p\left(\frac{m - m_0}{h}\right)\right).$$

Let us denote by X_i the set of vertices of G which are not covered by the i first independent sets $S_1, S_2, \ldots, S_i$ produced by the greedy algorithm. Let us denote by x_i the cardinality of X_i.

Let K be any fixed integer ≥ 9. We set

$$h = \lceil \log(c/K) \rceil,$$

which entails

$$ce^{-h} \geq 3.$$

We define, for $1 \leq k \leq h$,

$$A_k = \{i : ne^{-k} < x_i \leq ne^{1-k}\},$$
$$h_k = \#\{i : i \in A_k, |S_i| \geq (1 - \varepsilon)ne^{-1}(\log c - k)\},$$

where ε will be fixed later, and

$$g_k = |A_k| - h_k.$$

We have

$$h_k \leq 1 + \left\lceil \frac{c(e^{1-k} - e^{-k})}{(1 - \varepsilon)(\log c - k)} \right\rceil \leq \frac{c(e^{1-k} - e^{-k})}{(1 - \varepsilon)(\log c - k)} + 2.$$

We shall now derive a bound for g_k. Suppose that A_k is not empty, let us denote by I_0 the first index in A_k and let us write, for $j \geq 0$, $X_{k,j}$ for X_{I_0+j} and $x_{k,j}$ for x_{I_0+j}. Observe that the subgraph of G spanned by $X_{k,j}$ is a random graph with the same edge probability as G, which is, conditionally on $x_{k,j}$, independent of $x_{k,0}, x_{k,1}, \ldots, x_{k,j-1}$. Using Lemma 1, we obtain that the inequality

$$P(x_{k,j+1} - x_{k,j} \geq (1-\varepsilon)nc^{-1}(\log c - k) \mid x_{k,0}, \ldots, x_{k,j}) \geq 1 - \frac{3}{2n(1-3^{-\varepsilon})^2}$$

holds for $x_{k,j} \geq ne^{-k}$, that is to say for any value of j such that $I_0 + j$ belongs to A_k. This inequality enables us to apply Lemma 2 to the random variables $x_{k,j}$, $j = 0, 1, \ldots,$ with $m = ne^{1-k}$, $m_0 = ne^{-k}$, $h = n(1-\varepsilon)c^{-1}(\log c - k)$ and $p = (3/2)n^{-1}(1-3^{-\varepsilon})^{-2}$, which gives

$$Eg_k \leq \frac{c(e^{1-k} - e^{-k})}{(1-\varepsilon)(\log c - k)} \frac{3}{n(1-3^{-\varepsilon})^2} + 2. \tag{4}$$

Let $\varepsilon > 0$ be given. We set $k_0 = \lceil \log(1/\varepsilon) \rceil$, which implies $e^{-k_0} \leq \varepsilon$. Let us denote by n_ε the minimum value of n for which the following conditions are satisfied:

$$\frac{3}{n(1-3^{-\varepsilon})^2} \leq \varepsilon^2, \tag{5}$$

$$\log c \geq \frac{k_0}{\varepsilon}, \tag{6}$$

and

$$\frac{nc}{K} \geq N(\varepsilon, K), \tag{7}$$

where $N(\cdot, \cdot)$ is the function introduced in Lemma 2. We shall also require the conditions

$$\frac{\varepsilon c}{\log c} \geq 3 + 3K \tag{8}$$

and

$$\frac{\log^2 c}{c} \leq \varepsilon^2, \tag{9}$$

which are in fact implied by the preceding conditions if ε is sufficiently small.

In fact we shall stop the greedy algorithm precisely at the moment when the remaining graph is of density K. Let us denote by Q the chromatic number of this remaining graph. Then, because of (7) and (8), we have, using Lemma 3,

$$P\left(Q \leq \frac{\varepsilon c}{\log c}\right) \geq 1 - \varepsilon.$$

Obviously we have $\chi(G) \leq Q + L + H$, with

$$L = \sum_{k=1}^{h} g_k, \qquad H = \sum_{k=1}^{h} h_k.$$

We have, from (4) and (5),

$$EL \leq \frac{cS\varepsilon^2}{1-\varepsilon} + 2 \log c, \qquad (10)$$

where

$$S = \sum_{k=1}^{h} \frac{e^{1-k} - e^{-k}}{\log c - k}.$$

It is easily checked that S satisfies

$$S \leq \frac{1 + 2e^{-k_0}}{\log c - k_0}$$

for every $k_0 \leq h$. This inequality implies, with (6),

$$S \leq \frac{1 + 2\varepsilon}{(1-\varepsilon) \log c} \leq \frac{1 + 4\varepsilon}{\log c}.$$

By (10) and (9),

$$EL \leq \frac{\varepsilon^2 c(1 + 6\varepsilon)}{\log c} + \frac{2\varepsilon^2 c}{\log c} \leq \frac{4\varepsilon^2 c}{\log c},$$

so that

$$P\left(L \geq \frac{4\varepsilon c}{\log c}\right) \leq \varepsilon.$$

We have also

$$H \leq \frac{cS}{1-\varepsilon} + 2 \log c \leq \frac{c(1 + 4\varepsilon)}{(1-\varepsilon) \log c} \leq \frac{c(1 + 7\varepsilon)}{\log c}.$$

Putting together our bounds for Q, L and H, we get, for $n \geq n_\varepsilon$,

$$P\left(\chi(G) \leq \frac{c}{\log c}(1 + 12\varepsilon)\right) \geq 1 - 2\varepsilon,$$

which is equivalent to (3). $\square$

3. Proof of the lemmas

PROOF OF LEMMA 1   For convenience we will work with the random graph $G_{\omega,p}$ with point set the natural numbers, in which each edge occurs with probability $p = c/n$. $G_{n,p}$ coincides in distribution with the subgraph of $G_{\omega,p}$ spanned by the n first vertices.

Let $I_0 = 0, I_1, \ldots, I_k, \ldots$ be the vertices of the independent set of $G_{\omega,p}$ obtained by the greedy algorithm. We set $T_k = I_k - I_{k-1}$. Clearly the T_ks are independent random variables and the distribution of T_k coincides with the distribution of the rank of the first successful trial in a sequence of Bernouilli trials with individual success probabilities all equal to $(1 - c/n)^k$. Using the identities of Wald (see [7]) one gets

$$ET_k = (1 - c/n)^{-k} \leqslant e^{ck/n}$$

and

$$\text{Var } T_k = (1 - (1 - c/n)^k)(1 - c/n)^{-2k} \leqslant (1 - c/n)^{-2k} \leqslant e^{2ck/n}.$$

Then

$$EI_k = \sum_{i=1}^{k} ET_i \leqslant \frac{e^{ck/n}}{c}$$

and

$$\text{Var } I_k = \sum_{i=1}^{k} \text{Var } T_i \leqslant \frac{ne^{2ck/n}}{2c}.$$

The first independent set determined by the greedy algorithm applied to $G_{n,p}$ will be of cardinality at least $m + 1$ if and only if $I_m \leqslant n - 1$ holds. Taking $m = \lfloor (1 - \varepsilon)nc^{-1} \log c \rfloor$, we get

$$EI_m \leqslant nc^{-\varepsilon}, \qquad \text{Var } I_m \leqslant \tfrac{1}{2}nc^{1-2\varepsilon}.$$

Tchebycheff's inequality then gives

$$P(I_m \geqslant n) \leqslant \frac{(n/2)c^{1-2\varepsilon}}{n^2(1 - c^{-\varepsilon})^2} \leqslant \frac{c}{2n(1 - c^{-\varepsilon})^2}. \quad \square$$

PROOF OF LEMMA 2   For $m = n/3c$ we obtain, from (11),

$$EI_m \leqslant \frac{ne}{3c}, \qquad \text{Var } I_m \leqslant \frac{ne^2}{2c}.$$

These inequalities imply, using again Tchebycheff's inequality, for $c \geqslant 1$,

$$P(I_m \geqslant n) \leqslant \frac{\lambda}{nc}, \qquad \lambda = \frac{e^2}{2(1 - e/3)^2}.$$

Thus, for $c \geqslant 1$, the greedy algorithm will give, with probability $1 - O(1/n)$, an independent set of cardinality not less than $n/3c$. Lemma 2 is now easily deduced from this fact. If $c < 1$, then each component of G contains at most one cycle [2, Theorem 5e] and has therefore chromatic number not more than 3. Let us consider now a $c \geq 1$. Then, putting aside an independent set of cardinality $n/3c$ leaves a graph of density $c - 1/3$ and the lemma follows by induction. $\square$

PROOF OF LEMMA 3   Obviously we can define a sequence (Y_i) which coincides with (X_i) until the first time that X_i exceeds m and such that the condition of the lemma holds for every i. It will suffice to prove the lemma for (Y_i). Let us denote by I_j the indicator function of the event $Y_j - Y_{j-1} \geq h$ and by T' the first index k for which the inequality

$$\sum_{j=1}^{k} I_j \geq \frac{m - m_0}{h}$$

holds. Then "almost surely" $T \leq T'$ and

$$\sum_{j=1}^{\infty} 1_{\{T' \geq j\}} I_j \leq \frac{m - m_0}{h} + 1.$$

By the hypothesis of the lemma we have $P(I_j = 1 \mid T' \geq j) \geq 1 - p$, so that, taking expectations in the preceding inequality, we obtain

$$(1 - p) \sum_{j=1}^{\infty} E 1_{\{T' \geq j\}} \leq (1 - p) ET' \leq \frac{m - m_0}{h} + 1,$$

that is

$$ET' \leq \frac{1}{1 - p} \left(1 + \frac{m - m_0}{h} \right) \leq 2 + (1 + 2p) \frac{m - m_0}{h}.$$

We have $N \leq T' - (m - m_0)/h$, which implies, with the preceding inequality,

$$EN \leq ET' - (m - m_0)/h \leq 2 \left(1 + p \frac{m - m_0}{h} \right). \quad \square$$

References

1. B. Bollobás and P. Erdös, Cliques in random graphs. *Math. Proc. Camb. Phil. Soc.* **80** (1976), 419–427.
2. P. Erdös and A. Rényi, On the evolution of random graphs. *Publ. Math. Inst. Hungar. Acad. Sci.* **5A** (1960), 17–61.
3. P. Erdös and J. Spencer, *Probabilistic Methods in Combinatorics*, Academic Press, New York, 1974.

4. W. Fernandez de la Vega, Random graphs almost optimally colourable in polynomial time. To appear.
5. G. R. Grimmet and C. J. H. McDiarmid, On colouring random graphs. *Math. Proc. Camb. Phil. Soc.* **77** (1975), 313–324.
6. A. D. Koršunov, The chromatic number of n-vertex graphs [in Russian]. *Diskret Analiz*, **35** (1980), 15–44.
7. A. N. Shiryayev, *Optimal Stopping Rules*, Springer Verlag, New York, 1978.